Mathematics for Econom

Phoebus J. Dhrymes

Mathematics for Econometrics

Fourth Edition

 Springer

Phoebus J. Dhrymes
Department of Economics
Columbia University
New York, USA

ISBN 978-1-4614-8144-7   ISBN 978-1-4614-8145-4 (eBook)
DOI 10.1007/978-1-4614-8145-4
Springer New York Heidelberg Dordrecht London

Library of Congress Control Number: 2013944568

Printed on acid-free paper

Springer is part of Springer Science+Business Media (www.springer.com)

Preface

The fourth edition differs significantly from the third edition, in that it has undergone considerable expansion and revision.

The major expansion involves a more complete coverage of basic aspects of mathematics that have continued to play an increasingly significant role in the literature of econometrics. Thus, the chapter on difference equations has been expanded to include enhanced treatment of lag operators (backward shift operators in the statistical literature) that are important not only in the context of the dynamic simultaneous equation GLSEM (general linear structural econometric model), but also time series analysis.

In addition, a chapter on the basic mathematics underlying the analytics of probability theory has been added, as well as a chapter on laws of large numbers and central limit theorems that form the probabilistic basis of classical econometrics. Moreover, there is an informative but not exhaustive discussion of stationary time series analysis, including discussions of the taxonomy of time series, issues of causality and invertibility, with a limited treatment of certain non-linearities such as those found in the popular ARCH (autoregressive conditional heteroskedasticity) model, which together with its many variants has found extensive applications in the literature of financial econometrics. However, there is no discussion of non-stationary time series, which is the subject of the author's *Time Series Unit Roots and Cointegration*, Academic Press, 1998.

Finally, this edition contains two fairly extensive chapters on applications to the GLM (general linear model), GLSEM and time series analysis which treat issues relevant to their underlying theoretical bases, estimation and forecasting.

New York, USA   Phoebus J. Dhrymes

Preface to the Third Edition

The third edition differs from the second edition in several respects. The coverage of matrix algebra has been expanded. For example, the topic of inverting partitioned matrices in this edition deals explicitly with a problem that arises in estimation under (linear) constraints. Often this problem forces us to deal with a block portioned matrix whose (1,1) and (2,2) blocks are singular matrices. The standard method for inverting such matrices fails; unless the problem is resolved, explicit representation of estimators and associated Lagrange multipliers is not available. An important application is in estimating the parameters of the general linear structural econometric model, when the identifying restrictions are imposed by means of Lagrange multipliers. This formulation permits a near effortless test of the validity of such (overidentifying) restrictions.

This edition also contains a treatment of the vector representation of restricted matrices such as symmetric, triangular, diagonal and the like. The representation is in terms of restricted linear subspaces. Another new feature is the treatment of permutation matrices and the vec operator, leading to an explicit representation of the relationship between $A \otimes B$ and $B \otimes A$.

In addition, it contains three new chapters, one on asymptotic expansions and two on applications of the material covered in this volume to the general linear model and the general linear structural econometric model, respectively. The salient features of the estimation problems in these two topics are discussed rigorously and succinctly.

This version should be useful to students and professionals alike as a ready reference to mathematical tools and results of general applicability in econometrics. The two applications chapters should also prove useful to noneconomist professionals who are interested in gaining some understanding of certain topics in econometrics.

New York, USA   Phoebus J. Dhrymes

Preface to the Second Edition

The reception of this booklet has encouraged me to prepare a second edition.

The present version is essentially the original, but adds a number of very useful results in terms of inverses and other features of partitioned matrices, a discussion of the singular value decomposition for rectangular matrices, issues of stability for the general linear structural econometric model, and similar topics.

I would like to take this opportunity to express my thanks to many of my students and others for pointing out misprints and incongruities in the first edition.

New York, USA   Phoebus J. Dhrymes

Preface to the First Edition

This book began as an Appendix to *Introductory Econometrics*. As it progressed, requirements of consistency and completeness of coverage seemed to make it inordinately long to serve merely as an Appendix, and thus it appears as a work in its own right.

Its purpose is not to give rigorous instruction in mathematics. Rather it aims at filling the gaps in the typical student's or professional's mathematical training, to the extent relevant for the study of econometrics.

Thus, it contains a collection of mathematical results employed at various stage of *Introductory Econometrics*. More generally, however, it could serve as a useful adjunct and reference to students of econometrics, no matter what text is being employed.

In the vast majority of cases, proofs are provided and there is a modicum of verbal discussion of certain mathematical results, the objective being to reinforce the student's understanding of the formalities. In certain instances, however, when proofs are too cumbersome, or complex, or when they are too obvious, they are omitted.

New York, USA   Phoebus J. Dhrymes

Contents

Chapter 1

Vectors and Vector Spaces

In nearly all of the discussion in this volume, we deal with the set of **real** numbers. Occasionally, however, we deal with **complex** numbers as well. In order to avoid cumbersome repetition, we shall denote the set we are dealing with by $\mathcal{F}$ and let the context elucidate whether we are speaking of real or complex numbers, or both.

1.1 Complex Numbers and Vectors

For the sake of completeness, we begin with a brief review of complex numbers, although it is assumed that the reader is at least vaguely familiar with the subject.

A **complex** number, say z, is denoted by

$$z = x + iy,$$

where x and y are **real** numbers and the symbol i is defined by

$$i^2 = -1. \tag{1.1}$$

All other properties of the entity denoted by i are derivable from the basic definition in Eq. (1.1). For example,

$$i^4 = (i^2)(i^2) = (-1)(-1) = 1.$$

P.J. Dhrymes, *Mathematics for Econometrics*,
DOI 10.1007/978-1-4614-8145-4_1, © The Author 2013

Similarly,

$$i^3 = (i^2)(i) = (-1)i = -i,$$

and so on.

It is important for the reader to grasp, and bear in mind, that a complex number is describable in terms of an ordered pair of real numbers.

Let

$$z_j = x_j + iy_j, \qquad j = 1, 2,$$

be two complex numbers. We say

$$z_1 = z_2$$

if and only if

$$x_1 = x_2 \quad \text{and} \quad y_1 = y_2.$$

Operations with complex numbers are as follows.

Addition:

$$z_1 + z_2 = (x_1 + x_2) + i(y_1 + y_2).$$

Multiplication by a real scalar:

$$cz_1 = (cx_1) + i(cy_1).$$

Multiplication of two complex numbers:

$$z_1 z_2 = (x_1 x_2 - y_1 y_2) + i(x_1 y_2 + x_2 y_1).$$

Addition and multiplication are, evidently, associative and commutative; i.e. for complex z_j, $j = 1,2,3$

$$z_1 + z_2 + z_3 = (z_1 + z_2) + z_3 \quad \text{and} \quad z_1 z_2 z_3 = (z_1 z_2) z_3,$$

$$z_1 + z_2 = z_2 + z_1 \quad \text{and} \quad z_1 z_2 = z_2 z_1.$$

and so on.

The **conjugate** of a complex number z is denoted by $\bar{z}$ and is defined by

$$\bar{z} = x - iy.$$

Associated with each complex number is its **modulus** or **length** or **absolute value**, which is a **real** number often denoted by $|z|$ and defined by

$$|z| = (z\bar{z})^{1/2} = (x^2 + y^2)^{1/2}.$$

For the purpose of carrying out multiplication and division (an operation which we have not, as yet, defined) of complex numbers, it is convenient to express them in polar form.

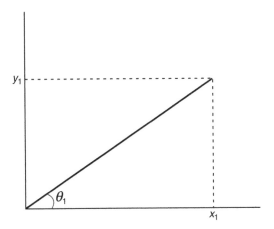

Figure 1.1.

1.1.1 Polar Form of Complex Numbers

Let z_1, a complex number, be represented in Fig. 1.1 by the point (x_1, y_1), its **coordinates**.

It is easily verified that the length of the line from the origin to the point (x_1, y_1) represents the modulus of z_1, which for convenience we denote by r_1. Let the angle described by this line and the abscissa be denoted by θ_1. As is well known from elementary trigonometry, we have

$$\cos \theta_1 = \frac{x_1}{r_1}, \qquad \sin \theta_1 = \frac{y_1}{r_1}. \tag{1.2}$$

We may thus write the complex number as

$$z_1 = x_1 + iy_1 = r_1 \cos \theta_1 + ir_1 \sin \theta_1 = r_1(\cos \theta_1 + i \sin \theta_1).$$

Further, we may define the quantity

$$e^{i\theta_1} = \cos \theta_1 + i \sin \theta_1, \tag{1.3}$$

and, consequently, write the complex number in the standard **polar form**

$$z_1 = r_1 e^{i\theta_1}. \tag{1.4}$$

In the representation above, r_1 is the **modulus** and θ_1 the **argument** of the complex number z_1. It may be shown that the quantity $e^{i\theta_1}$ as defined in Eq. (1.3) has all the properties of real exponentials insofar as the operations of multiplication and division are concerned. If we confine the **argument**

of a complex number to the range $[0, 2\pi)$, we have a unique correspondence
between the (x, y) coordinates of a complex number and the modulus and
argument needed to specify its polar form. Thus, for any complex number z,
the representations

$$z = x + iy, \qquad z = re^{i\theta},$$

where

$$r = (x^2 + y^2)^{1/2}, \quad \cos\theta = \frac{x}{r}, \quad \sin\theta = \frac{y}{r},$$

are completely equivalent.

In polar form, multiplication and division of complex numbers are
extremely simple operations. Thus,

$$z_1 z_2 = (r_1 r_2)e^{i(\theta_1 + \theta_2)}$$

$$\frac{z_1}{z_2} = \left(\frac{r_1}{r_2}\right)e^{i(\theta_1 - \theta_2)},$$

provided $z_2 \neq 0$.

We may extend our discussion to **complex vectors**, i.e. ordered n-tuples
of complex numbers. Thus

$$z = x + iy$$

is a complex vector, where x and y are n-element (real) vectors (a concept
to be defined immediately below). As in the scalar case, two complex vectors
z_1, z_2 are equal if and only if

$$x_1 = x_2, \qquad y_1 = y_2,$$

where now x_i, y_i, $i = 1, 2$, are n-element (column) vectors. The complex
conjugate of the vector z is given by

$$\bar{z} = x - iy,$$

and the modulus of the complex vector is defined by

$$(z'\bar{z})^{1/2} = [(x + iy)'(x - iy)]^{1/2} = (x'x + y'y)^{1/2},$$

the quantities $x'x$, $y'y$ being ordinary scalar products of two vectors.
Addition and multiplication of complex vectors are defined by

$$z_1 + z_2 = (x_1 + x_2) + i(y_1 + y_2),$$
$$z_1'z_2 = (x_1'x_2 - y_1'y_2) + i(x_1'y_2 + x_2'y_1),$$

where x_i, y_i, $i = 1, 2$, are real n-element column vectors. The notation
for example x_1', or y_2' means that the vectors are written in **row form**,

rather than the customary column form. Thus, $x_1 x_2'$ is a matrix, while $x_1' x_2$ is a scalar. These concepts (vector, matrix) will be elucidated below. It is somewhat awkward to introduce them now; still, it is best to set forth at the beginning what we need regarding complex numbers.

1.2 Vectors

Definition 1.1. Let[1] $a_i \in \mathcal{F}$, $i = 1, 2, \ldots, n$; then the ordered n-tuple

$$a = \begin{pmatrix} a_1 \\ a_2 \\ \vdots \\ a_n \end{pmatrix}$$

is said to be an n-**dimensional vector**. If $\mathcal{F}$ is the field of real numbers, it is termed an n-**dimensional real vector**.

Remark 1.1. Notice that a scalar is a trivial case of a vector whose dimension is $n = 1$.

Customarily we write vectors as **columns**, so strictly speaking we should use the term **column vectors**. But this is cumbersome and will not be used unless required for clarity.

If the elements of a vector, a_i, $i = 1, 2, \ldots, n$, belong to $\mathcal{F}$, we denote this by writing

$$a \in \mathcal{F}.$$

Definition 1.2. If $a \in \mathcal{F}$ is an n-dimensional column vector, its **transpose** is the n-dimensional **row** vector denoted by

$$a' = (a_1, \quad a_2, \quad a_3, \quad \ldots, \quad a_n).$$

If a, b are two n-dimensional vectors and $a, b \in \mathcal{F}$, we define their **sum** by

$$a + b = \begin{pmatrix} a_1 + b_1 \\ \vdots \\ a_n + b_n \end{pmatrix}.$$

---

[1]The symbol $\mathcal{F}$ is, in this discussion, a primitive and simply denotes the collection of objects we are dealing with.

If c is a scalar and $c \in \mathcal{F}$, we define

$$ca = \begin{pmatrix} ca_1 \\ ca_2 \\ \vdots \\ ca_n \end{pmatrix}.$$

If a, b are two n-dimensional vectors with elements in $\mathcal{F}$, their **inner product** (which is a scalar) is defined by

$$a'b = a_1 b_1 + a_2 b_2 + \cdots + a_n b_n.$$

The inner product of two vectors is also called their **scalar product**, and its square root is often referred to as the **length** or the **modulus** of the vector.

Definition 1.3. If $a, b \in \mathcal{F}$ are n-dimensional column vectors, they are said to be **orthogonal** if and only if $a'b = 0$. If, in addition, $a'a = b'b = 1$, they are said to be **orthonormal**.

Definition 1.4. Let $a_{(i)}$, $i = 1, 2, \ldots, k$, be n-dimensional vectors whose elements belong to $\mathcal{F}$. Let c_i, $i = 1, 2, \ldots, k$, be scalars such that $c_i \in \mathcal{F}$. If

$$\sum_{i=1}^{k} c_i a_{(i)} = 0$$

implies that

$$c_i = 0, \qquad i = 1, 2, \ldots, k,$$

the vectors $\{a_{(i)} : i = 1, 2, \ldots, k\}$ are said to be **linearly independent** or to constitute a **linearly independent set**. If there exist scalars c_i, $i = 1, 2, \ldots, k$, not all of which are zero, such that $\sum_{i=1}^{k} c_i a_{(i)} = 0$, the vectors $\{a_{(i)} : i = 1, 2, \ldots, k\}$ are said to be **linearly dependent** or to constitute a **linearly dependent set**.

Remark 1.2. Notice that if a set of vectors is **linearly dependent**, this means that one or more such vectors can be expressed as a linear combination of the remaining vectors. On the other hand if the set is **linearly independent** this is not possible.

Remark 1.3. Notice, further, that if a set of n-dimensional (non-null) vectors $a_{(i)} \in \mathcal{F}$, $i = 1, 2, \ldots, k$, are **mutually orthogonal**, i.e. for any $i \neq j$ $a'_{(i)} a_{(j)} = 0$ then they are linearly **independent**. The proof of this is quite straightforward.

Suppose not; then there exist constants $c_i \in \mathcal{F}$, not all of which are zero such that

$$0 = \sum_{i=1}^{k} c_i a_{(i)}.$$

Pre-multiply sequentially by $a'_{(s)}$ to obtain

$$0 = c_s a'_{(s)} a_{(s)}, \quad s = 1, 2, \ldots, k.$$

Since for all s, $a'_{(s)} a_{(s)} > 0$ we have a contradiction.

1.3 Vector Spaces

First we give a formal definition and then apply it to the preceding discussion.

Definition 1.5. A nonempty collection of elements $\mathcal{V}$ is said to be a linear space (or a vector space, or a linear vector space) over the set (of real or complex numbers) $\mathcal{F}$, if and only if there exist two functions, $+$, called vector addition, and $\cdot$, called scalar multiplication, such that the following conditions hold for all $x, y, z \in \mathcal{V}$ and $c, d \in \mathcal{F}$:

   i. $x + y = y + x$, $x + y \in \mathcal{V}$;

   ii. $(x + y) + z = x + (y + z)$;

   iii. There exists a unique zero element in $\mathcal{V}$ denoted by 0, and termed the zero vector, such that for all $x \in \mathcal{V}$,

$$x + 0 = x;$$

   iv. Scalar multiplication is distributive over vector addition, i.e. for all $x, y \in \mathcal{V}$ and $c, d \in \mathcal{F}$,

$$c \cdot (x + y) = c \cdot x + c \cdot y, \quad (c + d) \cdot x = c \cdot x + d \cdot x, \quad \text{and } c \cdot x \in \mathcal{V};$$

   v. Scalar multiplication is associative, i.e. for all $c, d \in \mathcal{F}$ and $x \in \mathcal{V}$,

$$(cd) \cdot x = c \cdot (d \cdot x);$$

   vi. For the zero and unit elements of $\mathcal{F}$, we have, for all $x \in \mathcal{V}$,

$$0 \cdot x = 0 \text{ (the zero vector of iii)}, \quad 1 \cdot x = x.$$

The elements of $\mathcal{V}$ are often referred to as **vectors**.

Remark 1.4. The notation $\cdot$, indicating scalar multiplication, is often suppressed, and one simply writes $c(x + y) = cx + cy$, the context making clear that c is a scalar and x, y are vectors.

Example 1.1. Let $\mathcal{V}$ be the collection of ordered n-tuplets with elements in $\mathcal{F}$ considered above. The reader may readily verify that **over the set** $\mathcal{F}$ such n-tuplets satisfy conditions i through vi of Definition 1.5. Hence, they constitute a **linear vector space**. If $\mathcal{F} = R$, where R is the collection of real numbers, the resulting n-dimensional vector space is denoted by R^n. Thus, if

$$a = \begin{pmatrix} a_1 & a_2 & a_3 & \ldots, a_n \end{pmatrix}',$$

we may use the notation $a \in R^n$, to denote the fact that a is an element of the n-dimensional Euclidean (vector) space. The concept, however, is much wider than is indicated by this simple representation.

1.3.1   Basis of a Vector Space

Definition 1.6 (Span of a vector space). Let V_n denote a generic n-dimensional vector space over $\mathcal{F}$, and suppose

$$a_{(i)} \in V_n, \qquad i = 1, 2, \ldots, m, \ m \geq n.$$

If any vector in V_n, say b, can be written as

$$b = \sum_{i=1}^{m} c_i a_{(i)}, \quad c_i \in \mathcal{F},$$

we say that the set $\{a_{(i)} : i = 1, 2, \ldots, m\}$ **spans** the vector space V_n.

Definition 1.7. A **basis** for a vector space V_n is a span of the space with minimal dimension, i.e. a **minimal** set of linearly independent vectors that span V_n.

Example 1.2. For the vector space $V_n = R^n$ above, it is evident that the set

$$\{e_{\cdot i} : \quad i = 1, 2, \ldots, n\}$$

forms a **basis**, where $e_{\cdot i}$ is an n-dimensional (column) vector all of whose elements are zero save the ith, which is unity. Such vectors are typically called **unit vectors**. Notice further that this is an **orthonormal** set in the sense that such vectors are mutually orthogonal and their length is unity.

Remark 1.5. It is clear that if V_n is a vector space and

$$A = \{a_{(i)} \colon a_{(i)} \in V_n \; i = 1, 2, \ldots, m, \; m \geq n\}$$

is a subset that spans V_n then there exists a subset of A that forms a basis for V_n. Moreover, if $\{a_{(i)} \colon i = 1, 2, \ldots, k, \; k < m\}$ is a linearly independent subset of A we can choose a basis that contains it. This is done by noting that since A spans V_n then, if it is linearly independent, it is a basis and we have the result. If it is not, then we simply eliminate some of its vectors that can be expressed as linear combinations of the remaining vectors. Because the remaining subset is linearly independent, it can be made part of the basis.

A basis is not unique, but all bases for a given vector space contain the same number of vectors. This number is called the **dimension** of the vector space V_n and is denoted by

$$\dim(V_n).$$

Suppose $\dim(V_n) = n$. Then, it may be shown that **any** $n + i$ vectors in V_n are **linearly dependent** for $i \geq 1$, and that no set containing less than n vectors can span V_n.

1.4 Subspaces of a Vector Space

Let V_n be a vector space and P_n a subset of V_n in the sense that $b \in P_n$ implies that $b \in V_n$. If P_n is also a vector space, then it is said to be a **subspace of** V_n, and all discussion regarding spanning, basis sets, and dimension applies to P_n as well.

Finally, notice that if $\{a_{(i)} \colon i = 1, 2, \ldots, n\}$ is a basis for a vector space V_n, every vector in V_n, say b, is uniquely expressible in terms of this basis. Thus, suppose we have two representations, say

$$b = \sum_{i=1}^{n} b_i^{(1)} a_{(i)} = \sum_{i=1}^{n} b_i^{(2)} a_{(i)},$$

where $b_i^{(1)}, b_i^{(2)}$, $i = 1, 2, \ldots, m$ are appropriate sets of scalars. This implies

$$0 = \sum_{i=1}^{n} (b_i^{(1)} - b_i^{(2)}) a_{(i)}.$$

But a basis is a linearly independent set; hence, we conclude

$$b_i^{(1)} = b_i^{(2)}, \qquad i = 1, 2, \ldots, n,$$

which shows uniqueness of representation.

Example 1.3. In the next chapter, we introduce matrices more formally. For the moment, let us deal with the rectangular array

$$A = \begin{bmatrix} a_{11} & a_{12} \\ a_{21} & a_{22} \end{bmatrix},$$

with elements $a_{ij} \in \mathcal{F}$, which we shall call a matrix. If we agree to look upon this matrix as the vector[2]

$$a = \begin{pmatrix} a_{11} \\ a_{21} \\ a_{12} \\ a_{22} \end{pmatrix},$$

we may consider the matrix A to be an element of the vector space R^4, for the case where $\mathcal{F} = R$. Evidently, the collection of unit vectors

$$e_{\cdot 1} = (1,0,0,0)', \; e_{\cdot 2} = (0,1,0,0)', \; e_{\cdot 3} = (0,0,1,0)', \; e_{\cdot 4} = (0,0,0,1)'$$

is a basis for this space because for arbitrary a_{ij} we can always write

$$a = a_{11}e_{\cdot 1} + a_{21}e_{\cdot 2} + a_{12}e_{\cdot 3} + a_{22}e_{\cdot 4},$$

which is equivalent to the display of A above.

Now, what if we were to specify that, in the matrix above, we must always have $a_{12} = a_{21}$? Any such matrix is still representable by the 4-dimensional vector a, except that now the elements of a have to satisfy the condition $a_{12} = a_{21}$, i.e. the second and third elements **must be the same.** Thus, a satisfies $a \in R^4$, with the additional restriction that its third and second elements are identical, and it is clear that this must be a subset of R^4. Is this subset a subspace? Clearly, if a, b satisfy the condition that their second and third elements are the same, the same is true of $a + b$, as well as $c \cdot a$, for any $c \in R$.

What is the basis of this subspace? A little reflection will show that it is

$$e_{\cdot 1} = (1,0,0,0)', \; e_{\cdot 4} = (0,0,0,1)', \; e^* = (0,1,1,0)'.$$

These three vectors are **mutually orthogonal,** but not orthonormal; moreover, if A is the special matrix

$$A = \begin{bmatrix} a_{11} & \alpha \\ \alpha & a_{22} \end{bmatrix},$$

---

[2]This is an instance of the vectorization of a matrix, a topic we shall discuss at length at a later chapter.

the corresponding vector is $a = (a_{11}, \alpha, \alpha, a_{22})'$ and we have the unique representation

$$a = a_{11}e_{.1} + \alpha e^* + a_{22}e_{.4}.$$

Because the basis for this vector space has three elements, the dimension of the space is three. Thus, these special matrices constitute a 3-dimensional subspace of R^4.

Chapter 2

Matrix Algebra

2.1 Basic Definitions

Definition 2.1. Let $a_{ij} \in \mathcal{F}$, $i = 1, 2, \ldots, m$, $j = 1, 2, \ldots, n$, where $\mathcal{F}$ is a suitable space, such as the one-dimensional Euclidean or complex space. Then, the ordered rectangular array

$$A = \begin{bmatrix} a_{11} & a_{12} & \cdots & a_{1n} \\ a_{21} & a_{22} & \cdots & a_{2n} \\ \vdots & \vdots & & \vdots \\ a_{m1} & a_{m2} & \cdots & a_{mn} \end{bmatrix} = [a_{ij}]$$

is said to be a matrix of dimension $m \times n$.

Remark 2.1. Note that the first subscript locates the **row** in which the typical element lies, whereas the second subscript locates the **column**. For example, a_{ks} denotes the element lying in the kth row and sth column of the matrix A. When writing a matrix, we usually write its typical element as well as its dimension. Thus,

$$A = (a_{ij}), \qquad i = 1, 2, \ldots, m, \ j = 1, 2, \ldots, n,$$

denotes a matrix whose typical element is a_{ij} and which has m rows and n columns.

Convention 2.1. Occasionally, we have reason to refer to the columns or rows of the matrix individually. If A is a matrix we shall denote its jth column by $a_{.j}$, i.e.

$$a_{.j} = \begin{pmatrix} a_{1j} \\ a_{2j} \\ \vdots \\ a_{mj} \end{pmatrix},$$

P.J. Dhrymes, *Mathematics for Econometrics*,
DOI 10.1007/978-1-4614-8145-4_2, © The Author 2013

and its i th row by

$$a_{i\cdot} = (a_{i1}, a_{i2}, \ldots, a_{in}).$$

Definition 2.2. Let A be a matrix as in Definition 2.1. Its transpose, denoted by A', is defined to be the $n \times m$ matrix

$$A' = [a_{ji}], \qquad j = 1, 2, \ldots, n, \ i = 1, 2, \ldots, m,$$

i.e. it is obtained by interchanging rows and columns.

Definition 2.3. Let A be as in Definition 2.1. If $m = n$, A is said to be a **square matrix**.

Definition 2.4. If A is a square matrix, it is said to be **symmetric** if and only if

$$A' = A.$$

If A is a square matrix with, say, n rows and n columns, it is said to be a **diagonal matrix** if and only if

$$a_{ij} = 0, \qquad i \neq j.$$

In this case, it is denoted by

$$A = \mathrm{diag}(a_{11}, a_{22}, \ldots, a_{nn}).$$

Remark 2.2. If A is square matrix, then, evidently, it is not necessary to refer to the number of its rows and columns separately. If it has, say n rows and n columns, we say that A is of **dimension** (or **order**) n.

Definition 2.5. Let A be a square matrix of order n. It is said to be an **upper triangular matrix** if and only if

$$a_{ij} = 0, \qquad i > j.$$

It is said to be a **lower triangular matrix** if and only if

$$a_{ij} = 0, \qquad i < j.$$

Remark 2.3. As the terms imply, for a **lower triangular matrix** all elements above the main diagonal must be zero, while for an **upper triangular matrix** all elements below the main diagonal must be zero.

Definition 2.6. The identity matrix of order n, denoted by I_n,[1] is a diagonal matrix all of whose non-null elements are unity.

---

[1]In most of the literature, the subscript is typically omitted. In this volume we shall include it more often than not for the greater clarity it brings to the discussion.

Definition 2.7. The **null** matrix of dimension $m \times n$ is a matrix all of whose elements are null (zeros).

Definition 2.8. Let A be a square matrix of order n. It is said to be an **idempotent** matrix if and only if

$$AA = A.$$

Usually, but not necessarily, idempotent matrices encountered in econometrics are also **symmetric**.

2.2 Basic Operations

Let A, B be two $m \times n$ matrices with elements in $\mathcal{F}$, and let c be a scalar in $\mathcal{F}$. Then, we have:

   i. Scalar multiplication:

$$cA = [ca_{ij}].$$

  ii. Matrix addition:

$$A + B = [a_{ij} + b_{ij}].$$

Remark 2.4. Note that while scalar multiplication is defined for every matrix, matrix addition for A and B **is not defined unless both have the same dimensions**.

Let A be $m \times n$ and B be $q \times r$, both with elements in $\mathcal{F}$; then, we have:

  iii. Matrix multiplication:

$$AB = \left[\sum_{s=1}^{n} a_{is} b_{sj} \right] \quad \text{provided } n = q;$$

$$BA = \left[\sum_{k=1}^{r} b_{ik} a_{kj} \right] \quad \text{provided } r = m.$$

Remark 2.5. Notice that matrix multiplication is not defined for any arbitrary two matrices A, B. They must satisfy certain conditions of **dimensional conformability**. Notice further that if the product

$$AB$$

is defined, the product

$$BA$$

need not be defined, and if it is, it is not **generally** true that

$$AB = BA.$$

Remark 2.6. If two matrices are such that a given operation between them is defined, we say that they are **conformable** with respect to that operation. Thus, for example, if A is $m \times n$ and B is $n \times r$ we say that A and B are **conformable** with respect to the operation of right multiplication, i.e. multiplying A on the right by B. If A is $m \times n$ and B is $q \times m$ we shall say that A and B are **conformable** with respect to the operation of left multiplication, i.e. multiplying A on the left by B. Or if A and B are both $m \times n$ we shall say that A and B are **conformable** with respect to matrix addition. Because being precise is rather cumbersome, we often merely say that two matrices are conformable, and we let the context define precisely the sense in which conformability is to be understood.

An immediate consequence of the preceding definitions is

Proposition 2.1. Let A be $m \times n$, and B be $n \times r$. The jth column of

$$C = AB$$

is given by

$$c_{\cdot j} = \sum_{s=1}^{n} a_{\cdot s} b_{sj}, \qquad j = 1, 2, \ldots, r.$$

Proof: Obvious from the definition of matrix multiplication.

Proposition 2.2. Let A be $m \times n$, B be $n \times r$. The ith row of

$$C = AB$$

is given by

$$c_{i \cdot} = \sum_{q=1}^{n} a_{iq} b_{q \cdot}, \qquad i = 1, 2, \ldots, m.$$

Proof: Obvious from the definition of matrix multiplication.

Proposition 2.3. Let A, B be $m \times n$, and $n \times r$, respectively. Then,

$$C' = B'A',$$

where

$$C = AB.$$

Proof: The typical element of C is given by

$$c_{ij} = \sum_{s=1}^{n} a_{is} b_{sj}.$$

By definition, the typical (i, j) element of C', say c'_{ij}, is given by

$$c'_{ij} = c_{ji} = \sum_{s=1}^{n} a_{js} b_{si}.$$

But

$$a_{js} = a'_{sj}, \qquad b_{si} = b'_{is},$$

i.e. a_{js} is the (s, j) element of A', say a'_{sj}, and b_{si} is the (i, s) element of B', say b'_{is}. Consequently,

$$c'_{ij} = c_{ji} = \sum_{s=1}^{n} a_{js} b_{si} = \sum_{s=1}^{n} b'_{is} a'_{sj},$$

which shows that the (i, j) element of C' is the (i, j) element of $B'A'$.

q.e.d.

2.3 Rank and Inverse of a Matrix

Definition 2.9. Let A be $m \times n$. The **column rank** of A is the maximum number of linearly independent columns it contains. The **row rank** of A is the maximum number of linearly independent rows it contains.

Remark 2.7. It may be shown—but not here—that the row rank of A is **equal** to its column rank. Hence, the concept of rank is unambiguous, and we denote by

$$r(A)$$

the rank of A. Thus, if we are told that A is $m \times n$ we can immediately conclude that

$$r(A) \leq \min(m, n).$$

Definition 2.10. Let A be $m \times n$, $m \leq n$. We say that A is of **full rank** if and only if

$$r(A) = m.$$

Definition 2.11. Let A be a square matrix of order m. We say that A is **nonsingular** if and only if

$$r(A) = m.$$

Remark 2.8. An example of a nonsingular matrix is the diagonal matrix

$$A = \text{diag}(a_{11}, a_{22}, \ldots, a_{mm})$$

for which

$$a_{ii} \neq 0, \qquad i = 1, 2, \ldots, m.$$

We are now in a position to define a matrix operation that corresponds to division for scalars. For example, if $c \in \mathcal{F}$ and $c \neq 0$, we know that for any $a \in \mathcal{F}$

$$\frac{a}{c}$$

means the operation of defining

$$\frac{1}{c}$$

(the "inverse" of c) and multiplying that by a. The "inverse" of a scalar, say c, is another scalar, say b, such that

$$bc = cb = 1.$$

We have a similar operation for square matrices.

2.3.1   Matrix Inversion

Let A be a square matrix of order m. Its inverse, say B, is another square matrix of order m such that B, if it exists, is defined by the property

$$AB = BA = I_m$$

Definition 2.12. Let A be a square matrix of order m. If its inverse exists it is denoted by A^{-1}, and the matrix A is said to be **invertible**.

Remark 2.9. The terms **invertible, nonsingular,** and **of full rank** are synonymous for square matrices. This is made clear below.

Proposition 2.4. Let A be a square matrix of order m. Then A is invertible if and only if

$$r(A) = m.$$

Proof: Necessity: Suppose A is invertible; then there exists a square matrix B (of order m) such that

$$AB = I_m. \tag{2.1}$$

Let $c \neq 0$ be any m-element vector and note that Eq. (2.1) implies

$$ABc = c.$$

Since $c \neq 0$ we must have that

$$Ad = c, \qquad d = Bc \neq 0.$$

But this means that if c is any m-dimensional vector it can be expressed as a linear combination of the columns of A, which in turn means that the columns of A span the vector space V_m consisting of all m-dimensional vectors with elements in $\mathcal{F}$. Because the dimension of this space is m, it follows that the (m) columns of A are linearly independent; hence, its rank is m.

Sufficiency: Conversely, suppose that

$$r(A) = m.$$

Then, its columns form a basis for V_m. The unit vectors (see Chap. 1) $\{e_{\cdot i}: i = 1, 2, \ldots, m\}$ all belong to V_m. Thus, we can write

$$e_{\cdot i} = Ab_{\cdot i} = \sum_{s=1}^{m} a_{\cdot s} b_{si}, \qquad i = 1, 2, \ldots, m.$$

The matrix

$$B = [b_{si}]$$

has the property[2]

$$AB = I_m.$$

q.e.d.

Corollary 2.1. Let A be a square matrix of order m. If A is invertible then the following is true for its inverse B: B is of rank m and thus B also is invertible; the inverse of B is A.

Proof: Obvious from the definition of the inverse and the proposition.

It is useful here to introduce the following definition

Definition 2.13. Let A be $m \times n$. The **column space** of A, denoted by $C(A)$, is the set of m-dimensional (column) vectors

$$C(A) = \{\xi : \xi = Ax\},$$

where x is n-dimensional with elements in $\mathcal{F}$. Similarly, the **row space** of A, $R(A)$, is the set of n-dimensional (row) vectors

$$R(A) = \{\zeta : \zeta = yA\},$$

where y is a row vector of dimension m with elements in $\mathcal{F}$.

---

[2]Strictly speaking, we should also provide an argument based on the rows of A and on $BA = I_m$, but this is repetitious and is omitted for the sake of simplicity.

Remark 2.10. It is clear that the column space of A is a vector space and that it is **spanned** by the columns of A. Moreover, the dimension of this vector space is simply the rank of A, i.e. $\dim C(A) = \mathrm{r}(A)$. Similarly, the row space of A is a vector space spanned by its rows, and the dimension of this space is also equal to the rank of A because the row rank of A is equal to its column rank.

Definition 2.14. Let A be $m \times n$. The (column) **null space** of A, denoted by $N(A)$, is the set
$$N(A) = \{x: \quad Ax = 0\}.$$

Remark 2.11. A similar definition can be made for the (row) null space of A.

Definition 2.15. Let A be $m \times n$, and consider its null space $N(A)$. This is a vector space; its dimension is termed the **nullity** of A and is denoted by
$$n(A).$$

We now have an important relation between the column space and column null space of any matrix.

Proposition 2.5. Let A be $p \times q$. Then,
$$r(A) + n(A) = q.$$

Proof: Suppose the nullity of A is $n(A) = n \le q$, and let $\{\xi_i : \ i = 1, 2, \ldots, n\}$ be a basis for $N(A)$. Note that each ξ_i is a q-dimensional (column) vector with elements in $\mathcal{F}$. We can extend this to a basis for V_q, the vector space containing all q-dimensional vectors with elements in $\mathcal{F}$; thus, let
$$\{\xi_1, \xi_2, \ldots, \xi_n, \zeta_1, \zeta_2, \ldots, \zeta_{q-n}\}$$

be such a basis. If x is any q-dimensional vector, we can write, uniquely,
$$x = \sum_{i=1}^{n} c_i \xi_i + \sum_{j=1}^{q-n} f_j \zeta_j.$$

Now, define
$$y = Ax \in C(A)$$

and note that
$$y = \sum_{i=1}^{n} c_i A\xi_i + \sum_{j=1}^{q-n} f_j A\zeta_j = \sum_{j=1}^{q-n} f_j (A\zeta_j). \tag{2.2}$$

This is so since

$$A\xi_i = 0, \qquad i = 1, 2, \ldots, n,$$

owing to the fact that the ξ's are a basis for the null space of A.

But Eq. (2.2) means that the vectors

$$\{A\zeta_j : j = 1, 2, \ldots, q - n\}$$

span $C(A)$, since x and (hence) y are arbitrary. We show that these vectors are linearly independent, and hence a basis for $C(A)$. Suppose not. Then, there exist scalars, g_j, $j = 1, 2, \ldots, q - n$, not all of which are zero, such that

$$0 = \sum_{j=1}^{q-n} (A\zeta_j) g_j = A \left(\sum_{j=1}^{q-n} \zeta_j g_j \right). \tag{2.3}$$

Equation (2.3) implies that

$$\zeta = \sum_{j=1}^{q-n} \zeta_j g_j \tag{2.4}$$

lies in the null space of A, because it states $A\zeta = 0$. As such, $\zeta \in V_q$ and has a unique representation in terms of the basis of that vector space, say

$$\zeta = \sum_{i=1}^{n} d_i \xi_i + \sum_{j=1}^{q-n} k_j \zeta_j. \tag{2.5}$$

Moreover, since $\zeta \in N(A)$, we know that in Eq. (2.5)

$$k_j = 0, \qquad j = 1, 2, \ldots, q - n.$$

But Eqs. (2.5) and (2.4) give two dissimilar representations of ζ in terms of a single basis for V_q, which is a contradiction, unless

$$g_j = 0, \qquad j = 1, 2, \ldots, q - n,$$
$$d_i = 0, \qquad i = 1, 2, \ldots, n.$$

This shows that Eq. (2.3) can be satisfied only by null g_j, $j = 1, 2, \ldots, q - n$; hence, the set $\{A\zeta_j : j = 1, 2, \ldots, q - n\}$ is linearly independent and, consequently, a basis for $C(A)$. Therefore, since the dimension of $C(A) = r(A)$, by Remark 2.10, we have

$$\dim[C(A)] = r(A) = q - n.$$

q.e.d.

Another useful result is the following.

Proposition 2.6. Let A be $p \times q$, let B be a nonsingular matrix of order q, and put $D = AB$. Then

$$r(D) = r(A).$$

Proof: We shall show that $C(A) = C(D)$, which, by the discussion in the proof of Proposition 2.5, is equivalent to the claim of the proposition.

Suppose $y \in C(A)$. Then, there exists a vector $x \in V_q$ such that $y = Ax$. Since B is nonsingular, define the vector $\xi = B^{-1}x$. We note $D\xi = ABB^{-1}x = y$, which shows that

$$C(A) \subset C(D). \tag{2.6}$$

Conversely, suppose $z \in C(D)$. This means there exists a vector $\xi \in V_q$ such that $z = D\xi$. Define the vector $x = B\xi$ and note that

$$Ax = AB\xi = D\xi = z;$$

this means that $z \in C(A)$, which shows

$$C(D) \subset C(A). \tag{2.7}$$

But Eqs. (2.6) and (2.7) together imply $C(A) = C(D)$.

q.e.d.

Finally, we have

Proposition 2.7. Let A be $p \times q$ and B $q \times r$, and put

$$D = AB.$$

Then

$$r(D) \leq \min[r(A), r(B)].$$

Proof: Since $D = AB$, we note that if $x \in N(B)$ then $x \in N(D)$; hence, we conclude

$$N(B) \subset N(D),$$

and thus that

$$n(B) \leq n(D). \tag{2.8}$$

But from

$$r(D) + n(D) = r,$$
$$r(B) + n(B) = r,$$

we find, in view of Eq. (2.8),

$$r(D) \leq r(B). \tag{2.9}$$

Next, suppose that $y \in C(D)$. This means that there exists a vector, say, $x \in V_r$, such that $y = Dx$ or $y = ABx = A(Bx)$, so that $y \in C(A)$. But this means that

$$C(D) \subset C(A),$$

or that

$$r(D) \leq r(A). \tag{2.10}$$

Together Eqs. (2.9) and (2.10) imply

$$r(D) \leq \min[r(A), r(B)].$$

q.e.d.

Remark 2.12. The preceding results can be stated in the following useful form: multiplying two (and therefore any finite number of) matrices results in a matrix whose rank cannot exceed the rank of the lowest ranked factor. The product of nonsingular matrices is nonsingular. Multiplying a matrix by a nonsingular matrix does not change its rank.

2.4 Hermite Forms and Rank Factorization

We begin with a few elementary aspects of matrix operations.

Definition 2.16. Let A be $m \times n$; any one of the following operations is said to be an **elementary transformation** of A:

 i. Interchanging two rows (or columns);

 ii. Multiplying the elements of a row (or column) by a (nonzero) scalar c;

 iii. Multiplying the elements of a row (or column) by a (nonzero) scalar c and adding the result to another row (or column).

The operations above are said to be **elementary row (or column) operations**.

Remark 2.13. The matrix performing operation i is the matrix obtained from the identity matrix by interchanging the two rows (or columns) in question.

The matrix performing operation ii is obtained from the identity matrix by multiplying the corresponding row (or column) by the scalar c.

Finally, the matrix performing operation iii is obtained from the identity matrix as follows: if it is desired to add c times the k th row to the i th row of a given matrix A, simply insert the scalar c in the (i, k) position of the appropriate identity matrix and use the resulting matrix to multiply A on the left.

Such matrices are termed **elementary matrices**. An elementary **row** operation is performed on A by multiplying A on the left by the corresponding elementary matrix, E, i.e. EA.

An elementary **column** operation is performed (*mutatis mutandis*) by AE.

Example 2.1. Let

$$A = \begin{bmatrix} a_{11} & a_{12} & a_{13} \\ a_{21} & a_{22} & a_{23} \\ a_{31} & a_{32} & a_{33} \end{bmatrix}$$

and suppose we want to interchange the position of the first and third rows (columns). Define

$$E_1 = \begin{bmatrix} 0 & 0 & 1 \\ 0 & 1 & 0 \\ 1 & 0 & 0 \end{bmatrix}.$$

Then,

$$E_1 A = \begin{bmatrix} a_{31} & a_{32} & a_{33} \\ a_{21} & a_{22} & a_{23} \\ a_{11} & a_{12} & a_{13} \end{bmatrix} ; \qquad AE_1 = \begin{bmatrix} a_{13} & a_{12} & a_{11} \\ a_{23} & a_{22} & a_{21} \\ a_{33} & a_{32} & a_{31} \end{bmatrix}.$$

Suppose we wish to multiply the second row (column) of A by the scalar c. Define

$$E_2 = \begin{bmatrix} 1 & 0 & 0 \\ 0 & c & 0 \\ 0 & 0 & 1 \end{bmatrix}.$$

Then,

$$E_2 A = \begin{bmatrix} a_{11} & a_{12} & a_{13} \\ ca_{21} & ca_{22} & ca_{23} \\ a_{31} & a_{32} & a_{33} \end{bmatrix}, \qquad AE_2 = \begin{bmatrix} a_{11} & ca_{12} & a_{13} \\ a_{21} & ca_{22} & a_{23} \\ a_{31} & ca_{32} & a_{33} \end{bmatrix}.$$

Finally, suppose we wish to add c times the first row (column) to the third row (column). Define

$$E_3 = \begin{bmatrix} 1 & 0 & 0 \\ 0 & 1 & 0 \\ c & 0 & 1 \end{bmatrix}, \quad \text{and note that}$$

$$E_3 A = \begin{bmatrix} a_{11} & a_{12} & a_{13} \\ a_{21} & a_{22} & a_{23} \\ ca_{11} + a_{31} & ca_{12} + a_{32} & ca_{13} + a_{33} \end{bmatrix}$$

$$AE_3 = \begin{bmatrix} a_{11} + ca_{13} & a_{12} & a_{13} \\ a_{21} + ca_{23} & a_{22} & a_{23} \\ a_{31} + ca_{33} & a_{32} & a_{33} \end{bmatrix}.$$

The result below follows immediately.

Proposition 2.8. Every elementary transformation matrix is nonsingular, and its inverse is a matrix of the same type.

Proof: For matrices of type E_1 it is clear that $E_1 E_1 = I$. The inverse of a matrix of type E_2 is of the same form but with c replaced by $1/c$. Similarly, the inverse of a matrix of type E_3 is of the same form but with c replaced by $-c$.

q.e.d.

Definition 2.17. An $m \times n$ matrix C is said to be an (**upper**) **echelon matrix** if:

i. It can be partitioned:
$$C = \begin{pmatrix} C_1 \\ 0 \end{pmatrix},$$
where C_1 is $r \times n$ $(r \le n)$ and there is no row in C_1 consisting entirely of zeros;

ii. The first nonzero element appearing in each row of C_1 is unity and, if the first nonzero element in row i is c_{ij} then all other elements in column j are zero, i.e. $c_{ij} = 0$ for $j > i$;

iii. When the first nonzero element in the kth row of C_1 is c_{kj_k}, then $j_1 < j_2 < j_3 < \cdots < j_k$.

An immediate consequence of the definition is

Proposition 2.9. Let A be $m \times n$; there exists a nonsingular $(m \times m)$ matrix B such that
$$BA = C$$
and C is an (upper) echelon matrix.

Proof: Consider the first column of A, and suppose it contains a nonzero element (if not, consider the second column, etc.). Without loss of generality, we suppose this to be a_{11} (if not, simply interchange rows so that it does become the first element). Multiply the first row by $1/a_{11}$. This is accomplished through multiplication on the left by a matrix of type E_2. Next, multiply the first row of the resulting matrix by $-a_{s1}$ and add to the s th row. This is accomplished through multiplication on the left by a matrix of type E_3. Continuing in this fashion, we make all elements in the first column zero except the first, which is unity. Repeat this for the second column, third column, and, in general, for all other columns of A. In the end some rows may consist entirely of zeros. If they do not all occur at the end (of the rows of the matrix) interchange rows so that all zero rows occur at the end. This can be done through multiplication on the left by a matrix of type E_1. The resulting matrix is, thus, in upper echelon form, and has been obtained through multiplication on the left by a number of elementary matrices. Because the latter are nonsingular, we have

$$BA = C,$$

where B is nonsingular and C is in upper echelon form.

q.e.d.

Proposition 2.10. Let A be $m \times n$, and suppose it can be reduced to an upper echelon matrix

$$BA = \begin{pmatrix} C_1 \\ 0 \end{pmatrix} = C$$

such that C_1 is $r \times n$. Then,

$$\mathrm{r}(A) = r.$$

Proof: By construction, the rows of C_1 are linearly independent; thus,

$$\mathrm{r}(C_1) = \mathrm{r}(C) = r \quad \text{and} \quad \mathrm{r}(C) \leq \mathrm{r}(A).$$

We also have $A = B^{-1}C$. Hence

$$\mathrm{r}(A) \leq \mathrm{r}(C),$$

which shows

$$\mathrm{r}(A) = \mathrm{r}(C) = r.$$

q.e.d.

Definition 2.18. An $n \times n$ matrix H^* is said to be in (**upper**) **Hermite form** if and only if:

  i. H^* is (upper) triangular;

 ii. The elements along the main diagonal of H^* are either zero or one;

iii. If a main diagonal element of H^* is zero, all elements in the **row** in which the null diagonal element occurs are zero;

iv. If a main diagonal element of H^* is unity then all other elements in the **column** in which the unit element occurs are zero.

Definition 2.19. An $n \times m$ matrix H is said to be in (**upper**) **Hermite canonical** form if and only if

$$H = \begin{bmatrix} I & H_1 \\ 0 & 0 \end{bmatrix}.$$

Proposition 2.11. Every matrix in Hermite form can be put in Hermite canonical form by elementary row and column operations.

Proof: Let H^* be a matrix in Hermite form; by interchanging rows we can put all zero rows at the end so that for some nonsingular matrix B_1 we have

$$B_1 H^* = \begin{pmatrix} H_1^* \\ 0 \end{pmatrix},$$

where the first nonzero element in each row of H_1^* is unity and H_1^* contains no zero rows. By interchanging columns, we can place the (unit) first nonzero elements of the rows of H_1^* along the main diagonal, so that there exists a nonsingular matrix B_2 for which

$$B_1 H^* B_2 = \begin{bmatrix} I & H_1 \\ 0 & 0 \end{bmatrix}.$$

q.e.d.

Proposition 2.12. The rank of a matrix H^*, in Hermite form, is equal to the dimension of the identity block in its Hermite canonical form.

Proof: Obvious.

Proposition 2.13. Every (square) matrix H^* in Hermite form is idempotent, although it is obviously not necessarily symmetric.

Proof: We have to show that

$$H^* H^* = H^*.$$

Because H^* is **upper triangular**, we know that H^*H^* is also **upper triangular**, and thus we need only determine its (i,j) element for $i \leq j$. Now, the (i,j) element of H^*H^* is

$$(H^*H^*)_{ij} = \sum_{k=1}^{n} h_{ik}^* h_{kj}^* = \sum_{k=i}^{j} h_{ik}^* h_{kj}^*.$$

If $h_{ii}^* = 0$ then $h_{ij}^* = 0$ for all j; hence $(H^*H^*)_{ij} = 0$ for all j. If $h_{ii}^* = 1$ then

$$(H^*H^*)_{ij} = h_{ij}^* + h_{i,i+1}^* h_{i+1,j}^* + \cdots + h_{ij}^* h_{jj}^*.$$

Now, $h_{i+1,i+1}^*$ is either zero or one; if zero then $h_{i+1,j}^* = 0$ for all j, and hence the second term on the right side in the equation above is zero. If $h_{i+1,i+1}^* = 1$ then $h_{i,i+1}^* = 0$, so that again the second term is null. Similarly, if $h_{i+2,i+2}^* = 0$, then the third term is null; if $h_{i+2,i+2}^* = 1$ then $h_{i,i+2}^* = 0$, so that again the third term on the right side of the equation defining $(H^*H^*)_{ij}$ is zero. Finally, if $h_{jj}^* = 1$ then $h_{ij}^* = 0$, and if $h_{jj}^* = 0$ then again $h_{ij}^* = 0$. Consequently, it is always the case that

$$(H^*H^*)_{ij} = h_{ij}^*$$

and thus

$$H^*H^* = H^*.$$

q.e.d.

2.4.1 Rank Factorization

Proposition 2.14. Let A be $n \times n$ of rank $r \leq n$. There exist nonsingular matrices Q_1, Q_2 such that

$$Q_1^{-1} A Q_2^{-1} = \begin{bmatrix} I_r & 0 \\ 0 & 0 \end{bmatrix}.$$

Proof: By Proposition 2.9, there exists a nonsingular matrix Q_1 such that

$$Q_1^{-1} A = \begin{pmatrix} A_1^* \\ 0 \end{pmatrix},$$

i.e. $Q_1^{-1}A$ is an (upper) echelon matrix. By transposition of columns, we obtain

$$\begin{pmatrix} A_1^* \\ 0 \end{pmatrix} B_1 = \begin{bmatrix} I_r & A_1 \\ 0 & 0 \end{bmatrix},$$

which in upper Hermite canonical form.

By elementary column operations, we can eliminate A_1, i.e.

$$\begin{bmatrix} I_r & A_1 \\ 0 & 0 \end{bmatrix} B_2 = \begin{bmatrix} I_r & 0 \\ 0 & 0 \end{bmatrix}.$$

Take $Q_2^{-1} = B_1 B_2$, and note that we have

$$Q_1^{-1} A Q_2^{-1} = \begin{bmatrix} I_r & 0 \\ 0 & 0 \end{bmatrix}.$$

q.e.d.

Proposition 2.15 (Rank Factorization). Let A be $m \times n$ $(m \leq n)$ of rank $r \leq m$. There exists an $m \times r$ matrix C_1 of rank r and an $r \times n$ matrix C_2 of rank r such that

$$A = C_1 C_2.$$

Proof: Let

$$A_0 = \begin{pmatrix} A \\ 0 \end{pmatrix},$$

where A_0 is $n \times n$ of rank r. By Proposition 2.14, there exist nonsingular matrices Q_1, Q_2 such that

$$A_0 = Q_1 \begin{bmatrix} I_r & 0 \\ 0 & 0 \end{bmatrix} Q_2.$$

Partition

$$Q_1 = \begin{bmatrix} C_1 & C_{11} \\ C_{21} & C_{22} \end{bmatrix}, \qquad Q_2 = \begin{pmatrix} C_2 \\ C_* \end{pmatrix}$$

so that C_1 is $m \times r$ and C_2 is $r \times n$ (of rank r). Thus,

$$A_0 = \begin{pmatrix} C_1 C_2 \\ C_{21} C_2 \end{pmatrix};$$

hence,

$$A = C_1 C_2.$$

Since

$$r = r(A) \leq \min[r(C_1), r(C_2)] = \min[r(C_1), r],$$

we must have that

$$r(C_1) = r.$$

q.e.d.

Remark 2.14. Proposition 2.15 is the so-called **rank factorization theorem**.

2.5   Trace and Determinants

Associated with square matrices are two important scalar functions, the **trace** and the **determinant**.

Definition 2.20. Let A be a square matrix of order m. Its trace is denoted by $\mathrm{tr}(A)$ and is defined by

$$\mathrm{tr}(A) = \sum_{i=1}^{m} a_{ii}.$$

An immediate consequence of the definition is

Proposition 2.16. Let A, B be two square matrices of order m. Then,

$$\begin{aligned}
\mathrm{tr}(A + B) &= \mathrm{tr}(A) + \mathrm{tr}(B), \\
\mathrm{tr}(AB) &= \mathrm{tr}(BA).
\end{aligned}$$

Proof: By definition, the typical element of $A + B$ is $a_{ij} + b_{ij}$. Hence,

$$\mathrm{tr}(A + B) = \sum_{i=1}^{m}(a_{ii} + b_{ii}) = \sum_{i=1}^{m} a_{ii} + \sum_{i=1}^{m} b_{ii} = \mathrm{tr}(A) + \mathrm{tr}(B).$$

Similarly, the typical element of AB is

$$\sum_{k=1}^{m} a_{ik}b_{kj}.$$

Hence,

$$\mathrm{tr}(AB) = \sum_{i=1}^{m} \sum_{k=1}^{m} a_{ik}b_{ki}.$$

The typical element of BA is

$$\sum_{i=1}^{m} b_{ki}a_{ij}.$$

Thus,

$$\mathrm{tr}(BA) = \sum_{k=1}^{m} \sum_{i=1}^{m} b_{ki}a_{ik} = \sum_{i=1}^{m} \sum_{k=1}^{m} a_{ik}b_{ki},$$

which shows

$$\mathrm{tr}(AB) = \mathrm{tr}(BA).$$

q.e.d.

Definition 2.21. Let A be a square matrix of order m; its **determinant**, denoted by $|A|$ or by $\det A$, is given by

$$|A| = \sum(-1)^s a_{1j_1} a_{2j_2} \cdots a_{mj_m},$$

where $j_1, j_2, \ldots, j_m$ is a permutation of the numbers $1, 2, \ldots, m$, and s is zero or one depending on whether the number of transpositions required to restore $j_1, j_2, \ldots, j_m$ to the natural sequence $1, 2, 3, \ldots, m$ is even or odd; the sum is taken over all possible such permutations.

Example 2.2. Consider the matrix

$$A = \begin{bmatrix} a_{11} & a_{12} & a_{13} \\ a_{21} & a_{22} & a_{23} \\ a_{31} & a_{32} & a_{33} \end{bmatrix}.$$

According to the definition, its determinant is given by

$$\begin{aligned} |A| = \; & (-1)^{s_1} a_{11} a_{22} a_{33} + (-1)^{s_2} a_{11} a_{23} a_{32} \\ & + (-1)^{s_3} a_{12} a_{21} a_{33} + (-1)^{s_4} a_{12} a_{23} a_{31} \\ & + (-1)^{s_5} a_{13} a_{21} a_{32} + (-1)^{s_6} a_{13} a_{22} a_{31}. \end{aligned}$$

To determine s_1, we note that the second subscripts in the corresponding term are in natural order; hence $s_1 = 0$. For the second term, we note that one transposition restores the second subscripts to the natural order; hence $s_2 = 1$. For the third term, $s_3 = 1$. For the fourth term two transpositions are required; hence, $s_4 = 0$. For the fifth term two transpositions are required; hence, $s_5 = 0$. For the sixth term one transposition is required; hence, $s_6 = 1$.

Remark 2.15. It should be noted that although Definition 2.21 is stated with the **rows** in natural order, a completely equivalent definition is one in which the **columns** are in natural order. Thus, for example, we could just as well have defined

$$|A| = \sum(-1)^d a_{i_1 1} a_{i_2 2} \cdots a_{i_m m}$$

where d is zero or one, according as the number of transpositions required to restore $i_1, i_2, \ldots, i_m$ to the natural order $1, 2, 3, \ldots, m$ is even or odd.

Example 2.3. Consider the matrix A of Example 2.2 and obtain the determinant in accordance with Remark 2.15. Thus

$$\begin{aligned} |A| = \; & (-1)^{d_1} a_{11} a_{22} a_{33} + (-1)^{d_2} a_{11} a_{32} a_{23} \\ = \; & +(-1)^{d_3} a_{21} a_{12} a_{33} + (-1)^{d_4} a_{21} a_{32} a_{13} \\ & (-1)^{d_5} a_{31} a_{12} a_{23} + (-1)^{d_6} a_{31} a_{22} a_{13}. \end{aligned}$$

It is easily determined that $d_1 = 0$, $d_2 = 1$, $d_3 = 1$, $d_4 = 0$, $d_5 = 0$, $d_6 = 1$. Noting, in comparison with Example 2.2, that $s_1 = d_1$, $s_2 = d_2$, $s_3 = d_3$, $s_4 = d_5$, $s_5 = d_4$, $s_6 = d_6$, we see that we have exactly the same terms.

An immediate consequence of the definition is the following proposition.

Proposition 2.17. Let A be a square matrix of order m. Then

$$|A'| = |A|.$$

Proof: Obvious from Definition 2.21 and Remark 2.15.

Proposition 2.18. Let A be a square matrix of order m, and consider the matrix B that is obtained by interchanging the kth and rth rows of A $(k \leq r)$. Then,

$$|B| = -|A|.$$

Proof: By definition,

$$|B| = \sum (-1)^s b_{1j_1} b_{2j_2} \cdots b_{mj_m},$$

$$|A| = \sum (-1)^s a_{1j_1} a_{2j_2} \cdots a_{mj_m}. \tag{2.11}$$

However, each term in $|B|$ (except possibly for sign) is exactly the same as in $|A|$ but for the interchange of the kth and rth rows. Thus, for example, we can write

$$|B| = \sum (-1)^s a_{1j_1} \cdots a_{k-1j_{k-1}} a_{rj_k} a_{k+1j_{k+1}}$$

$$\cdots a_{r-1j_{r-1}} a_{kr_r} a_{r+1j_{r+1}} \cdots a_{mj_m}. \tag{2.12}$$

Now, if we restore to their natural order the first subscripts in Eq. (2.12), we will have an expression like the one for $|A|$ in Eq. (2.11), except that in Eq. (2.12) we would require an odd number of additional transpositions to restore the second subscripts to the natural order $1, 2, \ldots, m$. Hence, the sign of each term in Eq. (2.12) is exactly the opposite of the corresponding term in Eq. (2.11). Consequently,

$$|B| = -|A|.$$

q.e.d.

Proposition 2.19. Let A be a square matrix of order m, and suppose it has two identical rows. Then

$$|A| = 0.$$

Proof: Let B be the matrix obtained by interchanging the two identical rows. Then by Proposition 2.18,

$$|B| = -|A|. \tag{2.13}$$

Since these two rows are identical $B = A$, and thus

$$|B| = |A|. \tag{2.14}$$

But Eqs. (2.13) and (2.14) imply

$$|A| = 0.$$

q.e.d.

Proposition 2.20. Let A be a square matrix of order m, and suppose all elements in its r th row are zero. Then,

$$|A| = 0.$$

Proof: By the definition of a determinant, we have

$$|A| = \sum (-1)^s a_{1j_1} a_{2j_2} a_{3j_3} \cdots a_{mj_m},$$

and it is clear that every term above contains an element from the i th row, say a_{ij_i}. Hence, all terms vanish and thus

$$|A| = 0.$$

q.e.d.

Remark 2.16. It is clear that, in any of the propositions regarding determinants, we may substitute "column" for "row" without disturbing the conclusion. This is clearly demonstrated by Remark 2.15 and the example following. Thus, while most of the propositions are framed in terms of rows, an equivalent result would hold in terms of columns.

Proposition 2.21. Let A be a square matrix of order m. Let B be the matrix obtained when we multiply the i th row by a scalar k. Then

$$|B| = k|A|.$$

Proof: By Definition 2.21,

$$|B| = \sum (-1)^s b_{1j} b_{2j_1} \cdots b_{mj_m}$$
$$= k \sum (-1)^s a_{1j_m} a_{2j_m} \cdots a_{mj_m} = k|A|.$$

This is so because

$$b_{sj_s} = a_{sj_s} \quad \text{for } s \neq i,$$
$$= ka_{sj_s} \quad \text{for } s = i.$$

q.e.d.

Proposition 2.22. Let A be a square matrix of order m. Let B be the matrix obtained when to the r th row of A we add k times its s th row. Then,

$$|B| = |A|.$$

Proof: By Definition 2.21,

$$|B| = \sum(-1)^s b_{1j_1} b_{2j_2} \cdots b_{mj_m}$$
$$= \sum(-1)^s a_{1j_1} \cdots a_{r-1j_{r-1}} (a_{rj_r} + ka_{sj_r}) \cdots a_{mj_m}$$
$$= \sum(-1)^s a_{1j_1} \cdots a_{r-1j_{r-1}} a_{rj_r} \cdots a_{mj_n} + k\sum(-1)^s a_{1j_1}$$
$$\cdots a_{r-1j_{r-1}} a_{sj_r} \cdots a_{mj_m}.$$

The first term on the rightmost member of the equation above gives $|A|$, and the second term represents k times the determinant of a matrix having two identical rows. By Proposition 2.19, that determinant is zero. Hence,

$$|B| = |A|.$$

q.e.d.

Remark 2.17. It is evident, by a simple extension of the argument above, that if we add a linear combination of any number of the remaining rows (or columns) to the r th row (or column) of A, we do not affect the determinant of A.

Remark 2.18. While Definition 2.21 is intuitively illuminating and, indeed, leads rather easily to the derivation of certain important properties of the determinant, it is not particularly convenient for computational purposes. We give below a number of useful alternatives for evaluating determinants.

Definition 2.22. Let A be a square matrix of order m and let B_{ij} be the matrix obtained by deleting from A its i th row and j th column. The quantity

$$A_{ij} = (-1)^{i+j} |B_{ij}|$$

is said to be the **cofactor** of the element a_{ij} of A. The matrix B_{ij} is said to be an $(m-1)$-**order minor** of A.

Proposition 2.23. (Expansion by cofactors) Let A be a square matrix of order m. Then,

$$|A| = \sum_{j=1}^{m} a_{ij} A_{ij}, \qquad |A| = \sum_{i=1}^{m} a_{ij} A_{ij}.$$

Proof: For definiteness, we shall prove this for a specific value of i (for the expansion by cofactors in a given row). By the definition of a determinant, $|A| = \sum (-1)^s a_{1j_1} a_{2j_2} \cdots a_{mj_m}$. This can also be written more suggestively as follows:

$$|A| = a_{11} \sum (-1)^s a_{2j_2} \cdots a_{mj_m} + a_{12} \sum (-1)^s a_{2j_2} \cdots a_{mj_m}$$

$$+ \cdots + a_{1m} \sum (-1)^s a_{2j_2} \cdots a_{mj_m}$$

$$= a_{11} f_{11} + a_{12} f_{12} + a_{1m} f_{1m}, \tag{2.15}$$

where, for example, $f_{1r} = \sum (-1)^s a_{2j_2} \cdots a_{mj_m}$, and the numbers $j_2, j_3, \ldots, j_m$ represent some arrangement (permutation) of the integers $1, 2, \ldots, m$, excluding the integer $r \le m$. But it is clear that, except (possibly) for sign, f_{1r} is simply the determinant of an $(m-1)$-order minor of A obtained by deleting its first row and rth column. In that determinant, s would be zero or one depending on whether the number of transpositions required to restore $j_2, j_3, \ldots, j_m$ to the natural order $1, 2, \ldots, r-1$, $r+1, \ldots, m$ is even or odd. In Eq. (2.15), however, the corresponding s would be zero or one depending on whether the number of transpositions required to restore r, $j_2, j_3, \ldots, j_m$ to the natural order $1, 2, \ldots, r-1$, r, $r+1, \ldots, m$ is even or odd. But for $r > 1$, this would be exactly $r-1$ more than before, and would be exactly the same if $r = 1$. Thus,

$$f_{11} = |B_{11}|, \quad f_{12} = (-1)|B_{12}|, \quad f_{13} = (-1)^2 |B_{13}|, \quad \ldots, \quad f_{1m} = (-1)^{m-1} |B_{1m}|.$$

Noting that $A_{1j} = (-1)^{j+1} |B_{1j}|$, and $(-1)^{j+1} = (-1)^{j-1}$, we conclude that $f_{1j} = A_{1j}$. A similar argument can be made for the expansion along any row—and not merely the first—as well as expansion along any column.

q.e.d.

Remark 2.19. Expansion by cofactors is a very useful way of evaluating a determinant; it is also the one most commonly used in actual computations. For certain instances, however, another method—the Laplace expansion—is preferable. Its proof, however, is cumbersome and will be omitted.

Definition 2.23. Let A be a square matrix of order m. Let P be an n-order $(n < m)$ minor formed by the rows $i_1, i_2, \ldots, i_n$ and the columns $j_1, j_2, \ldots, j_n$ of A, and let Q be the $(m - n)$-order minor formed by taking the remaining rows and columns. Then Q is said to be the **complementary minor of** P (and conversely P is said to be the complementary minor of Q.) Moreover,

$$M = [(-1)^{\sum_{i=1}^{n}(i_r + j_r)}]|Q|$$

is said to be the **complementary cofactor of** P.

Proposition 2.24 (Laplace expansion). Let A be a square matrix of order m. Let $P(i_1, i_2, \ldots, i_n \mid j_1, j_2, \ldots, j_n)$ be an n-order minor of A formed by rows $i_1, i_2, \ldots, i_n$ and columns $j_1, j_2, \ldots, j_n$, $n < m$. Let M be its associated complementary cofactor. Then

$$|A| = \sum_{j_1 < j_2 < \cdots < j_n} |P(i_1, i_2, \ldots, i_n \mid j_1, j_2, \ldots, j_n)|M,$$

where the sum is taken over all possible choices of n columns of P, the number of which is in fact

$$\binom{m}{n};$$

similarly,

$$|A| = \sum_{i_1 < i_2 < \cdots < i_n} |P(i_1, i_2, \ldots, i_n \mid j_1, j_2, \ldots, j_n)|M,$$

the sum now chosen over all $\binom{m}{n}$ ways in which n of the rows of P may be chosen.

Proof: The proof, while conceptually simple, is rather cumbersome and not particularly instructive. The interested reader is referred to Hadley (1961).

Remark 2.20. The first representation in Proposition 2.24 refers to an expansion by n columns, the second to an expansion by n rows. It is simple to see that this method is a generalization of the method of expansion by cofactors. The usefulness of the Laplace expansion lies chiefly in the evaluation of determinants of partitioned matrices, a fact that will become apparent in later discussion.

In dealing with determinants, it is useful to establish rules for evaluating such quantities for sums or products of matrices. We have

Proposition 2.25. Let A, B be two square matrices of order m. Then, in general,

$$|A + B| \neq |A| + |B|,$$

in the sense that any of the following three relations is possible:

$$|A + B| = |A| + |B|,$$
$$|A + B| > |A| + |B|,$$
$$|A + B| < |A| + |B|.$$

Proof: We establish the validity of the proposition by a number of examples. Thus, for

$$A = \begin{bmatrix} 2 & 0 \\ 0 & 3 \end{bmatrix}, \qquad B = \begin{bmatrix} -\frac{2}{3} & 0 \\ 0 & 1 \end{bmatrix},$$

we have $|A| = 6$, $|B| = -\frac{2}{3}$, $|A + B| = 5\frac{1}{3}$, and we see that

$$|A + B| = |A| + |B|.$$

For the matrices

$$A = \begin{bmatrix} 1 & 0 \\ 0 & 1 \end{bmatrix}, \qquad B = \begin{bmatrix} 1 & 0 \\ 0 & 1 \end{bmatrix},$$

we find $|A| = 1$, $|B| = 1$, $|A + B| = 4$. Thus, $|A + B| > |A| + |B|$.

For the matrices

$$A = \begin{bmatrix} 1 & 0 \\ 0 & 1 \end{bmatrix}, \qquad B = \begin{bmatrix} -1 & 0 \\ 0 & -1 \end{bmatrix},$$

we find $|A| = 1$, $|B| = 1$, $|A + B| = 0$. Thus, $|A + B| < |A| + |B|$.

q.e.d.

Proposition 2.26. Let A, B be two square matrices of order m. Then,

$$|AB| = |A| |B|.$$

Proof: Define the $2m \times 2m$ matrix

$$C = \begin{bmatrix} A & 0 \\ -I & B \end{bmatrix}.$$

Multiply the last m rows of C on the left by A (i.e. take m linear combinations of such rows) and add them to the first rows. The resulting matrix is

$$C^* = \begin{bmatrix} 0 & AB \\ -I & B \end{bmatrix}.$$

By Proposition 2.22 and Remark 2.17, we have

$$|C| = |C^*|. \tag{2.16}$$

Expand $|C^*|$ by the method of Proposition 2.24 and note that, using m-order minors involving the last m rows, their associated complementary cofactors will vanish (since they involve the determinant of a matrix containing a zero column), except for the one corresponding to $-I$. The complementary cofactor for that minor is

$$[(-1)^{\sum_{i=1}^{m}(i+m+i)}]|AB| = (-1)^{m^2+m^2+m}|AB|.$$

Moreover,

$$|-I| = (-1)^m.$$

Hence,

$$|C^*| = (-1)^{2m^2+2m}|AB| = |AB|. \qquad (2.17)$$

Similarly, expand $|C|$ by the same method using m-order minors involving the first m rows. Notice now that all m-order minors involving the first m rows of C have a zero determinant except for the one corresponding to A, whose determinant is, evidently, $|A|$. Its associated complementary cofactor is

$$[(-1)^{\sum_{i=1}^{m}(i+i)}]|B| = (-1)^{2[m(m+1)/2]}|B| = |B|.$$

Hence, we have $|C| = |A||B|$. But this result, together with Eqs. (2.16) and (2.17), implies $|AB| = |A||B|$.

q.e.d.

Corollary 2.2. Let A be an invertible matrix of order m. Then,

$$|A^{-1}| = \frac{1}{|A|}.$$

Proof: By definition, $|AA^{-1}| = |I| = 1$ —the last equality following immediately from the fundamental definition of the determinant. Since $|AA^{-1}| = |A||A^{-1}|$, we have $|A^{-1}| = |A|^{-1}$.

q.e.d.

We conclude this section by introducing

Definition 2.24. Let A be a square matrix of order m. Let A_{ij} be the cofactor of the (i,j) element of A, a_{ij}, and define

$$B = (A_{ij}), \qquad i,j = 1,2,\ldots,m.$$

The **adjoint** of A, denoted by adj A, is defined by

$$\text{adj } A = B'.$$

2.6 Computation of the Inverse

We defined earlier the inverse of a matrix, say A, to be another matrix, say B, having the properties $AB = BA = I$.

Although this describes the essential property of the inverse, it does not provide a useful way to determine the elements of B. In the preceding section, we have laid the foundation for providing a practicable method for determining the elements of the inverse of a given matrix. To this end we have the following proposition.

Proposition 2.27. Let A be an invertible square matrix of order m. Its inverse, denoted by A^{-1}, is given by

$$A^{-1} = \frac{\text{adj } A}{|A|}.$$

Proof: In the standard notation for the inverse, denote the (i, j) element of A^{-1} by a^{ij}; then, the proposition asserts that

$$a^{ij} = \frac{A_{ji}}{|A|},$$

where A_{ji} is the cofactor of the element in the jth row and ith column of A. Let us now verify the validity of the assertion by determining the typical element of AA^{-1}. It is given by

$$\sum_{k=1}^{m} a_{ik} a^{kj} = \frac{1}{|A|} \sum_{k=1}^{m} a_{ik} A_{jk}. \tag{2.18}$$

Now for $i = j$ we have the expansion by cofactors along the ith row of A. Hence all diagonal elements of AA^{-1} are unity. For $i \neq j$, we may evaluate the quantity in Eq. (2.18) as follows. Strike out the jth row of A and replace it by the ith row. The resulting matrix has two identical rows and as such its determinant is zero. Now, expand by cofactors along the jth row. The cofactors of such elements are plainly A_{jk} because the other rows of A have not been disturbed. Thus, expanding by cofactors along the jth row we conclude $\sum_{k=1}^{m} a_{ik} A_{jk} = 0$. This is so because above we have a representation of the determinant of a matrix with two identical rows. Thus, $AA^{-1} = I$. Similarly, consider the typical element of $A^{-1}A$, namely

$$\sum_{k=1}^{m} a^{ik} a_{kj} = \frac{1}{|A|} \sum_{k=1}^{m} a_{kj} A_{ki}. \tag{2.19}$$

Again for $j = i$ we have, in the right-hand summation, the determinant of A evaluated by an expansion along the ith column. Hence, all diagonal elements of $A^{-1}A$ are unity. For $i \neq j$, consider the matrix obtained when we strike out the ith column of A and replace it by its jth column. The resulting matrix has two identical columns and hence its determinant is zero. Evaluating its determinant by expansion along the ith column, we note that the cofactors are given by A_{ki} because the other columns of A have not been disturbed. But then we have

$$\sum_{k=1}^{m} a_{kj} A_{ki} = 0, \qquad i \neq j,$$

and thus we conclude that $A^{-1}A = I_m$.

q.e.d.

Proposition 2.28. Let A, B be two invertible matrices of order m. Then, $(AB)^{-1} = B^{-1}A^{-1}$.

Proof: We verify

$$(AB)(B^{-1}A^{-1}) = A(BB^{-1})A^{-1} = AA^{-1} = I_m,$$
$$(B^{-1}A^{-1})(AB) = B^{-1}A^{-1}AB = B^{-1}B = I_m.$$

q.e.d.

Remark 2.21. For any two conformable and invertible matrices, A, B, we have

$$(A + B)^{-1} \neq A^{-1} + B^{-1},$$

and, indeed, $A + B$ need not be invertible. For example, suppose $B = -A$. Then, even though A^{-1}, B^{-1} exist, $A + B = 0$, which is evidently not invertible since its determinant is zero.

2.7   Partitioned Matrices

Frequently, we find it convenient to deal with partitioned matrices. In this section, we derive certain useful results that will facilitate operations with such matrices. Let A be $m \times n$ and write

$$A = \begin{bmatrix} A_{11} & A_{12} \\ A_{21} & A_{22} \end{bmatrix},$$

where A_{11} is $m_1 \times n_1$, A_{22} is $m_2 \times n_2$, A_{12} is $m_1 \times n_2$, and A_{21} is $m_2 \times n_1$, where ($m_1 + m_2 = m$, and $n_1 + n_2 = n$). The above is said to be a **partition** of the matrix A.

Now, let B be $m \times n$, and partition it conformably with A, i.e. put

$$B = \begin{bmatrix} B_{11} & B_{12} \\ B_{21} & B_{22} \end{bmatrix},$$

where B_{11} is $m_1 \times n_1$, B_{22} is $m_2 \times n_2$, and so on.

Addition of (conformably) partitioned matrices is defined by

$$A + B = \begin{bmatrix} A_{11} + B_{11} & A_{12} + B_{12} \\ A_{21} + B_{21} & A_{22} + B_{22} \end{bmatrix}.$$

If A is $m \times n$, C is $n \times q$, and A is partitioned as above, let

$$C = \begin{bmatrix} C_{11} & C_{12} \\ C_{21} & C_{22} \end{bmatrix},$$

where C_{11} is $n_1 \times q_1$, C_{22} is $n_2 \times q_2$, and so on.

Multiplication of two (conformably) partitioned matrices is defined by

$$AC = \begin{bmatrix} A_{11} & A_{12} \\ A_{21} & A_{22} \end{bmatrix} \begin{bmatrix} C_{11} & C_{12} \\ C_{21} & C_{22} \end{bmatrix}$$

$$= \begin{bmatrix} A_{11}C_{11} + A_{12}C_{21} & A_{11}C_{12} + A_{12}C_{22} \\ A_{21}C_{11} + A_{22}C_{21} & A_{21}C_{12} + A_{22}C_{22} \end{bmatrix}.$$

In general, and for either matrix addition or matrix multiplication, readers will not commit an error if, upon (conformably) partitioning two matrices, they proceed to regard the partition blocks as ordinary (scalar) elements and apply the usual rules except for division. Thus, for example, consider

$$A = \begin{bmatrix} A_{11} & A_{12} & \cdots & A_{1s} \\ A_{21} & A_{22} & \cdots & A_{2s} \\ \vdots & \vdots & & \vdots \\ A_{s1} & A_{s2} & \cdots & A_{ss} \end{bmatrix},$$

where A_{ij} is $m_i \times n_j$, the matrix A is $m \times n$, and

$$\sum_{i=1}^{s} m_i = m, \qquad \sum_{j=1}^{s} n_j = n.$$

Similarly, consider

$$B = \begin{bmatrix} B_{11} & B_{12} & \cdots & B_{1s} \\ B_{21} & B_{22} & \cdots & B_{2s} \\ \vdots & \vdots & & \vdots \\ B_{s1} & B_{s2} & \cdots & B_{ss} \end{bmatrix},$$

where B_{ij} is $m_i \times n_j$ as above. Then A and B are conformably partitioned with respect to matrix addition and their sum is simply

$$A + B = \begin{bmatrix} A_{11} + B_{11} & \cdots & A_{1s} + B_{1s} \\ A_{21} + B_{21} & \cdots & A_{2s} + B_{2s} \\ \vdots & & \vdots \\ A_{s1} + B_{s1} & \cdots & A_{ss} + B_{ss} \end{bmatrix},$$

If, instead of being $m \times n$, B is $n \times q$ and its partition blocks B_{ij} are $n_i \times q_j$ matrices such that

$$\sum_{i=1}^{s} n_i = n, \qquad \sum_{j=1}^{s} q_j = q,$$

then A and B are conformably partitioned with respect to multiplication, and their product is given by

$$AB = \begin{bmatrix} \sum_{r=1}^{s} A_{1r} B_{r1} & \cdots & \sum_{r=1}^{s} A_{1r} B_{rs} \\ \sum_{r=1}^{s} A_{2r} B_{r1} & \cdots & \sum_{r=1}^{s} A_{2r} B_{rs} \\ \vdots & & \vdots \\ \sum_{r=1}^{s} A_{sr} B_{r1} & \cdots & \sum_{r=1}^{s} A_{sr} B_{rs} \end{bmatrix},$$

where the (i, j) block of AB, namely

$$\sum_{r=1}^{s} A_{ir} B_{rj},$$

is a matrix of dimension $m_i \times q_j$.

For inverses and determinants of partitioned matrices, we may prove certain useful results.

Proposition 2.29. Let A be a square matrix of order m. Partition

$$A = \begin{bmatrix} A_{11} & A_{12} \\ A_{21} & A_{22} \end{bmatrix}$$

and let A_{ij} be $m_i \times m_j$, $i, j = 1, 2$, $m_1 + m_2 = m$. Also, let

$$A_{21} = 0.$$

Then

$$|A| = |A_{11}| \, |A_{22}|.$$

Proof: This follows immediately from the Laplace expansion by noting that if we expand along the last m_2 rows the only $m_2 \times m_2$ minor with non-vanishing determinant is A_{22}. Its complementary cofactor is

$$[(-1)^{\sum_{i=m_1+1}^{m_1+m_2}(i+i)}]|A_{11}| = |A_{11}|.$$

Consequently,

$$|A| = |A_{11}||A_{22}|.$$

q.e.d.

Corollary 2.3. If, instead, we had assumed

$$A_{12} = 0,$$

then

$$|A| = |A_{11}||A_{22}|.$$

Proof: Obvious from the preceding.

Definition 2.25. A matrix of the form

$$A = \begin{bmatrix} A_{11} & A_{12} \\ 0 & A_{22} \end{bmatrix}$$

(as in Proposition 2.29) is said to be an **upper block triangular** matrix. A matrix of the form

$$A = \begin{bmatrix} A_{11} & 0 \\ A_{21} & A_{22} \end{bmatrix}$$

is said to be a **lower block triangular** matrix.

Definition 2.26. Let A be as in Proposition 2.29, but suppose

$$A_{12} = 0, \qquad A_{21} = 0,$$

i.e. A is of the form

$$A = \begin{bmatrix} A_{11} & 0 \\ 0 & A_{22} \end{bmatrix}.$$

Then, A is said to be a **block diagonal** matrix and is denoted by

$$A = \text{diag}(A_{11}, A_{22}).$$

Corollary 2.4. Let A be a block diagonal matrix as above. Then,

$$|A| = |A_{11}||A_{22}|.$$

Proof: Obvious.

Remark 2.22. Note that, in the definition of block triangular (or block diagonal) matrices, the blocks A_{11}, A_{22} need not be triangular (or diagonal) matrices.

Proposition 2.30. Let A be a partitioned square matrix of order m,

$$A = \begin{bmatrix} A_{11} & A_{12} \\ A_{21} & A_{22} \end{bmatrix},$$

where the A_{ii} are nonsingular square matrices of order m_i, $i = 1, 2,$, $m_1 + m_2 = m$. Then,

$$|A| = |A_{22}| \, |A_{11} - A_{12}A_{22}^{-1}A_{21}|, \quad \text{and} \quad |A| = |A_{11}| \, |A_{22} - A_{21}A_{11}^{-1}A_{12}|.$$

Proof: Consider the matrix

$$A_* = \begin{bmatrix} I_{m_1} & -A_{12}A_{22}^{-1} \\ 0 & I_{m_2} \end{bmatrix} A = \begin{bmatrix} A_{11} - A_{12}A_{22}^{-1}A_{21} & 0 \\ A_{21} & A_{22} \end{bmatrix}.$$

By Proposition 2.29 and Corollary 2.3,

$$\det \begin{bmatrix} I_{m_1} & -A_{12}A_{22}^{-1} \\ 0 & I_{m_2} \end{bmatrix} = 1;$$

thus, we conclude $|A_*| = |A|$. Again by Proposition 2.29, the determinant of A_* may be evaluated as

$$|A_*| = |A_{22}| \, |A_{11} - A_{12}A_{22}^{-1}A_{21}|.$$

Hence, we conclude

$$|A| = |A_{22}| \, |A_{11} - A_{12}A_{22}^{-1}A_{21}|.$$

Similarly, consider

$$A^* = \begin{bmatrix} I_{m_1} & 0 \\ -A_{21}A_{11}^{-1} & I_{m_2} \end{bmatrix} A = \begin{bmatrix} A_{11} & A_{12} \\ 0 & A_{22} - A_{21}A_{11}^{-1}A_{12} \end{bmatrix}$$

and thus conclude

$$|A| = |A_{11}| \, |A_{22} - A_{21}A_{11}^{-1}A_{12}|.$$

q.e.d.

Next, we turn to the determination of inverses of partitioned matrices. We have

Proposition 2.31. Let A be a square nonsingular matrix of order m, and partition

$$A = \begin{bmatrix} A_{11} & A_{12} \\ A_{21} & A_{22} \end{bmatrix}$$

such that the A_{ii}, $i = 1, 2$, are nonsingular matrices of order m_i, $i = 1, 2$, respectively $(m_1 + m_2 = m)$. Then,

$$B = A^{-1} = \begin{bmatrix} B_{11} & B_{12} \\ B_{21} & B_{22} \end{bmatrix}$$

where

$$B_{11} = (A_{11} - A_{12}A_{22}^{-1}A_{21})^{-1}, \quad B_{12} = -A_{11}^{-1}A_{12}(A_{22} - A_{21}A_{11}^{-1}A_{12})^{-1},$$

$$B_{21} = -A_{22}^{-1}A_{21}(A_{11} - A_{12}A_{22}^{-1}A_{21})^{-1}, \quad B_{22} = (A_{22} - A_{21}A_{11}^{-1}A_{12})^{-1}.$$

Proof: By definition of the inverse B, we have

$$AB = \begin{bmatrix} A_{11}B_{11} + A_{12}B_{21} & A_{11}B_{12} + A_{12}B_{22} \\ A_{21}B_{11} + A_{22}B_{21} & A_{21}B_{12} + A_{22}B_{22} \end{bmatrix} = \begin{bmatrix} I & 0 \\ 0 & I \end{bmatrix},$$

which implies

$$A_{11}B_{11} + A_{12}B_{21} = I_{m_1}, \qquad A_{11}B_{12} + A_{12}B_{22} = 0,$$
$$A_{21}B_{11} + A_{22}B_{21} = 0, \qquad A_{21}B_{12} + A_{22}B_{22} = I_{m_2}.$$

Solving these equations by substitution, we have the proposition.

q.e.d.

The result above may be utilized to obtain the inverse of certain types of matrices that occur frequently in econometrics.

Proposition 2.32. Let A be $m \times n$, B be $n \times m$, and suppose $I_m + AB$, $I_n + BA$ are nonsingular matrices. Then,

$$(I_m + AB)^{-1} = I_m - A(I_n + BA)^{-1}B.$$

Proof: Observe that

$$\begin{bmatrix} I_n & -B \\ 0 & I_m \end{bmatrix} \begin{bmatrix} I_n & B \\ -A & I_m \end{bmatrix} = \begin{bmatrix} I_n + BA & 0 \\ -A & I_m \end{bmatrix}.$$

Consequently,

$$\begin{bmatrix} (I_n + BA)^{-1} & 0 \\ A(I_n + BA)^{-1} & I_m \end{bmatrix} = \begin{bmatrix} (I_n + BA)^{-1} & -(I_n + BA)^{-1}B \\ A(I_n + BA)^{-1} & (I_m + AB)^{-1} \end{bmatrix} \begin{bmatrix} I_n & B \\ 0 & I_m \end{bmatrix},$$

which implies, in particular, $(I_m + AB)^{-1} + A(I_n + BA)^{-1}B = I_m$, which further implies $(I_m + AB)^{-1} = I_m - A(I_n + BA)^{-1}B$.

q.e.d.

Corollary 2.5. Let C, D be square nonsingular matrices of order m and n, respectively; let X be $m \times n$, Y be $n \times m$, and suppose the nonsingularity conditions of Proposition 2.32 hold. Then,

$$(C + XDY)^{-1} = C^{-1} - C^{-1}X(D^{-1} + YC^{-1}X)^{-1}YC^{-1}.$$

Proof: Note that

$$(C + XDY)^{-1} = [C(I_m + C^{-1}XDY)]^{-1} = (I_m + C^{-1}XDY)^{-1}C^{-1}.$$

Let $C^{-1}X$, DY be, respectively, the matrices A, B of Proposition 2.32. Then,

$$\begin{aligned}(I_m + C^{-1}XDY)^{-1} &= I_m - C^{-1}X(I_n + DYC^{-1}X)^{-1}DY \\ &= I_m - C^{-1}X(D^{-1} + YC^{-1}X)^{-1}Y.\end{aligned}$$

Consequently,

$$\begin{aligned}(C + XDY)^{-1} &= (I_m + C^{-1}XDY)^{-1}C^{-1} \\ &= C^{-1} - C^{-1}X(D^{-1} + YC^{-1}X)^{-1}YC^{-1}.\end{aligned}$$

q.e.d.

Remark 2.23. A certain special case occurs sufficiently frequently in econometrics to deserve special notice. Precisely, let $n = 1$ so that D is now a **scalar**, say d. Let x, y be two m-element column vectors, so that with

$$n = 1, \quad X = x, \quad Y = y', \quad D = d,$$

the result of the corollary becomes

$$[C + dxy']^{-1} = C^{-1} - \alpha C^{-1}xy'C^{-1}, \quad \alpha = \frac{d}{1 + dy'C^{-1}x}.$$

The determinant of such matrices may also be expressed in relatively simple form. In order to do so we present a very useful result on determinants, which we will rederive later as a by-product of more general considerations.

Proposition 2.33. Let A be $m \times n$, and B be $n \times m$; then,

$$|I_m + AB| = |I_n + BA|.$$

Proof: Note that

$$\begin{bmatrix} I_m & A \\ -B & I_n \end{bmatrix} \begin{bmatrix} I_m & 0 \\ B & I_n \end{bmatrix} = \begin{bmatrix} I_m + AB & A \\ 0 & I_n \end{bmatrix},$$

$$\begin{bmatrix} I_m & 0 \\ B & I_n \end{bmatrix} \begin{bmatrix} I_m & A \\ -B & I_n \end{bmatrix} = \begin{bmatrix} I_m & A \\ 0 & I_n + BA \end{bmatrix}.$$

Using Proposition 2.29, and the results above, we conclude

$$|I_n + BA| = \det \begin{bmatrix} I_m & A \\ -B & I_n \end{bmatrix} = |I_m + AB|.$$

q.e.d.

We immediately have

Corollary 2.6. Let C, D, X, and Y be as in Corollary 2.5. Then,

$$|C + XDY| = |C|\,|D|\,|D^{-1} + YC^{-1}X|.$$

Proof: Since $C + XDY = C(I_m + C^{-1}XDY)$, we find, by Proposition 2.26, $|C + XDY| = |C|\,|I_m + C^{-1}XDY|$. By Proposition 2.33,

$$|I_m + C^{-1}XDY| = |I_m + DYC^{-1}X| = |D|\,|D^{-1} + YC^{-1}X|$$

and consequently $|C + XDY| = |C|\,|D|\,|D^{-1} + YC^{-1}X|$.

q.e.d.

Remark 2.24. Again, the special case when $n = 1$ and thus D is a **scalar**, say, d, deserves special mention. Thus, let x, y be m-element column vectors and

$$D = d, \quad X = x, \quad Y = y'.$$

The result of Corollary 2.6, for the special case $n = 1$, is rendered as

$$|C + dxy'| = |C|(1 + dy'C^{-1}x).$$

Remark 2.25. In view of Proposition 2.33, we see that in the statement of Proposition 2.32, it is not necessary to explicitly assume that both

$$I_m + AB \quad \text{and} \quad I_n + BA$$

are nonsingular, because if one is, the other must be as well.

In the discussion above we examined the question of obtaining the inverse of a partitioned matrix

$$A = \begin{bmatrix} A_{11} & A_{12} \\ A_{21} & A_{22} \end{bmatrix},$$

where A_{11} and A_{22} were assumed to be nonsingular.

In many instances, the situation arises that such matrices are singular, while A is nonsingular. In this context Proposition 2.31 is inapplicable; instead, the following proposition is applicable.

Proposition 2.31a. Let A be a nonsingular matrix, and partition it as

$$A = \begin{bmatrix} A_{11} & A_{12} \\ A_{21} & A_{22} \end{bmatrix}.$$

Let A_{11}, A_{22} be singular, and suppose the inverses

$$V_{11} = (A_{11} + A_{12}A_{21})^{-1},$$
$$V_{22} = (A_{22} - A_{21}V_{11}A_{12} - A_{21}V_{11}A_{12}A_{22})^{-1}$$

exist. Then,

$$A^{-1} = \begin{bmatrix} B_{11} & B_{12} \\ B_{21} & B_{22} \end{bmatrix},$$

where

$$B_{11} = V_{11} + V_{11}A_{12}(I + A_{22})V_{22}A_{21}V_{11},$$

$$B_{12} = V_{11}A_{12} + V_{11}A_{12}(I + A_{22})V_{22}A_{21}V_{11}A_{12} - V_{11}A_{12}(I + A_{22})V_{22}$$

$$B_{21} = -V_{22}A_{21}V_{11},$$

$$B_{22} = V_{22} - V_{22}A_{21}V_{11}A_{12}.$$

Proof: Obvious by direct verification.

Remark 2.26. One arrives at the result of Proposition 2.31a by multiplying A, on the left, first by the matrix

$$C_1 = \begin{bmatrix} I & A_{12} \\ 0 & I \end{bmatrix}$$

and then, also on the left, by

$$C_2 = \begin{bmatrix} I & 0 \\ -A_{21}V_{11} & I \end{bmatrix}$$

to obtain the block triangular matrix

$$\begin{bmatrix} V_{11}^{-1} & A_{12}(I + A_{22}) \\ 0 & V_{22}^{-1} \end{bmatrix}.$$

Inverting this matrix and multiplying on the **right** by C_2C_1 gives the desired result.

Some simplification of the representation of the inverse may be possible, but it hardly seems worthwhile without additional assumptions on the nature of the matrix A.

In the case of symmetric matrices, we have the following very useful result.

Corollary 2.7. Let A be as in Proposition 2.31a, and suppose, in addition, that:

i. A is symmetric;

ii. The matrices $(I + A_{22})$, $I - A_{21}V_{11}A_{12}$ are nonsingular.

Then,

$$A^{-1} = \begin{bmatrix} B_{11} & B_{12} \\ B_{21} & B_{22} \end{bmatrix},$$

where

$$B_{11} = V_{11} + V_{11}A_{12}(I + A_{22})V_{22}A_{21}V_{11},$$

$$B_{12} = -V_{11}A_{12}V_{22}',$$

$$B_{21} = -V_{22}A_{21}V_{11},$$

$$B_{22} = V_{22} - V_{22}A_{21}V_{11}A_{12}.$$

Proof: We note that since A is symmetric $A_{21}' = A_{12}$. Moreover,

$$(I + A_{22})V_{22} = (V_{22}^{-1}(I + A_{22})^{-1})^{-1},$$

$$A_{22}(I + A_{22})^{-1} = I - (I + A_{22})^{-1}.$$

Hence,

$$(I + A_{22})V_{22} = (I - A_{21}V_{11}A_{12} - (I + A_{22})^{-1})^{-1},$$

which shows first that

$$(I + A_{22})V_{22} = V_{22}'(I + A_{22})$$

i.e. that the matrix above is symmetric, and second, since V_{11} is evidently a symmetric matrix, that B_{11} is also symmetric, as required.

We note further the identity

$$(I - A_{21}V_{11}A_{12})V_{22}' = (V_{22}'^{-1}(I - A_{21}V_{11}A_{12})^{-1})^{-1}$$
$$= (I + A_{22} - (I - A_{21}V_{11}A_{12})^{-1})^{-1},$$

which shows that the matrix

$$B_{22} = V_{22} - V_{22}A_{21}V_{11}A_{12}$$

is symmetric, as is also required. To complete the proof, we need only show that, in this case, the representation of B_{12} in Proposition 2.31a reduces to that given in this corollary.

Regrouping the elements in B_{12} of Proposition 2.31a, we find

$$B_{12} = V_{11}A_{12} - V_{11}A_{12}(I + A_{22})V_{22}(I - A_{21}V_{11}A_{12})$$

and, using the identity above, we conclude

$$(I + A_{22})V_{22}(I - A_{21}V_{11}A_{12}) = (I - (I - A_{21}V_{11}A_{12})^{-1}(I + A_{22})^{-1})^{-1}.$$

Finally, using the identity in Corollary 2.5, with

$$C = I, \quad D = I, \quad X = -(I - A_{21}V_{11}A_{12})^{-1}, \quad Y = (I + A_{22})^{-1},$$

we conclude that

$$(I + A_{22})V_{22}(I - A_{21}V_{11}A_{12}) = I + V'_{22}.$$

Consequently,

$$B_{12} = -V_{11}A_{12}V'_{22}.$$

q.e.d.

Remark 2.27. In some econometric problems, namely those involving consumer expenditure systems, we need to minimize a certain function subject to a set of constraints, involving a matrix of restrictions, R, which is of full **row** rank.

In order to solve the system, and thus obtain estimators for the parameters and the relevant Lagrange multipliers, we would need to invert the matrix

$$A = \begin{bmatrix} A_{11} & A_{12} \\ A_{21} & A_{22} \end{bmatrix}$$

with $A_{12} = R'$, $A_{21} = R$, $A_{22} = 0$ and singular (positive semidefinite) A_{11}.

Since R is of full row rank, and A is nonsingular, it follows that the matrices

$$V_{11}^{-1} = A_{11} + R'R, \qquad RV_{11}R'$$

are nonsingular. Moreover, in this case

$$I + A_{22} = I, \qquad V_{22} = -(RV_{11}R')^{-1}$$

and, consequently, we have that the required inverse is

$$\begin{bmatrix} V_{11} - V_{11}R'(RV_{11}R')^{-1}RV_{11} & V_{11}R'(RV_{11}R')^{-1} \\ \\ (RV_{11}R')^{-1}RV_{11} & I - (RV_{11}R')^{-1} \end{bmatrix}.$$

2.8 Kronecker Products of Matrices

Definition 2.27. Let A be $m \times n$, and B be $p \times q$. The **Kronecker product** of the two matrices, denoted by

$$A \otimes B,$$

is defined by

$$A \otimes B = \begin{bmatrix} a_{11}B & a_{12}B & \cdots & a_{1n}B \\ a_{21}B & a_{22}B & \cdots & a_{2n}B \\ \vdots & \vdots & & \vdots \\ a_{m1}B & a_{m2}B & \cdots & a_{mm}B \end{bmatrix}.$$

Often, it is written more compactly as

$$A \otimes B = [a_{ij}B]$$

and is a matrix of dimension $(mp) \times (nq)$.

One operates with Kronecker products as follows.

Matrix addition. Let A_1, A_2 be matrices of dimension $m \times n$ and B_1, B_2, be matrices of dimension $p \times q$, and put

$$D_i = (A_i \otimes B_1), \qquad i = 1, 2.$$

Then,

$$D_1 + D_2 = (A_1 + A_2) \otimes B_1.$$

Similarly, if

$$E_i = (A_1 \otimes B_i), \qquad i = 1, 2$$

then

$$E_1 + E_2 = A_1 \otimes (B_1 + B_2).$$

Scalar multiplication. Let

$$C_i = (A_i \otimes B_i), \qquad i = 1, 2,$$

with the A_i, B_i as previously defined, and let α be a scalar. Then,

$$\alpha C_i = (\alpha A_i \otimes B_i) = (A_i \otimes \alpha B_i).$$

Matrix multiplication. Let C_i, $i = 1, 2$, be two Kronecker product matrices,

$$C_i = A_i \otimes B_i, \qquad i = 1, 2,$$

and suppose A_1 is $m \times n$, A_2 is $n \times r$, B_1 is $p \times q$, and B_2 is $q \times s$. Then,

$$C_1 C_2 = A_1 A_2 \otimes B_1 B_2.$$

Matrix inversion. Let C be a Kronecker product

$$C = A \otimes B$$

and suppose A, B are invertible matrices of order m and n, respectively. Then,

$$C^{-1} = A^{-1} \otimes B^{-1}.$$

All of the above can be verified directly either from the rules for operating with partitioned matrices or from other appropriate definitions.

We have

Proposition 2.34. Let A, B be square matrices of orders m, and n, respectively. Then,

$$\text{tr}(A \otimes B) = \text{tr}(A)\text{tr}(B).$$

Proof: Since $A \otimes B$ is a **block** matrix, its trace is the sum of the traces of the diagonal blocks. Hence,

$$\text{tr}(A \otimes B) = \sum_{i=1}^{m} (\text{tr} a_{ii} B) = \left(\sum_{i=1}^{m} a_{ii} \right) \text{tr}(B) = \text{tr}(A)\text{tr}(B).$$

q.e.d.

Proposition 2.35. Let A, and B be nonsingular matrices of orders m, and n, respectively. Then,

$$|A \otimes B| = |A|^n |B|^m.$$

Proof: Denote by A_{i+1} the matrix obtained when we suppress the first i rows and columns of A. Similarly, denote by $a_{\cdot i}^{i}$ the ith row of A after its first i elements are suppressed, and by $a_{\cdot i}^{(i)}$ the ith column of A after its first i elements have been suppressed.

Now, partition

$$A = \begin{bmatrix} a_{11} & a_{1\cdot}^{(1)} \\ a_{\cdot 1}^{(1)} & A_2 \end{bmatrix}$$

and write the Kronecker product in the partitioned form

$$A \otimes B = \begin{bmatrix} a_{11}B & a_{1\cdot} \otimes B \\ a_{\cdot 1} \otimes B & A_2 \otimes B \end{bmatrix}.$$

Apply Proposition 2.30 to the partitioned matrix above to obtain

$$|A \otimes B| = |a_{11}B - (a_{1\cdot} \otimes B)(A_2 \otimes B)^{-1}(a_{\cdot 1} \otimes B)| \, |A_2 \otimes B|.$$

Because A is nonsingular, we assume that A_2 is also nonsingular,[3] as well as $A_3, A_4, \ldots,$ etc. Thus, we may evaluate

$$(a_{1\cdot} \otimes B)(A_2 \otimes B)^{-1}(a_{\cdot 1} \otimes B) = a_{1\cdot}A_2^{-1}a_{\cdot 1} \otimes B,$$

where, evidently, $a_{1\cdot}^{(1)} A_2^{-1} a_{\cdot 1}^{(1)}$ is a scalar. Consequently,

$$\left| a_{11}B - \left(a_{1\cdot}^{(1)} \otimes B \right) (A_2 \otimes B)^{-1} \left(a_{\cdot 1}^{(1)} \otimes B \right) \right| = \left| \left(a_{11} - a_{1\cdot}^{(1)} A_2^{-1} a_{\cdot 1}^{(1)} \right) \otimes B \right|$$
$$= (a_{11} - a_{1\cdot}^{(1)} A_2^{-1} a_{\cdot 1}^{(1)})^n |B|.$$

Applying Proposition 2.30 to the partition of A above, we note that

$$|A| = \left(a_{11} - a_{1\cdot}^{(1)} A_2^{-1} a_{\cdot 1}^{(1)} \right) |A_2|,$$

and so we find

$$|A \otimes B| = |A|^n |A_2|^{-n} |B| \, |A_2 \otimes B|.$$

Applying the same procedure, we also find

$$|A_2 \otimes B| = |A_2|^n |A_3|^{-n} |B| \, |A_3 \otimes B|,$$

and thus

$$|A \otimes B| = |A|^n |A_3|^{-n} |B|^2 |A_3 \otimes B|.$$

Continuing in this fashion $m - 1$ times, we have

$$|A \otimes B| = |A|^n |A_m|^{-n} |B|^{m-1} |A_{m-1} \otimes B|.$$

But

$$A_m = a_{mm}$$

---

[3]This involves some loss of generality but makes a proof by elementary methods possible. The results stated in the proposition are valid without these restrictive assumptions.

and

$$|A_m \otimes B| = |a_{mm}B| = a_{mm}^n|B|.$$

Since

$$|A_m| = a_{mm},$$

we conclude

$$|A \otimes B| = |A|^n|B|^m.$$

q.e.d.

2.9 Characteristic Roots and Vectors

Definition 2.28. Let A be a square matrix of order m; let λ, x be, respectively, a scalar and an m-element non-null vector. If

$$Ax = \lambda x,$$

λ is said to be a **characteristic root** of A and x its associated **characteristic vector**.

Remark 2.28. Characteristic vectors are evidently not unique. If x is a characteristic vector and c a non-null scalar, cx is also a characteristic vector. We render characteristic vectors unique by imposing the requirement, or convention, that their length be unity, i.e. that $x'x = 1$.

Proposition 2.36. Let A be a square matrix of order m and let Q be an invertible matrix of order m. Then,

$$B = Q^{-1}AQ$$

has the same characteristic roots as A, and if x is a characteristic vector of A then $Q^{-1}x$ is a characteristic vector of B.

Proof: Let (λ, x) be any pair of characteristic root and associated characteristic vector of A. They satisfy

$$Ax = \lambda x.$$

Pre-multiply by Q^{-1} to obtain

$$Q^{-1}Ax = \lambda Q^{-1}x.$$

But we may also write

$$Q^{-1}A = Q^{-1}AQQ^{-1}.$$

Thus, we obtain

$$Q^{-1}AQ(Q^{-1}x) = \lambda(Q^{-1}x),$$

which shows that the pair $(\lambda, Q^{-1}x)$ is a characteristic root and associated characteristic vector of B.

q.e.d.

Remark 2.29. If A and Q are as above, and

$$B = Q^{-1}AQ,$$

B and A are said to be **similar** matrices.

Moreover, it is clear that if there exists a matrix P such that

$$P^{-1}AP = D,$$

where D is a diagonal matrix, then P must be the matrix of the characteristic vectors of A, and D the matrix of its characteristic roots, provided the columns of P have unit length. This is so because the equation above implies

$$AP = PD$$

and the columns of this relation read

$$Ap_{\cdot i} = d_i p_{\cdot i}, \qquad i = 1, 2, \ldots, m,$$

thus defining the pair $(d_i, p_{\cdot i})$ as a characteristic root and its associated characteristic vector.

We now investigate the conditions under which a matrix A is similar to a diagonal matrix.

Proposition 2.37. Let A be a square matrix of order m, and suppose

$$r_i, \qquad i = 1, 2, \ldots, n, \ n \leq m,$$

are the **distinct** characteristic roots of A. If

$$\{x_{\cdot i} : i = 1, 2, \ldots, n\}$$

is the set of associated characteristic vectors, then it is a linearly independent set.

Proof: Put

$$X = (x_{\cdot 1}, x_{\cdot 2}, \ldots, x_{\cdot n})$$

and note that X is $m \times n$, and $n \leq m$. Suppose the columns of X are not linearly independent. Then, there exists a non-null vector

$$b = (b_1, b_2, \ldots, b_n)'$$

such that

$$Xb = 0. \tag{2.20}$$

Let

$$R = \operatorname{diag}(r_1, r_2, \ldots, r_n)$$

and note that, since

$$AX = XR,$$

multiplying Eq. (2.20) (on the left) by A we have

$$0 = AXb = XRb.$$

Repeating this j times, we find

$$XR^j b = 0, \qquad j = 1, 2, \ldots, n-1, \tag{2.21}$$

where

$$R^j = \operatorname{diag}(r_1^j, r_2^j, \ldots, r_n^j).$$

Consider now the matrix whose jth column is $XR^j b$, with the understanding that for $j = 0$ we have Eq. (2.20). In view of Eq. (2.21) this is the null matrix. But note also that

$$0 = (Xb, XRb, \ldots, XR^{n-1}b) = XBV, \tag{2.22}$$

where

$$B = \operatorname{diag}(b_1, b_2, \ldots, b_n)$$

and V is the so-called **Vandermonde** matrix

$$V = \begin{bmatrix} 1 & r_1 & r_1^2 & \cdots & r_1^{n-1} \\ 1 & r_2 & r_2^2 & \cdots & r_2^{n+1} \\ \vdots & \vdots & \vdots & & \\ 1 & r_n & r_n^2 & \cdots & r_n^{n-1} \end{bmatrix}.$$

It may be shown (see Proposition 2.38) that if the r_i are distinct, V is nonsingular. Hence, from Eq. (2.22) we conclude

$$XB = 0.$$

But this means

$$b_i x_{\cdot i} = 0, \qquad i = 1, 2, \ldots, n.$$

Thus, unless

$$b_i = 0, \qquad i = 1, 2, \ldots, n,$$

we must have, for some i, say i_0,

$$x_{\cdot i_0} = 0.$$

This is a contradiction and shows that Eq. (2.20) cannot hold for non-null b; hence, the characteristic vectors corresponding to distinct characteristic roots are linearly independent.

q.e.d.

Proposition 2.38. Let

$$V = \begin{bmatrix} 1 & r_1 & r_1^2 & \cdots & r_1^{n-1} \\ 1 & r_2 & r_2^2 & \cdots & r_2^{n-1} \\ \vdots & \vdots & \vdots & & \vdots \\ 1 & r_n & r_n^2 & \cdots & r_n^{n-1} \end{bmatrix}$$

and suppose the r_i, $i = 1, 2, \ldots, n$, are distinct. Then

$$|V| \neq 0.$$

Proof: Expand $|V|$ by cofactors along the first row to obtain

$$|V| = a_0 + a_1 r_1 + a_2 r_1^2 + \cdots + a_{n-1} r_1^{n-1},$$

where a_i is the cofactor of r_1^i. This shows $|V|$ to be a polynomial of degree $n-1$ in r_1; it is immediately evident that $r_2, r_3, \ldots, r_n$ are its roots since if for r_1 we substitute r_i, $i \geq 2$, we have the determinant of a matrix with two identical rows. From the fundamental theorem of algebra, we can thus write

$$|V| = a_{n-1} \prod_{j=2}^{n} (r_1 - r_j).$$

But

$$a_{n-1} = (-1)^{n+1} |V_1|,$$

where V_1 is the matrix obtained by striking out the first row and nth column of V. Hence, we can also write

$$|V| = |V_1| \prod_{j=2}^{n} (r_j - r_1).$$

But V_1 is of exactly the same form as V except that it is of dimension $n-1$ and does not contain r_1.

Applying a similar procedure to V_1, we find

$$|V_1| = |V_2| \prod_{j=3}^{n} (r_j - r_2),$$

where V_2 is evidently the matrix obtained when we strike out the first and second rows as well as columns n and $n-1$ of V. Continuing in this fashion we find

$$|V| = \sum_{i=1}^{n-1} \prod_{j_i=i+1}^{n} (r_{j_i} - r_i).$$

Since $r_{j_i} \neq r_i$ it is evident that

$$|V| \neq 0.$$

q.e.d.

An immediate consequence of Proposition 2.37 is

Proposition 2.39. Let A be a square matrix of order m, and suppose all its roots are distinct. Then, A is similar to a diagonal matrix.

Proof: Let $(\lambda_i, x_{\cdot i})$, $i = 1, 2, \ldots, m$, be the characteristic roots and associated characteristic vectors of A. Let

$$\Lambda = \text{diag}(\lambda_1, \lambda_2, \ldots, \lambda_m), \qquad X = (x_{\cdot 1}, x_{\cdot 2}, \ldots, x_{\cdot m}),$$

and note that the relationship

$$Ax_{\cdot i} = \lambda_i x_{\cdot i}, \qquad i = 1, 2, \ldots, m,$$

between A and its characteristic roots and vectors may be written compactly as

$$AX = X\Lambda.$$

By Proposition 2.37, X is nonsingular; hence,

$$X^{-1}AX = \Lambda.$$

q.e.d.

The usefulness of this proposition is enhanced by the following approximation result.

Proposition 2.40. Let A be a square matrix of order m. Then, there exists a square matrix of order m, say B, such that B has distinct roots and

$$\sum_{i,j=1}^{m} |a_{ij} - b_{ij}| < \varepsilon,$$

where ε is any arbitrary preassigned positive quantity however small.

Proof: The proof of this result lies entirely outside the scope of this volume. The interested reader is referred to Bellman (1960), pp. 199 ff.

In the preceding, we have established a number of properties regarding the characteristic roots and their associated characteristic vectors without explaining how such quantities may be obtained. It is thus useful to deal with these aspects of the problem before we proceed.

By the definition of characteristic roots and vectors of a square matrix A, we have

$$Ax = \lambda x,$$

or more revealingly

$$(\lambda I - A)x = 0, \tag{2.23}$$

where λ is a characteristic root and x the associated characteristic vector. We recall that, for a characteristic vector, we require

$$x \neq 0, \quad x'x = 1. \tag{2.24}$$

Clearly, Eq. (2.23) together with Eq. (2.24) implies that the columns of $\lambda I - A$ are linearly dependent. Hence, we can find all λ's for which Eqs. (2.23) and (2.24) are satisfied for appropriate x's, by finding the λ's for which

$$|\lambda I - A| = 0. \tag{2.25}$$

Definition 2.29. Let A be a square matrix of order m. The relation in Eq. (2.25) regarded as an equation in λ is said to be the **characteristic equation** of the matrix A.

From the basic definition of a determinant, we easily see that Eq. (2.25) represents a polynomial of degree m in λ. This is so because in evaluating a determinant we take the sum of all possible products involving the choice of one element from each row and column. In this case, the largest power of λ occurs in the term involving the choice of the diagonal elements of $\lambda I - A$. This term is

$$(\lambda - a_{11})(\lambda - a_{22}) \cdots (\lambda - a_{mm}),$$

and we easily see that the highest power of λ occurring in the characteristic equation is λ^m, and its coefficient is unity. Moreover, collecting terms involving λ^j, $j = 0, 1, 2, \ldots, m$, we can write the characeristic equation as

$$0 = |\lambda I - A| = \lambda^m + b_{m-1}\lambda^{m-1} + b_{m-2}\lambda^{m-2} + \cdots + b_0. \qquad (2.26)$$

It is also clear from this discussion that

$$b_0 = |-A| = (-1)^m |A|. \qquad (2.27)$$

The fundamental theorem of algebra assures us that, over the field of complex numbers, the polynomial of degree m in Eq. (2.26) has m roots. These may be numbered, say, in order of decreasing magnitude: $\lambda_1,\ \lambda_2,\ \lambda_3, \ldots, \lambda_m$. The characteristic equation of Eq. (2.26) can also be written as

$$0 = |\lambda I - A| = \prod_{i=1}^{m}(\lambda - \lambda_i). \qquad (2.28)$$

The roots of the characteristic equation of A, as exhibited Eq. (2.26) or Eq. (2.28) are said to be the **characteristic roots** of A. If λ_i is one of the characteristic roots of A, the columns of $\lambda_i I - A$ are linearly dependent; it follows, therefore, that there exists at least one non-null vector, say $x_{\cdot i}$, such that

$$(\lambda_i I - A)x_{\cdot i} = 0.$$

But this means that the pair $(\lambda_i, x_{\cdot i})$ represents a characteristic root and its associated characteristic vector, provided the latter is normalized so that its length is unity.

Thus, obtaining the characteristic roots of a matrix involves solving a polynomial equation of degree m; obtaining the characteristic vectors involves solving a system of m linear equations. An immediate consequence of the preceding discussion is

Proposition 2.41. Let A be a square matrix of order m. Let $\lambda_1, \lambda_2, \ldots, \lambda_m$ be its characteristic roots. Then,

$$|A| = \prod_{i=1}^{m}\lambda_i.$$

Proof: If in Eq. (2.28) we compute the constant term of the polynomial in the right member, we find

$$\prod_{i=1}^{m}(-\lambda_i) = (-1)^m \prod_{i=1}^{m}\lambda_i.$$

From Eq. (2.27), we see that

$$b_0 = |-A| = (-1)^m |A|.$$

Because Eqs. (2.28) and (2.26) are two representations of the same polynomial, we conclude

$$|A| = \prod_{i=1}^{m} \lambda_i.$$

q.e.d.

Remark 2.30. The preceding proposition implies that if A is a singular matrix, then at least one of its roots is zero. It also makes clear the terminology **distinct** and **repeated** characteristic roots. In particular, let $s < m$ and suppose Eq. (2.28) turns out to be of the form

$$|\lambda I - A| = \prod_{j=1}^{s} (\lambda - \lambda_{(j)})^{m_j},$$

where

$$\sum_{j=1}^{s} m_j = m, \qquad \lambda_{(j)} \neq \lambda_{(i)} \quad \text{for } i \neq j.$$

Then, we say that A has s **distinct** roots, viz. the roots $\lambda_{(1)}, \lambda_{(2)}, \ldots, \lambda_{(s)}$, and that the root $\lambda_{(i)}$ is **repeated** m_i times, since the factor corresponding to it in the factorization of the characteristic equation is raised to the m_i power.

Remark 2.31. It may further be shown, but will not be shown here, that if A is a square matrix of order m and rank $r \leq m$ then it has r nonzero roots and $m - r$ zero roots, i.e. the zero root is repeated $m - r$ times or, alternatively, its characteristic equation is of the form

$$|\lambda I - A| = \lambda^{m-r} f(\lambda), \tag{2.29}$$

where

$$f(\lambda_i) = 0, \qquad i = 1, 2, \ldots, r,$$

and

$$\lambda_i \neq 0, \qquad i = 1, 2, \ldots, r.$$

From the method for obtaining characteristic roots, we easily deduce

Proposition 2.42. Let A be a square matrix of order m and let

$$\lambda_i, \qquad i = 1, 2, \ldots, m,$$

be its characteristic roots. Then,

   i. The characteristic roots of A' are exactly those of A, and

   ii. If A is nonsingular, the characteristic roots of A^{-1} are given by

$$\mu_i = \frac{1}{\lambda_i}, \qquad i = 1, 2, \ldots, m.$$

Proof: The characteristic roots of A are simply the solution of $|\lambda I - A| = 0$. The characteristic roots of A' are obtained by solving $|vI - A'| = 0$. Since $vI - A' = (vI - A)'$, Proposition 2.17 implies that the determinant of $(vI - A)'$ is exactly the same as the determinant of $vI - A$. Hence, if by v_i, $i = 1, 2, \ldots, m$, we denote the characteristic roots of A', we conclude

$$v_i = \lambda_i, \qquad i = 1, 2, \ldots, m,$$

which proves part i.

   For part ii,

$$|\mu I - A^{-1}| = 0$$

is the characteristic equation for A^{-1}, and moreover

$$\mu I - A^{-1} = A^{-1}(\mu A - I) = -\mu A^{-1}\left(\frac{1}{\mu}I - A\right).$$

Thus,

$$|\mu I - A^{-1}| = (-1)^m \mu^m |A^{-1}|\,|\lambda I - A|, \qquad \lambda = \frac{1}{\mu},$$

and we see that since $\mu = 0$ is not a root,

$$|\mu I - A^{-1}| = 0$$

if and only if

$$|\lambda I - A| = 0,$$

where

$$\lambda = \frac{1}{\mu}.$$

Hence, if μ_i are the roots of A^{-1}, we must have

$$\mu_i = \frac{1}{\lambda_i}, \qquad i = 1, 2, \ldots, m.$$

q.e.d.

Another important result that may be derived by using the characteristic equation is

Proposition 2.43. Let A, B be two square matrices of order m. Then, the characteristic roots of AB are exactly the characteristic roots of BA.

Proof: The characteristic roots of AB and BA are, respectively, the solutions of

$$|\lambda I - AB| = 0, \qquad |\lambda I - BA| = 0.$$

We show that

$$|\lambda I - AB| = |\lambda I - BA|,$$

thus providing the desired result. For some square matrix C of order m, consider

$$\psi(t) = |\lambda I + tC|,$$

where t is an indeterminate. Quite clearly, $\psi(t)$ is a polynomial of degree m. As such it may be represented by a Taylor series expansion about $t = 0$. If the expansion contains $m + 1$ terms, the resulting representation will be exact. Expanding, we find

$$\psi(t) = \psi(0) + \psi'(0)t + \tfrac{1}{2}\psi''(0)t^2 + \cdots + \frac{1}{m!}\psi^{(m)}(0)t^m.$$

By the usual rules for differentiating determinants (see Sect. 4.3), we easily find that

$$\psi(0) = \lambda^m, \qquad \psi'(0) = \lambda^{m-1}\operatorname{tr} C$$

and, in general,

$$\frac{1}{j!}\psi^{(j)}(0) = \lambda^{m-j}h_j(C),$$

where $h_j(C)$ depends only on tr C, tr $C^2, \ldots,$ tr C^j. Evaluating ψ at $t = -1$, gives the characteristic equation for C as

$$0 = \psi(-1) = |\lambda I - C| = \lambda^m - \lambda^{m-1}\operatorname{tr} C + \lambda^{m-2}h_2(C)$$
$$-\lambda^{m-3}h_3(C) + \cdots + (-1)^m h_m(C).$$

Let

$$C_1 = AB, \qquad C_2 = BA$$

and note that

$$\operatorname{tr} C_1 = \operatorname{tr} C_2, \qquad \operatorname{tr} C_1^2 = \operatorname{tr} C_2^2,$$

and, in general,

$$\operatorname{tr} C_1^j = \operatorname{tr} C_2^j.$$

This is so because

$$C_1^j = (AB)(AB) \cdots (AB),$$

$$C_2^j = (BA)(BA) \cdots (BA) = \overbrace{B(AB)(AB) \cdots (AB)}^{j-1 \text{ terms}} A = BC_1^{j-1}A.$$

Thus,

$$\operatorname{tr} C_2^j = \operatorname{tr} BC_1^{j-1}A = \operatorname{tr} C_1^{j-1}AB = \operatorname{tr} C_1^j.$$

Consequently, we see that

$$h_j(C_1) = h_j(C_2)$$

and, moreover,

$$|\lambda I - AB| = |\lambda I - BA|.$$

q.e.d.

Corollary 2.8. Let A and B be, respectively, $m \times n$ and $n \times m$ matrices, where $m \leq n$. Then, the characteristic roots of BA, an $n \times n$ matrix, consist of $n - m$ zeros and the m characteristic roots of AB, an $m \times m$ matrix.

Proof: Define the matrices

$$A_* = \begin{pmatrix} A \\ 0 \end{pmatrix}, \qquad B_* = (B, 0)$$

such that A_* and B_* are $n \times n$ matrices. By Proposition 2.43, the characteristic roots of A_*B_* are exactly those of B_*A_*. But

$$A_*B_* = \begin{bmatrix} AB & 0 \\ 0 & 0 \end{bmatrix}, \qquad B_*A_* = BA.$$

Thus,

$$\lambda I - A_*B_* = \begin{bmatrix} \lambda I - AB & 0 \\ 0 & \lambda I \end{bmatrix}$$

and, consequently, $|\lambda I - BA| = |\lambda I - B_*A_*| = |\lambda I - A_*B_*| = \lambda^{n-m}|\lambda I - AB|$.

q.e.d.

Corollary 2.9. Let A be a square matrix of order m, and let λ_i, $i = 1, 2, \ldots, m$, be its characteristic roots. Then,

$$\operatorname{tr} A = \sum_{i=1}^{m} \lambda_i.$$

Proof: From the proof of Proposition 2.43, we have that

$$|\lambda I - A| = \lambda^m - \lambda^{m-1} \operatorname{tr} A + \lambda^{m-2} h_2(A) + \cdots + (-1)^m h_m(A).$$

From the factorization of polynomials, we have

$$|\lambda I - A| = \prod_{i=1}^{m}(\lambda - \lambda_i) = \lambda^m - \lambda^{m-1}\left(\sum_{i=1}^{m}\lambda_i\right) + \cdots + (-1)^m \prod_{i=1}^{m}\lambda_i.$$

Equating the coefficients for λ^{m-1}, we find $\operatorname{tr} A = \sum_{i=1}^{m}\lambda_i$.

q.e.d.

Proposition 2.44. Let A be a square matrix of order m. Then A is diagonalizable, i.e. it is similar to a diagonal matrix, if and only if for each characteristic root λ of A the multiplicity of λ is equal to the nullity of $\lambda I - A$.

Proof: Suppose A is diagonalizable. Then, we can write

$$Q^{-1}AQ = \Lambda, \qquad \Lambda = \operatorname{diag}(\lambda_1, \lambda_2, \ldots, \lambda_m).$$

Now, suppose the **distinct** roots are $\lambda_{(i)}$, $i = 1, 2, \ldots, s$, $s \le m$. Let the multiplicity of $\lambda_{(i)}$ be m_i, where

$$\sum_{i=1}^{s} m_i = m.$$

It is clear that $\lambda_{(i)}I - \Lambda$ has m_i zeros on its diagonal and hence is of rank

$$r[\lambda_{(i)}I - \Lambda] = m - m_i = \sum_{j \ne i} m_j.$$

Since

$$\lambda_{(i)}I - A = \lambda_{(i)}I - Q\Lambda Q^{-1} = Q(\lambda_{(i)}I - \Lambda)Q^{-1},$$

it follows that

$$r(\lambda_{(i)}I - A) = r(\lambda_{(i)}I - \Lambda) = \sum_{j \ne i} m_j.$$

But $\lambda_{(i)}I - A$ is an $m \times m$ matrix, and by Proposition 2.5 its nullity obeys

$$n[\lambda_{(i)}I - A] = m - r[\lambda_{(i)}I - A] = m_i,$$

which is the multiplicity of $\lambda_{(i)}$.

Conversely, suppose that the nullity of $\lambda_{(i)}I - A$ is m_i and $\sum_{i=1}^{s} m_i = m$. Choose the basis

$$\xi_{\cdot 1}, \xi_{\cdot 2}, \ldots, \xi_{\cdot m_1}$$

for the null space of $\lambda_{(1)}I - A$,

$$\xi_{\cdot m_1 + 1}, \ldots, \xi_{\cdot m_1 + m_2}$$

for the null space of $\lambda_{(2)}I - A$, and so on until the null space of $\lambda_{(s)}I - A$. Thus, we have m, m-element, vectors

$$\xi_{\cdot 1}, \xi_{\cdot 2}, \ldots, \xi_{\cdot m},$$

and each (appropriate) subset of m_i vectors, $i = 1, 2, \ldots, s$, is linearly independent. We claim that the entire set of m vectors is linearly independent. Suppose not. Then, we can find a set of scalars a_i, not all of which are zero, such that

$$\sum_{k=1}^{m} \xi_{\cdot k} a_k = 0.$$

We can also write the equation above as

$$\sum_{i=1}^{s} \zeta_{\cdot i} = 0, \quad \zeta_{\cdot i} = \sum_{j=m_1 + \cdots + m_{i-1} + 1}^{m_1 + \cdots + m_i} \xi_{\cdot j} a_j, \quad i = 1, 2, \ldots, s, \qquad (2.30)$$

it being understood that $m_0 = 0$. Because of the way in which we have chosen the $\xi_{\cdot k}$, $k = 1, 2, \ldots, m$, the second equation of Eq. (2.30) implies that the $\zeta_{\cdot i}$ obey

$$(\lambda_{(i)}I - A)\zeta_{\cdot i} = 0,$$

i.e. that they are characteristic vectors of A corresponding to the distinct roots $\lambda_{(i)}$, $i = 1, 2, \ldots, s$. The first equation in Eq. (2.30) then implies that the $\zeta_{\cdot i}$ are linearly dependent. By Proposition 2.37, this is a contradiction. Hence,

$$a_k = 0, \quad k = 1, 2, \ldots, m,$$

and the $\xi_{\cdot i}$, $i = 1, 2, \ldots, m$ are a linearly independent set. Let

$$X = (\xi_{\cdot 1}, \xi_{\cdot 2}, \ldots, \xi_{\cdot m})$$

and arrange the (distinct) roots

$$|\lambda_{(1)}| > |\lambda_{(2)}| > \cdots > |\lambda_{(s)}|.$$

Putting

$$\Lambda = \text{diag}(\lambda_{(1)}I_{m_1}, \lambda_{(2)}I_{m_2}, \ldots, \lambda_{(s)}I_{m_s}),$$

we must have

$$AX = X\Lambda.$$

Because X is nonsingular, we conclude

$$X^{-1}AX = \Lambda.$$

q.e.d.

2.9.1 Kronecker Product Matrices

Although Kronecker product matrices were examined in an earlier section, and their characteristic roots and vectors may be determined by the preceding discussion, it is very useful here to explain and make explicit their connection to the corresponding entities of their constituent matrices.

Proposition 2.45. Let $D = A \otimes B$, where A is $m \times m$ and B is $n \times n$, with characteristic roots and vectors, respectively,

$$\{(\lambda_i, \ x_{\cdot i}) : \ i = 1, 2, \ldots, m\}, \quad \{(\mu_j, y_{\cdot j}) : \ j = 1, 2, \ldots, n\}.$$

The following statements are true:

i. The characteristic roots and associated characteristic vectors of D are given by

$$\{(\nu_{ij}, z_{\cdot ij}) : \ \nu_{ij} = \lambda_i \mu_j, \ z_{\cdot ij} = x_{\cdot i} \otimes y_{\cdot j}\},$$

for $i = 1, 2, \ldots, m, \ j = 1, 2, \ldots n.$

ii. $\mathrm{r}\,(D) \ = \mathrm{r}\,(A)\,\mathrm{r}\,(B).$

Proof: Since by hypothesis

$$Ax_{\cdot i} = \lambda_i x_{\cdot i}, \quad By_{\cdot j} = \mu_j y_{\cdot j}, \quad i = 1, 2, \ldots, m, \ j = 1, 2, \ldots, n,$$

we have

$$Dz_{ij} = Ax_{\cdot i} \otimes By_{\cdot j} = \lambda_i x_{\cdot i} \otimes \mu_j y_{\cdot j} = (\lambda_i \otimes \mu_j)(x_{\cdot i} \otimes y_{\cdot j}) = \nu_{ij} z_{ij},$$

which proves part i.

To prove part ii, we shall also assume that A, B are diagonalizable[4] and let $Q_1, \ Q_2$ be nonsingular matrices of order $m, \ n$, respectively, such that

$$Q_1^{-1}AQ_1 = \Lambda, \quad Q_2^{-1}BQ_2 = M,$$

---

[4]This is an assumption that simplifies the proof considerably. Strictly speaking, it is not necessary. The cost it imposes on the generality of the result is, in any event, miniscule in view of Proposition 2.40.

where Λ and M are the diagonal matrices containing the characteristic roots of A and B, respectively. Consequently,

$$(Q_1^{-1} \otimes Q_2^{-1})D(Q_1 \otimes Q_2) = \Lambda \otimes M.$$

Since, evidently, $r(A) = r(\Lambda)$, $r(B) = r(M)$, i.e. they are equal to the nonzero characteristic roots of A and B, respectively and, moreover, the number of nonzero roots of D is equal to the product of the nonzero roots of A and the nonzero roots of B, we conclude that

$$r(D) = r(A)r(B).$$

q.e.d.

2.10   Orthogonal Matrices

Although the term orthogonal was informally defined in Chap. 1, we repeat the formal definition for completeness.

Definition 2.30. Let a, b be two m-element vectors. They are said to be (mutually) **orthogonal** if and only if

$$a'b = 0. \tag{2.31}$$

They are said to be **orthonormal** if Eq. (2.31) holds and, in addition, $a'a = 1$, $b'b = 1$.

Definition 2.31. Let Q be a square matrix of order m. It is said to be **orthogonal** if and only if its columns are orthonormal.

An immediate consequence of the definition is the proposition below.

Proposition 2.46. Let Q be an orthogonal matrix of order m. Then, it is nonsingular.

Proof: It will suffice to shall show that its columns are linearly independent. Suppose there exist scalars c_i, $i = 1, 2, \ldots, m$, such that

$$\sum_{i=1}^{m} c_i q_{\cdot i} = 0, \tag{2.32}$$

the $q_{\cdot i}$ being the (orthonormal) columns of Q. Pre-multiply Eq. (2.32) by $q'_{\cdot j}$, $j = 1, 2, 3, \ldots, m$, **and note that we obtain**

$$c_j q'_{\cdot j} q_{\cdot j} = 0, \quad j = 1, 2, 3, \ldots, m.$$

But since

$$q'_{\cdot j} q_{\cdot j} = 1$$

we conclude that Eq. (2.32) implies

$$c_j = 0, \; j = 1, 2, \ldots, m.$$

q.e.d.

A further consequence is

Proposition 2.47. Let Q be an orthogonal matrix of order m. Then, $Q' = Q^{-1}$.

Proof: By the definition of an orthogonal matrix,

$$Q'Q = I_m.$$

By Proposition 2.46, its inverse exists. Multiplying on the right by Q^{-1} we find

$$Q' = Q^{-1}.$$

q.e.d.

Proposition 2.48. Let Q be an orthogonal matrix of order m. The following statements are true:

   i. $|Q| = 1$ or $|Q| = -1$;

  ii. If $\lambda_i, \; i = 1, 2, \ldots, m,$ are the characteristic roots of Q, $\lambda_i = \pm 1, \; i = 1, 2, \ldots, m.$

Proof: The validity of i follows immediately from $Q'Q = I_m$, which implies

$$|Q|^2 = 1, \qquad |Q| = \pm 1.$$

For ii, we note that, by Proposition 2.42, the characteristic roots of Q' are exactly those of Q, and the characteristic roots of Q^{-1} are $1/\lambda_i, \; i = 1, 2, \ldots, m,$ where the λ_i are the characteristic roots of Q. Because for an orthogonal matrix $Q' = Q^{-1}$, we conclude $\lambda_i = 1/\lambda_i$, which implies

$$\lambda_i = \pm 1.$$

q.e.d.

It is interesting that given a set of linearly independent vectors we can transform them into an orthonormal set. This procedure, known as Gram-Schmidt orthogonalization, is explained below.

Proposition 2.49 (Gram-Schmidt orthogonalization). If $\xi_{\cdot i}$, $i = 1, 2, \ldots, m$, is a set of m linearly independent, m-element **column** vectors, they can be transformed into a set of orthonormal vectors.

Proof: First, we transform the $\xi_{\cdot i}$ into an orthogonal set, and then divide each resulting vector by its modulus to produce the desired orthonormal set. To this end, define

$$y_{\cdot 1} = \xi_{\cdot 1}$$

$$y_{\cdot 2} = a_{12}\xi_{\cdot 1} + \xi_{\cdot 2}$$

$$y_{\cdot 3} = a_{13}\xi_{\cdot 1} + a_{23}\xi_{\cdot 2} + \xi_{\cdot 3}$$

$$\vdots$$

$$y_{\cdot m} = a_{1m}\xi_{\cdot 1} + a_{2m}\xi_{\cdot 2} \cdots + a_{m-1,m}\xi_{\cdot m-1} + \xi_{\cdot m}.$$

The condition for defining the a_{ij} is that

$$y_{\cdot i}' y_{\cdot j} = 0, \qquad i = 1, 2, \ldots, j - 1. \tag{2.33}$$

But since $y_{\cdot i}$ depends only on $\xi_{\cdot 1}, \xi_{\cdot 2}, \ldots, \xi_{\cdot i}$, a condition equivalent to Eq. (2.33) is

$$\xi_{\cdot i}' y_{\cdot j} = 0, \qquad i = 1, 2, \ldots, j - 1.$$

To make the notation compact, put

$$X_j = (\xi_{\cdot 1}, \xi_{\cdot 2}, \ldots, \xi_{\cdot j-1}), \qquad a_{\cdot j} = (a_{1j}, a_{2j}, \ldots, a_{j-1,j})',$$

and note that the y's may be written compactly as

$$y_{\cdot 1} = \xi_{\cdot 1}$$

$$y_{\cdot j} = X_j a_{\cdot j} + \xi_{\cdot j}, \qquad j = 2, \ldots, m.$$

We wish the $y_{\cdot j}$ to satisfy the condition

$$X_j' y_{\cdot j} = X_j' X_j a_{\cdot j} + X_j' \xi_{\cdot j} = 0. \tag{2.34}$$

The matrix $X_j' X_j$ is nonsingular because the columns of X_j are linearly independent.[5] Hence,

$$a_{\cdot j} = -(X_j' X_j)^{-1} X_j' \xi_{\cdot j}, \qquad j = 2, 3, \ldots, m, \tag{2.35}$$

---

[5]A simple proof of this is as follows. Suppose there exists a non-null vector c such that $X_j' X_j c = 0$. But $c' X_j' X_j c = 0$, implies $X_j c = 0$, which is a contradiction.

and we can define the desired orthogonal set by $y_{\cdot 1} = \xi_{\cdot 1}$
$y_{\cdot i} = \xi_{\cdot i} - X_i(X_i'X_i)^{-1}X_i'\xi_{\cdot i}, \quad i \geq 2$. Put

$$\zeta_{\cdot i} = \frac{y_{\cdot i}}{(y_{\cdot i}'y_{\cdot i})^{1/2}}, \qquad i = 1, 2, \ldots, m,$$

and note that $\zeta_{\cdot i}'\zeta_{\cdot i} = 1, \; i = 1, 2, \ldots, m$. The set

$$\{\zeta_{\cdot i} : i = 1, 2, \ldots, m\}$$

is the desired orthonormal set.

q.e.d.

A simple consequence is

Proposition 2.50. Let a be an m-element non-null **column** vector with unit length (modulus). Then, there exists an orthogonal matrix with a as the first column.

Proof: Given a, there certainly exist m-element vectors $\xi_{\cdot 2}, \; \xi_{\cdot 3}, \ldots, \xi_{\cdot m}$ such that the set

$$\{a, \xi_{\cdot 2}, \ldots, \xi_{\cdot m}\}$$

is linearly independent. The desired matrix is then obtained by applying Gram-Schmidt orthogonalization to this set.

q.e.d.

Remark 2.32. Evidently, Propositions 2.49 and 2.50 are applicable to row vectors.

2.11 Symmetric Matrices

In this section, we shall establish certain useful properties of symmetric matrices.

Proposition 2.51. Let S be a symmetric matrix of order m whose elements are real. Then, its characteristic roots are also real.

Proof: Let λ be any characteristic root of S and let z be its associated characteristic vector. Put

$$\lambda = \lambda_1 + i\lambda_2, \qquad z = x + iy,$$

so that we allow that λ, z may be complex. Since they form a pair of characteristic root and its associated characteristic vector, they satisfy

$$Sz = \lambda z. \tag{2.36}$$

Pre-multiply by $\bar{z}'$, $\bar{z}$ being the complex conjugate of z. We find

$$\bar{z}'Sz = \lambda\bar{z}'z. \tag{2.37}$$

We note that since z is a characteristic vector

$$\bar{z}'z = x'x + y'y > 0.$$

In Eq. (2.36) take the complex conjugate to obtain

$$S\bar{z} = \bar{\lambda}\,\bar{z}, \tag{2.38}$$

because the elements of S are real. Pre-multiply Eq. (2.38) by z' to find

$$z'S\bar{z} = \bar{\lambda}z'\bar{z}. \tag{2.39}$$

Since $\bar{z}'Sz$ is a scalar (a 1×1 "matrix"),

$$(z'S\bar{z}) = (z'S\bar{z})' = \bar{z}'Sz$$

and, moreover, $\bar{z}'z = z'\bar{z}$. Subtracting Eq. (2.39) from Eq. (2.37), we find $0 = (\lambda - \bar{\lambda})\bar{z}'z$. Since $\bar{z}'z > 0$, we conclude $\lambda = \bar{\lambda}$. But

$$\lambda = \lambda_1 + i\lambda_2, \qquad \bar{\lambda} = \lambda_1 - i\lambda_2,$$

which implies $\lambda_2 = -\lambda_2$ or $\lambda_2 = 0$. Hence, $\lambda = \lambda_1$, and the characteristic root is real.

q.e.d.

Another important property is

Proposition 2.52. Let S be a symmetric matrix of order m. Let its distinct roots be $\lambda_{(i)}$, $i = 1, 2, \ldots, s$, $s \leq m$, and let the multiplicity of $\lambda_{(i)}$ be m_i, $\sum_{i=1}^{s} m_i = m$. Then, corresponding to the root $\lambda_{(i)}$ there exist m_i **linearly independent** orthonormal characteristic vectors.

Proof: Since $\lambda_{(i)}$ is a characteristic root of S, let $q_{\cdot 1}$ be its associated characteristic vector (of unit length). By Proposition 2.49, there exist vectors

$$p_{\cdot j}^{(i)}, \qquad j = 2, 3, \ldots, m,$$

such that

$$Q_1 = \left(q_{\cdot 1}, p_{\cdot 2}^{(1)}, p_{\cdot 3}^{(1)}, \ldots, p_{\cdot m}^{(1)}\right)$$

is an orthogonal matrix. Consider

$$S_1 = Q_1'SQ_1 = \begin{bmatrix} \lambda_{(i)} & 0 \\ 0 & A_1 \end{bmatrix},$$

where A_1 is a matrix whose i, j element is

$$p_{\cdot i}^{(1)'} S p_{\cdot j}^{(1)}, \qquad i, j = 2, 3, \ldots, m.$$

But S and S_1 have exactly the same roots. Hence, if $m_i \geq 2$,

$$|\lambda I - S| = |\lambda I - S_1| \;=\; \begin{vmatrix} \lambda - \lambda_{(i)} & 0 \\ 0 & \lambda I_{m-1} - A_1 \end{vmatrix}$$

$$= (\lambda - \lambda_{(i)})|\lambda I_{m-1} - A_1| = 0$$

implies that $\lambda_{(i)}$ is also a root of

$$|\lambda I_{m-1} - A_1| = 0.$$

Hence, the nullity of $\lambda_{(i)} I - S$ is at least two, i.e.,

$$n(\lambda_{(i)} I - S) \geq 2,$$

and we can thus find another vector, say, $q_{\cdot 2}$, satisfying

$$(\lambda_{(i)} I - S) q_{\cdot 2} = 0$$

and such that $q_{\cdot 1}, q_{\cdot 2}$ are linearly independent and of unit length, and such that the matrix

$$Q_2 = \left[q_{\cdot 1}, q_{\cdot 2}, p_{\cdot 3}^{(2)}, p_{\cdot 4}^{(2)}, \ldots, p_{\cdot m}^{(2)} \right]$$

is orthogonal.

Define

$$S_2 = Q_2' S Q_2$$

and note that S_2 has exactly the same roots as S. Note further that

$$|\lambda I - S| = |\lambda I - S_2| = \begin{vmatrix} \lambda - \lambda_{(i)} & 0 & \\ 0 & \lambda - \lambda_{(i)} & 0 \\ & 0 & \lambda I_{m-2} - A_2 \end{vmatrix}$$

$$= (\lambda - \lambda_{(i)})^2 |\lambda I_{m-2} - A_2| = 0$$

and $m_i > 2$ implies

$$|\lambda_{(i)} I_{m-2} - A_2| = 0.$$

Hence, $n(\lambda_{(i)} I - S) \geq 3$ and consequently we can choose another characteristic vector, $q_{\cdot 3}$, of unit length orthogonal to $q_{\cdot 1}, q_{\cdot 2}$ and such that

$$Q_3 = \left[q_{\cdot 1}, q_{\cdot 2}, q_{\cdot 3}, p_{\cdot 4}^{(3)}, p_{\cdot 5}^{(3)}, \ldots, p_{\cdot m}^{(3)} \right]$$

is an orthogonal matrix.

Continuing in this fashion, we can choose m_i orthonormal vectors

$$q._1, q._2, \ldots, q._{m_i}$$

corresponding to $\lambda_{(i)}$ whose multiplicity is m_i. It is clear that we cannot choose more than m_i such vectors since, after the choice of $q._{m_i}$, we will be dealing with

$$|\lambda I - S| = |\lambda I - S_{m_i}| = \begin{vmatrix} (\lambda - \lambda_{(i)})I_{m_i} & 0 \\ 0 & \lambda I_{m_i^*} - A_{m_i} \end{vmatrix}$$
$$= (\lambda - \lambda_{(i)})^{m_i} |\lambda I_{m_i^*} - A_{m_i}| = 0,$$

where $m_i^* = m - m_i$. It is evident that

$$|\lambda I - S| = (\lambda - \lambda_{(i)})^{m_i} |\lambda I_{m_i^*} - A_{m_i}| = 0$$

implies

$$|\lambda_{(i)} I_{m_i^*} - A_{m_i}| \neq 0, \qquad (2.40)$$

for, if not, the multiplicity of $\lambda_{(i)}$ would exceed m_i. In turn Eq. (2.40) means that

$$r(\lambda_{(i)} I - S) = m - m_i$$

and thus

$$n(\lambda_{(i)} I - S) = m_i.$$

Because we have chosen m_i linearly independent characteristic vectors corresponding to $\lambda_{(i)}$, they form a basis for the null space of $\lambda_{(i)} I - S$ and thus a larger number of such vectors would form a linearly dependent set.

q.e.d.

Corollary 2.10. If S is as in Proposition 2.51, the multiplicity of the root $\lambda_{(i)}$ is equal to the nullity of

$$\lambda_{(i)} I - S.$$

Proof: Obvious from the proof of the proposition above.

q.e.d.

An important consequence of the preceding is

Proposition 2.53. Let S be a symmetric matrix of order m. Then, the characteristic vectors of S can be chosen to be an orthonormal set, i.e. there exists an orthogonal matrix Q such that

$$Q'SQ = \Lambda,$$

or equivalently S is orthogonally similar to a diagonal matrix.

Proof: Let the distinct characteristic roots of S be $\lambda_{(i)}$, $i = 1, 2, \ldots, s$, $s \leq m$, where $\lambda_{(i)}$ is of multiplicity m_i, and $\sum_{i=1}^{s} m_i = m$. By Corollary 2.10, the nullity of $\lambda_{(i)} I - S$ is equal to the multiplicity m_i of the root $\lambda_{(i)}$. By Proposition 2.52, there exist m_i orthonormal characteristic vectors corresponding to $\lambda_{(i)}$. By Proposition 2.37, characteristic vectors corresponding to distinct characteristic roots are linearly independent. Hence, the matrix

$$Q = (q_{\cdot 1}, q_{\cdot 2}, \ldots, q_{\cdot m}),$$

where the first m_1 columns are the characteristic vectors corresponding to $\lambda_{(1)}$, the next m_2 columns are those corresponding to $\lambda_{(2)}$, and so on, is an orthogonal matrix. Define

$$\Lambda = \mathrm{diag}(\lambda_{(1)} I_{m_1}, \lambda_{(2)} I_{m_2}, \ldots, \lambda_{(s)} I_{m_s})$$

and note that we have

$$SQ = Q\Lambda.$$

Consequently, $Q'SQ = \Lambda$.

q.e.d.

Proposition 2.54 (Simultaneous diagonalization). Let A, B be two symmetric matrices of order m; Then, there exists an orthogonal matrix Q such that

$$Q'AQ = D_1, \qquad Q'BQ = D_2,$$

where the D_i, $i = 1, 2$, are diagonal matrices if and only if

$$AB = BA.$$

Proof: Sufficiency: This part is trivial since if such an orthogonal matrix exists,

$$Q'AQQ'BQ = D_1 D_2,$$
$$Q'BQQ'AQ = D_2 D_1.$$

But the two equations above imply

$$AB = QD_1 D_2 Q',$$
$$BA = QD_2 D_1 Q',$$

which shows that

$$AB = BA \qquad (2.41)$$

because the diagonal matrices D_1, D_2 commute.

Necessity: Suppose Eq. (2.41) holds. Since A is symmetric, let Λ be the diagonal matrix containing its (real) characteristic roots and let Q_1 be the matrix of associated characteristic vectors. Thus,

$$Q_1' A Q_1 = \Lambda.$$

Define

$$C = Q_1' B Q_1$$

and note that

$$\Lambda C = Q_1' A Q_1 Q_1' B Q_1$$

$$= Q_1' A B Q_1$$

$$= Q_1' B A Q_1 = Q_1' B Q_1 Q_1' A Q_1 = C \Lambda. \tag{2.42}$$

If all the roots of A are distinct, we immediately conclude from Eq. (2.42) that

$$C = Q_1' B Q_1$$

is a diagonal matrix. Thus, taking

$$D_1 = \Lambda, \qquad D_2 = C,$$

the proof is completed. If not, let

$$\lambda_{(i)}, \qquad i = 1, 2, \dots, s,$$

be the distinct roots of A and let $\lambda_{(i)}$ be of multiplicity m_i, where $\sum_{i=1}^{s} m_i = m$. We may write

$$\Lambda = \begin{bmatrix} \lambda_{(1)} I_{m_1} & 0 & \cdots & 0 \\ & & \ddots & \cdots \\ 0 & \lambda_{(2)} I_{m_2} & & \\ \vdots & \ddots & 0 & \\ 0 & \cdots & 0 & \lambda_{(s)} I_{m_s} \end{bmatrix}.$$

Partition C conformably with Λ, i.e.

$$C = \begin{bmatrix} C_{11} & C_{12} & \cdots & C_{1s} \\ C_{21} & C_{22} & \cdots & C_{2s} \\ \vdots & \vdots & & \vdots \\ C_{s1} & C_{s2} & \cdots & C_{ss} \end{bmatrix},$$

so that C_{ij} is a matrix of dimension $m_i \times m_j$. From Eq. (2.42), we thus conclude

$$\lambda_{(i)} C_{ij} = \lambda_{(j)} C_{ij}. \tag{2.43}$$

But for $i \neq j$ we have

$$\lambda_{(i)} \neq \lambda_{(j)},$$

and Eq. (2.43) implies

$$C_{ij} = 0, \qquad i \neq j. \tag{2.44}$$

Thus, C is the block diagonal matrix

$$C = \operatorname{diag}(C_{11}, C_{22}, \ldots, C_{ss}).$$

Clearly, the C_{ii}, $i = 1, 2, \ldots, s$, are symmetric matrices. Thus, there exist orthogonal matrices, say,

$$Q_i^*, \qquad i = 1, 2, 3, \ldots, s,$$

that diagonalize them, i.e.

$$Q_i^{*'} C_{ii} Q_i^* = D_i^*, \qquad i = 1, 2, \ldots, m,$$

the D_i^* being diagonal matrices. Define

$$Q_2 = \operatorname{diag}(Q_1^*, Q_2^*, \ldots, Q_s^*)$$

and note that Q_2 is an orthogonal matrix such that

$$D_2 = Q_2' C Q_2 = Q_2' Q_1' B Q_1 Q_2$$

with

$$D_2 = \operatorname{diag}(D_1^*, D_2^*, \ldots, D_s^*).$$

Evidently, D_2 is a diagonal matrix. Define $Q = Q_1 Q_2$ and note:

i. $Q'Q = Q_2' Q_1' Q_1 Q_2 = Q_2' Q_2 = I_m$, so that Q is an orthogonal matrix;

ii. $Q'AQ = Q_2' \Lambda Q_2 = \Lambda$, which follows from the construction of Q_2;

iii. $Q'BQ = D_2$.

Taking $D_1 = \Lambda$, we see that

$$Q'AQ = D_1, \qquad Q'BQ = D_2.$$

q.e.d.

Corollary 2.11. Let A, B be two symmetric matrices of order m such that

$$AB = 0.$$

Then, there exists an orthogonal matrix Q such that $Q'AQ = D_1$, $Q'BQ = D_2$, and, moreover, $D_1 D_2 = 0$.

Proof: Since A, B are symmetric and $AB = 0$, we see that $0 = (AB)' = B'A' = BA = AB$.

By Proposition 2.54, there exists an orthogonal matrix Q such that

$$Q'AQ = D_1, \qquad Q'BQ = D_2.$$

Moreover,

$$D_1 D_2 = Q'AQQ'BQ = Q'ABQ = 0.$$

q.e.d.

We close this section by stating an interesting result that connects the rank of a matrix to the number of the latter's nonzero characteristic roots. This result holds for all matrices, as implied by Proposition 2.14; however, the discussion and proof are greatly simplified in the case of symmetric matrices. Thus, we have

Corollary 2.12. Let S be as in Proposition 2.53; then,

$$r(S) = r \leq m$$

if and only if the number of nonzero characteristic roots of S is r.

Proof: From Proposition 2.53, there exists an orthogonal matrix Q such that

$$S = Q\Lambda Q', \quad \Lambda = Q'SQ, \quad \Lambda = \text{diag}(\lambda_1, \lambda_2, \ldots, \lambda_m),$$

the λ_i, $i = 1, 2, \ldots, m$, being the characteristic roots of S. But the first two relations above imply

$$r(S) = r(\Lambda).$$

Now, suppose

$$r(S) = r.$$

Then, $(m - r)$ of the diagonal elements of Λ must be zero; hence, only r of the characteristic roots of S are nonzero. Conversely, if only r of the characteristic roots of S are nonzero, then

$$r(\Lambda) = r$$

and, consequently,

$$r(S) = r.$$

q.e.d.

2.12 Idempotent Matrices

We recall from Definition 2.8 that a square matrix A is said to be idempotent if and only if

$$AA = A.$$

An easy consequence of the definition is

Proposition 2.55. Let A be a square matrix of order m; suppose further that A is idempotent. Then, its characteristic roots are either zero or one.

Proof: Let λ, x be a pair consisting of a characteristic root and its associated (normalized) characteristic vector. Thus,

$$Ax = \lambda x. \tag{2.45}$$

Pre-multiplying by A, we find

$$Ax = AAx = \lambda Ax = \lambda^2 x. \tag{2.46}$$

But Eqs. (2.45) and (2.46) imply, after pre-multiplication by x',

$$\lambda = \lambda^2.$$

This condition is satisfied only by

$$\lambda = 0, \quad \text{or} \quad \lambda = 1.$$

q.e.d.

Remark 2.33. In idempotent matrices, we have a non-obvious example of a matrix with repeated roots. If A is a **symmetric** idempotent matrix it is **diagonalizable**, i.e. it is similar to a diagonal matrix. The result above is to be understood in the context of Propositions 2.37, 2.39, 2.40, and 2.44.

Example 2.4. An example of a (non-symmetric) matrix whose characteristic vectors are not linearly independent is

$$A = \begin{bmatrix} 1 & 1 \\ 0 & 1 \end{bmatrix}, \quad |\lambda I_2 - A| = (\lambda - 1)^2,$$

which has a repeated root, namely $\lambda = 1$, of multiplicity 2. The nullity of the matrix $\lambda I - A$ for $\lambda = 1$ is defined by the dimension of the (null) space of this matrix, i.e. by the dimension of the collection of vectors

$$\left(x : \begin{bmatrix} 0 & 1 \\ 0 & 0 \end{bmatrix} x = 0 \right).$$

This (null) space is generated (spanned) by the vector $(1,0)'$ and is thus of dimension 1; since the repeated root is of multiplicity 2, the characteristic vectors of A cannot be linearly **independent**. In fact, the equations for the characteristic vectors associated with the unit root (of multiplicity 2) are given by

$$x_1 + x_2 = x_1, \quad x_2 = x_2,$$

which implies that $x_2 = 0$, and x_1 is arbitrary. Thus, the characteristic vectors corresponding to the repeated root 1 are $(1,0)'$ and $(c,0)'$, where c is an arbitrary constant, and the matrix of characteristic vectors is **singular**. Consequently, A is non-diagonalizable.

Proposition 2.56. Let A be an idempotent matrix of order m and rank r. Then,

$$\text{tr } A = r(A).$$

Proof: From Corollary 2.9, we have

$$\text{tr } A = \sum_{i=1}^{m} \lambda_i.$$

By Proposition 2.55,

$$\lambda_i = 0 \quad \text{or} \quad \lambda_i = 1.$$

Hence,

$$\text{tr } A = \text{number of nonzero roots}$$

or

$$\text{tr } A = r(A).$$

$$\text{q.e.d.}$$

2.13   Semi-definite and Definite Matrices

Definition 2.32. Let A be a square matrix of order m and let x be an m-element vector. Then, A is said to be **positive semi-definite** if and only if for all vectors x

$$x'Ax \geq 0.$$

The matrix A is said to be **positive definite** if and only if for non-null x

$$x'Ax > 0.$$

Definition 2.33. Let A be a square matrix of order m. Then, A is said to be **negative (semi)definite** if and only if $-A$ is positive (semi)definite.

Remark 2.34. It is clear that we need only study the properties of positive (semi)definite matrices, since the properties of negative (semi)definite matrices can easily be derived from the latter.

Remark 2.35. A definite or semi-definite matrix B need not be symmetric. However, because the defining property of such matrices involves the **quadratic form** $x'Bx$, we see that if we put

$$A = \tfrac{1}{2}(B + B')$$

we have $x'Ax = x'Bx$, with A symmetric. Thus, whatever properties may be ascribed to B, by virtue of the fact that for any x, say

$$x'Bx \geq 0,$$

can also be ascribed to A. Thus, we sacrifice no generality if we always take definite or semi-definite matrices to be symmetric. In subsequent discussion, it should be understood that if we say that A is positive (semi) definite we also mean that A is symmetric as well.

Certain properties follow immediately from the definition of definite and semi-definite matrices.

Proposition 2.57. Let A be a square matrix of order m. If A is positive definite, it is also positive semi-definite. The converse, however, is not true.

Proof: The first part is obvious from the definition since if x is any m-element vector and A is positive definite, $x'Ax \geq 0$, so that A is also positive semi-definite.

That the converse is not true is established by an example. Take

$$A = \begin{bmatrix} 1 & 1 \\ 1 & 1 \end{bmatrix}.$$

For any vector $x = (x_1, x_2)'$, $x'Ax = (x_1 + x_2)^2 \geq 0$, so that A is positive semi-definite. For the choice of $x_1 = -x_2$, $x_2 \neq 0$, we have $x'Ax = 0$, which shows that A is not positive definite.

q.e.d.

Proposition 2.58. Let A be a square matrix of order m. Then,

i. If A is positive definite,

$$a_{ii} > 0, \qquad i = 1, 2, \ldots, m$$

ii. If A is only positive semi-definite,

$$a_{ii} \geq 0, \qquad i = 1, 2, \ldots, m.$$

Proof: Let $e_{\cdot i}$ be the m-element unit vector (all of whose elements are zero except the ith, which is unity). If A is positive definite, since $e_{\cdot i}$ is not the null vector, we must have

$$e'_{\cdot i} A e_{\cdot i} > 0, \qquad i = 1, 2, \ldots, m.$$

But

$$e'_{\cdot i} A e_{\cdot i} = a_{ii}, \qquad i = 1, 2, \ldots, m.$$

If A is positive semi-definite but not positive definite, then, repeating the argument above we find

$$a_{ii} = e'_{\cdot i} A e_{\cdot i} \geq 0, \qquad i = 1, 2, \ldots, m.$$

q.e.d.

Another interesting property is the following.

Proposition 2.59 (Triangular decomposition theorem). Let A be a positive definite matrix of order m. Then, there exists a lower triangular matrix T such that

$$A = TT'.$$

Proof: Let

$$T = \begin{bmatrix} t_{11} & 0 & & \cdots & & 0 \\ t_{21} & t_{22} & & \ddots & & \vdots \\ t_{31} & t_{32} & t_{33} & & & \\ & & & \ddots & & 0 \\ t_{m1} & t_{m2} & t_{m3} & & \cdots & t_{mm} \end{bmatrix}.$$

Setting

$$A = TT'$$

we obtain the equations (by equating the (i, j) elements of A and TT')

$$t_{11}^2 = a_{11}, \quad t_{11}t_{21} = a_{12}, \quad t_{11}t_{31} = a_{13}, \quad \ldots, \quad t_{11}t_{m1} = a_{1m}$$

$$t_{21}t_{11} = a_{21}, \qquad t_{21}^2 + t_{22}^2 = a_{22}, \qquad t_{21}t_{31} + t_{22}t_{32} = a_{23}, \quad \ldots,$$

$$t_{21}t_{m1} + t_{22}t_{m2} = a_{2m}$$

$$\vdots$$

$$t_{m1}t_{11} = a_{m1}, \qquad t_{m1}t_{21} + t_{m2}t_{22} = a_{m2}, \quad \ldots, \quad \sum_{i=1}^{m} t_{mi}^2 = a_{mm}.$$

In solving the equations as they are arranged, line by line, we see that we are dealing with a recursive system. From the first line, we have

$$t_{11} = \pm\sqrt{a_{11}}, \quad t_{21} = \frac{a_{12}}{t_{11}}, \quad t_{31} = \frac{a_{13}}{t_{11}}, \quad \ldots, \quad t_{m1} = \frac{a_{1m}}{t_{11}}.$$

From the second line, we have

$$t_{21} = \frac{a_{21}}{t_{11}}, \qquad t_{22} = \pm \left(\frac{a_{22}a_{11} - a_{21}^2}{a_{11}} \right)^{1/2},$$

and in general

$$t_{i2} = \frac{a_{2i} - t_{21}t_{i1}}{t_{22}}, \qquad i = 3, 4, \ldots, m.$$

Similarly, in the third line, we find

$$t_{33} = \pm \left(a_{33} - \frac{a_{31}^2}{t_{31}^2} - \frac{(a_{23} - t_{21}t_{31})^2}{t_{22}^2} \right)^{1/2},$$

$$t_{i3} = \frac{a_{3i} - t_{31}t_{i1} - t_{32}t_{i2}}{t_{33}}, \qquad i = 4, 5, \ldots, m,$$

and so on.

q.e.d.

Remark 2.36. Evidently, the lower triangular matrix above is not unique. In particular, we see that for t_{11} we have the choice

$$t_{11} = \sqrt{a_{11}} \quad \text{or} \quad t_{11} = -\sqrt{a_{11}}.$$

Similarly, for t_{22} we have the choice

$$t_{22} = \left(\frac{a_{22}a_{11} - a_{21}^2}{a_{11}} \right)^{1/2} \quad \text{or} \quad t_{22} = - \left(\frac{a_{22}a_{11} - a_{21}^2}{a_{11}} \right)^{1/2},$$

and so on. The matrix T can be rendered unique if we specify, say, that all diagonal elements must be positive.

Notice further that the same argument as in Proposition 2.59 can establish the existence of a unique **upper triangular** matrix T^* such that $A = T^*T^{*'}$.

In the literature of econometrics, the triangular decomposition of Proposition 2.59 is occasionally referred to as the **Choleski decomposition**.

The properties of characteristic roots of (semi)definite matrices are established in

Proposition 2.60. Let A be a symmetric matrix of order m and let λ_i, $i = 1, 2, \ldots, m$, be its (real) characteristic roots. If A is positive definite,

$$\lambda_i > 0, \qquad i = 1, 2, \ldots, m.$$

If it is only positive semi-definite,

$$\lambda_i \geq 0, \qquad i = 1, 2, \ldots, m.$$

Proof: Let $x._i$ be the normalized characteristic vector corresponding to the root λ_i of A. If A is positive definite,

$$x'._i A x._i = \lambda_i > 0, \qquad i = 1, 2, \ldots, m.$$

If A is merely positive semi-definite, we can only assert

$$x'._i A x._i = \lambda_i \geq 0, \qquad i = 1, 2, \ldots, m.$$

q.e.d.

By now, the reader should have surmised that positive definite matrices are nonsingular, and positive semi-definite matrices (which are not also positive definite) are singular matrices. This is formalized in

Proposition 2.61. Let A be a symmetric matrix of order m. If A is positive definite then

$$r(A) = m.$$

If A is merely positive semi-definite, i.e. it is not also positive definite, then

$$r(A) < m.$$

Proof: Since A is symmetric, let Λ denote the diagonal matrix of its (real) characteristic roots and Q the associated (orthogonal) matrix of characteristic vectors. We have

$$AQ = Q\Lambda.$$

By Propositions 2.51 and 2.44, Q is nonsingular so that A is similar to the diagonal matrix Λ, which is evidently of rank m. Thus $r(A) = m$ and its inverse exists and is given by

$$A^{-1} = Q\Lambda^{-1}Q',$$

which also shows that the characteristic roots of A^{-1} are the diagonal elements of Λ^{-1}, as exhibited below

$$\Lambda^{-1} = \text{diag}\left(\frac{1}{\lambda_1}, \frac{1}{\lambda_2}, \ldots, \frac{1}{\lambda_m}\right).$$

This establishes the first part of the proposition.

For the second part suppose A is **only** positive semi-definite. From Proposition 2.60, we merely know that $\lambda_i \geq 0$, $i = 1, 2, \ldots, m$. We now establish that at least one root must be zero, thus completing the proof of the proposition.

We have the representation

$$Q'AQ = \Lambda.$$

Consequently, for any vector y,

$$y'Q'AQy = \sum_{i=1}^{m} \lambda_i y_i^2.$$

Now, if x is any non-null vector by the semi-definiteness of A we have

$$0 \leq x'Ax = x'QQ'AQQ'x = x'Q\Lambda Q'x = \sum_{i=1}^{m} \lambda_i y_i^2, \qquad (2.47)$$

where now we have put

$$y = Q'x.$$

Since x is non-null, y is also non-null.

If none of the λ_i is zero, Eq. (2.47) implies that for any non-null x $x'Ax > 0$, thus showing A to be positive definite. Consequently, at least one of the λ_i, $i = 1, 2, \ldots, m$, must be zero, and there must exist at least one non-null x such that $x'Ax = \sum_{i=1}^{m} \lambda_i y_i^2 = 0$. But this shows that $r(A) < m$.

q.e.d.

Remark 2.37. Roughly speaking, positive definite and semi-definite matrices correspond to positive and nonnegative numbers in the usual number system. The reader's intuitive comprehension would be aided if he thinks of them as a sort of matrix generalization of positive and nonnegative real numbers. Just as a nonnegative number can always be written as the square of some other number, the same holds *mutatis mutandis* for definite and semi-definite matrices. In fact, this conceptualization leads to the concept of the square root for such matrices.

Proposition 2.62. Let A be a symmetric matrix of order m. Then A is positive definite if and only if there exists a matrix S of dimension $n \times m$ and rank m $(n \geq m)$ such that

$$A = S'S.$$

It is positive semi-definite if and only if

$$r(S) < m.$$

Proof: Sufficiency: If A is positive (semi)definite then, as in the proof of Proposition 2.61, we have the representation

$$A = Q\Lambda Q'.$$

Taking

$$S = \Lambda^{1/2}Q',$$

we have $A = S'S$. If A is positive definite, Λ is nonsingular, and thus

$$r(S) = m.$$

If A is merely positive semi-definite, $r(\Lambda) < m$, and hence

$$r(S) < m.$$

Necessity: Suppose $A = S'S$, where S is $n \times m$ $(n \geq m)$ of rank m. Let x be any non-null vector and note $x'Ax = x'S'Sx$. The right member of the equation above is a sum of squares and thus is zero if and only if

$$Sx = 0. \tag{2.48}$$

If A is positive definite, Eq. (2.48) can be satisfied only with null x. Hence the rank of S is m.

Evidently, for any x, $x'Ax = x'S'Sx \geq 0$, and if A is positive semi-definite (but not positive definite) there exists at least one non-null x such that $x'Ax = 0$; hence, for that x, $Sx = 0$ thus S is of rank less than m.

                                                                    q.e.d.

An obvious consequence of the previous discussion is

Corollary 2.13. If A is a positive definite matrix, $|A| > 0$, $\mathrm{tr}(A) > 0$.

Proof: Let λ_i, $i = 1, 2, \ldots, m$, be the characteristic roots of A. Since

$$|A| = \prod_{i=1}^{m} \lambda_i, \qquad \mathrm{tr}(A) = \sum_{i=1}^{m} \lambda_i,$$

the result follows immediately from Proposition 2.60.

                                                                    q.e.d.

Corollary 2.14. Let A be a positive semi-definite, but not a positive definite, matrix. Then $|A| = 0$, $\mathrm{tr}(A) \geq 0$; $\mathrm{tr}(A) = 0$ if and only if A is the null matrix.

Proof: From the representation

$$|A| = \prod_{i=1}^{m} \lambda_i$$

we conclude that $|A| = 0$ by Proposition 2.61.

For the second part, we note that

$$A = Q\Lambda Q' \tag{2.49}$$

and

$$\text{tr}(A) = 0$$

if and only if

$$\text{tr}(\Lambda) = 0.$$

But

$$\text{tr}(\Lambda) = \sum_{i=1}^{m} \lambda_i = 0, \qquad \lambda_i \geq 0 \text{ all } i$$

implies

$$\lambda_i = 0, \qquad i = 1, 2, \ldots, m.$$

If this holds, then Eq. (2.49) implies

$$A = 0.$$

Consequently, if A is not a null matrix,

$$\text{tr}(A) > 0.$$

q.e.d.

Corollary 2.15 (Square root of a positive definite matrix). Let A be a positive definite matrix of order m. Then, there exists a nonsingular matrix W such that

$$A = W'W.$$

Proof: Obvious from Propositions 2.59, 2.61, and 2.62. A particular choice of W may be $W = Q\Lambda^{1/2}Q'$, which is the usual expression for defining the **square root of a positive definite matrix** A.

q.e.d.

In previous discussions, when considering characteristic roots and characteristic vectors, we did so in the context of the characteristic equation

$$|\lambda I - A| = 0.$$

Often it is more convenient to broaden the definition of characteristic roots and vectors as follows.

Definition 2.34. Let A, B be two matrices of order m, where B is non-singular. The **characteristic roots of A in the metric of B**, and their associated characteristic vectors, are connected by the relation

$$Ax = \lambda Bx,$$

where λ is a characteristic root and x is the associated (non-null) characteristic vector.

Remark 2.38. It is evident that the **characteristic roots of A in the metric of B** are found by solving the polynomial equation

$$|\lambda B - A| = 0.$$

It is also clear that this is a simple generalization of the ordinary definition of characteristic roots where the role of B is played by the identity matrix.

Definition 2.34 is quite useful in dealing with differences of positive (semi)definite matrices, and particularly in determining whether such differences are positive (semi)definite or not. This is intimately connected with the question of relative efficiency in comparing two estimators. We have

Proposition 2.63. Let B be a positive definite matrix and let A be positive (semi)definite, both of order m. Then, the characteristic roots of A in the metric of B, say λ_i, obey

$$\lambda_i > 0, \qquad i = 1, 2, \ldots, m,$$

if A is positive definite, and

$$\lambda_i \geq 0, \qquad i = 1, 2, \ldots, m,$$

if A is positive semi-definite.

Proof: Consider

$$|\lambda B - A| = 0.$$

Since B is positive definite, by Corollary 2.15 there exists a nonsingular matrix P such that

$$B = P^{'-1}P^{-1}.$$

Consequently, by Proposition 2.26,

$$0 = |\lambda B - A| = |\lambda P^{'-1}P^{-1} - A| = |\lambda I - P'AP|\, |P|^{-2}.$$

Thus, the **characteristic roots of A in the metric of B** are simply the usual characteristic roots of $P'AP$, i.e. the solution of

$$|\lambda I - P'AP| = 0.$$

If A is positive definite, $P'AP$ is also positive definite; if A is only positive semi-definite, $P'AP$ is only positive semi-definite. Hence, in the former case

$$\lambda_i > 0, \qquad i = 1, 2, \ldots, m,$$

whereas in the latter case

$$\lambda_i \geq 0, \qquad i = 1, 2, \ldots, m.$$

q.e.d.

A very useful result in this context is

Proposition 2.64 (Simultaneous decomposition). Let B be a positive definite matrix and A positive (semi)definite, both of order m. Let

$$\Lambda = \text{diag}(\lambda_1, \lambda_2, \ldots, \lambda_m)$$

be the diagonal matrix of the characteristic roots of A in the metric of B. Then, there exists a nonsingular matrix W such that

$$B = W'W, \qquad A = W'\Lambda W.$$

Proof: From Proposition 2.63, we have that the λ_i are also the (ordinary) characteristic roots of $P'AP$, where P is such that

$$B = P'^{-1}P^{-1}.$$

Let Q be the (orthogonal) matrix of (ordinary) characteristic vectors of $P'AP$. Thus, we have

$$P'APQ = Q\Lambda. \tag{2.50}$$

From Eq. (2.50), we easily establish

$$A = P'^{-1}Q\Lambda Q'P^{-1}.$$

Putting $W = Q'P^{-1}$, we have

$$A = W'\Lambda W, \qquad B = W'W.$$

q.e.d.

From the preceding two propositions flow a number of useful results regarding differences of positive (semi)definite matrices. Thus,

Proposition 2.65. Let B be a positive definite matrix and A be positive (semi)definite. Then $B - A$ is positive (semi)definite if and only if

$$\lambda_i < 1 \qquad (\lambda_i \le 1),$$

respectively, where the λ_i are the characteristic roots of A in the metric of B, $i = 1, 2, \ldots, m$.

Proof: From Proposition 2.64, there exists a nonsingular matrix W such that

$$B = W'W, \qquad A = W'\Lambda W.$$

Hence,

$$B - A = W'(I - \Lambda)W.$$

Let x be any m-element vector, and note

$$x'(B - A)x = y'(I - \Lambda)y = \sum_{i=1}^{m}(1 - \lambda_i)y_i^2, \tag{2.51}$$

where $y = Wx$. If, for arbitrary non-null x, $x'(B - A)x > 0$, we must have $1 - \lambda_i > 0$, or

$$\lambda_i < 1, \qquad i = 1, 2, \ldots, m, \tag{2.52}$$

thus concluding the if part of the proof; conversely, if Eq. (2.52) holds, then y is **non-null**, for **arbitrary non-null** x; consequently, it follows from Eq. (2.51) that $B - A$ is positive definite, thus concluding the only if part for positive definite matrices.

If, on the other hand, $B - A$ is only positive semi-definite, then for at least one index i we must have

$$\lambda_i = 1,$$

and conversely.

q.e.d.

Another useful result easily obtained from the simultaneous decomposition of matrices is given in

Proposition 2.66. Let A, B be two positive definite matrices, both of order m. If $B - A$ is positive definite, $A^{-1} - B^{-1}$ is also positive definite. If $B - A$ is positive semidefinite, so is $A^{-1} - B^{-1}$.

Proof: We may write $B = W'W$, $A = W'\Lambda W$; by Proposition 2.65, the diagonal elements of Λ (i.e., the roots of A in the metric of B) are less than unity. Hence, $B^{-1} = W^{-1}W'^{-1}$, $A^{-1} = W^{-1}\Lambda^{-1}W'^{-1}$. Thus,

$$A^{-1} - B^{-1} = W^{-1}(\Lambda^{-1} - I_m)W'^{-1}.$$

The diagonal elements of $\Lambda^{-1} - I$ are given by

$$\frac{1}{\lambda_i} - 1 > 0, \qquad i = 1, 2, \ldots, m,$$

and thus

$$A^{-1} - B^{-1}$$

is positive definite by Proposition 2.61.

If $B - A$ is only positive semi-definite, then for at least one of the roots we have $\lambda_{i_0} = 1$. Hence,

$$A^{-1} - B^{-1} = W^{-1}(\Lambda^{-1} - I_m)W'^{-1} \geq 0,$$

owing to the fact that at least one of the diagonal elements of $\Lambda^{-1} - I_m$ is zero.

q.e.d.

Finally, we have

Proposition 2.67. Let B be positive definite, and A be positive (semi)definite. If $B - A$ is positive (semi)definite then

$$|B| > |A|, \quad (|B| \geq |A|), \quad \text{tr}(B) > \text{tr}(A), \quad (\text{tr}(B) \geq \text{tr}(A)).$$

Proof: As in Proposition 2.66, we can write $B = W'W$, $A = W'\Lambda W$, and by Proposition 2.65 we know that the diagonal elements of Λ, viz., the λ_i, obey $\lambda_i < 1$, for all i. Consequently, by Proposition 2.26,

$$|B| = |W|^2, \qquad |A| = |W|^2|\Lambda|.$$

Moreover, by Proposition 2.65, if $B - A$ is positive definite, $|\lambda_i| < 1$, for all i, and hence $|B| > |A|$.

Evidently, the inequality above is automatically satisfied if A itself is merely positive semi-definite. Moreover,

$$\text{tr}(B) - \text{tr}(A) = \text{tr}(B - A) > 0.$$

On the other hand, if $B - A$ is merely positive semi-definite then we can only assert $|\Lambda| \leq 1$, and hence we conclude that

$$|B| \geq |A|, \qquad \text{tr}(B) \geq \text{tr}(A).$$

q.e.d.

Corollary 2.16. In Proposition 2.67, the strict inequalities hold unless

$$B = A.$$

Proof: Since

$$\frac{|A|}{|B|} = |\Lambda| = \prod_{i=1}^{m} \lambda_i,$$

we see that $|A| = |B|$ implies $\lambda_i = 1$, $\quad i = 1, 2, \ldots, m$. Hence, $B = A$. Moreover, from the proof of Proposition 2.65,

$$\text{tr}(B) = \text{tr}(A)$$

implies

$$0 = \text{tr}(B - A) = \text{tr}[W'(I - \Lambda)W].$$

But this means

$$W'(I - \Lambda)W = 0,$$

which in turn implies

$$\Lambda = I_m,$$

and consequently

$$B = A.$$

q.e.d.

We now present the very useful **singular value decomposition** theorem for an arbitrary matrix A.

In Sect. 2.5 we examined the **rank factorization** theorem and showed that if A is $m \times n$ of rank $r \le n \le m$, then there exist matrices C_1, C_2, respectively, of dimension $m \times r$, $r \times n$ and both of rank r, such that

$$A = C_1 C_2.$$

The matrices C_1, C_2 are, of course, non-unique. The construction given in that section proceeds from first principles and essentially utilizes elementary row and column operations. Although conceptually simple and straightforward, that construction is not particularly useful for applied work. In view of the ready availability of computer software for obtaining the characteristic roots and vectors of symmetric matrices, the following result is perhaps more convenient.

Proposition 2.68 (Singular value decomposition theorem). Let A be $m \times n$ of rank r, $r \le n \le m$. Then, there exist matrices B_1, B_2 and a diagonal matrix, D, with positive diagonal elements, such that

$$A = B_1 D B_2.$$

Proof: Consider the matrices AA', $A'A$; both are of rank r and of dimension $m \times m$, $n \times n$, respectively.

By Proposition 2.52, we have the representation

$$AA' = Q\Lambda Q',$$

where Q is the (orthogonal) matrix of characteristic vectors and Λ is the (diagonal) matrix of the corresponding characteristic roots. Similarly,

$$A'A = RMR'$$

where again R, M are, respectively, the matrices of characteristic vectors and corresponding characteristic roots of $A'A$.

By Corollary 2.11, we conclude that since AA' and $A'A$ are both of rank r, only r of their characteristic roots are positive, the remaining being zero. Hence, we can write

$$\Lambda = \begin{bmatrix} \Lambda_r & 0 \\ 0 & 0 \end{bmatrix}, \qquad M = \begin{bmatrix} M_r & 0 \\ 0 & 0 \end{bmatrix}.$$

Partition Q, R conformably with Λ and M, respectively, i.e.

$$Q = (Q_r, Q_*), \qquad R = (R_r, R_*)$$

such that Q_r is $m \times r$, R_r is $n \times r$ and correspond, respectively, to the non-zero characteristic roots of AA' and $A'A$.

Take

$$B_1 = Q_r, \; B_2 = R'_r, \qquad \Lambda_r = D = \mathrm{diag}(\lambda_1, \lambda_2, \ldots, \lambda_r),$$

where λ_i, $i = 1, 2, \ldots, r$ are the positive characteristic roots of AA' and hence, by Corollary 2.8, those of $A'A$ as well.

Now, define

$$S = Q_r \Lambda_r^{1/2} R'_r = B_1 D B_2.$$

We show that $S = A$, thus completing the proof. We easily verify that

$$S'S = A'A, \qquad SS' = AA'.$$

From the first relation above we conclude that for an arbitrary orthogonal matrix, say P_1,

$$S = P_1 A$$

whereas from the second we conclude that for an arbitrary orthogonal matrix P_2 we must have

$$S = AP_2.$$

The preceding, however, implies that for arbitrary orthogonal matrices P_1, P_2 the matrix A satisfies

$$AA' = P_1 AA' P_1', \qquad A'A = P_2' A'AP_2,$$

which in turn implies that

$$P_1 = I_m, \qquad P_2 = I_n.$$

Thus,

$$A = S = Q_r \Lambda_r^{1/2} R_r' = B_1 D B_2.$$

q.e.d.

We close this chapter by formally defining the "square root" of a positive (semi)definite matrix.

Definition 2.35. Let A be a positive (semi)definite matrix, Λ the diagonal matrix of its (nonnegative) characteristic roots, and Q the matrix of its associated characteristic vectors. The **square root** of A, denoted by $A^{1/2}$, is defined by

$$A^{1/2} = Q\Lambda^{1/2}Q', \qquad \Lambda = \operatorname{diag}(\lambda_1^{1/2}, \lambda_2^{1/2}, \ldots, \lambda_m^{1/2})$$

Chapter 3

Systems of Linear Equations

3.1 Introduction

Consider the system of linear equations

$$Ax = b, \qquad (3.1)$$

where A is $m \times n$ and b is an m-element vector. The meaning of Eq. (3.1), as a system of equations, is that we seek an n-element vector x satisfying that system. If $m = n$ and if A is nonsingular, the equation system above has the unique solution

$$x^* = A^{-1}b. \qquad (3.2)$$

If A is singular, i.e. if $\mathrm{r}(A) < m$ or if $n > m$, it is clear that more than one solution may exist.

Moreover, if A is $m \times n$ but $m > n$, the system may be **inconsistent**, i.e. there may not exist a vector x satisfying all the conditions (equations) specified in Eq. (3.1). In such a case, we may wish to derive "approximate" solutions. In doing so, we are in effect defining "pseudoinverses" (literally fake inverses) that mimic some or all of the properties of the proper inverse.

By way of motivation, note that in the context of the general linear model (GLM) we may characterize estimation in the following terms. Find a vector b such that

$$y = Xb \qquad (3.3)$$

is satisfied approximately, where X is a $T \times (n+1)$ matrix of observations on the explanatory variables and y is a T-element vector of observations on the dependent variable. Typically

$$T > n + 1,$$

and the system in Eq. (3.3) is inconsistent, since no vector b exists strictly satisfying Eq. (3.3). It is such considerations that prompt us to study various types of pseudoinverses.

P.J. Dhrymes, *Mathematics for Econometrics*,   95
DOI 10.1007/978-1-4614-8145-4_3, © The Author 2013

3.2   The c-, s-, and g-Inverses

Definition 3.1. Let A be $m \times n$ $(m \leq n)$; the $n \times m$ matrix A_c is said to be the **conditional inverse**, generally also referred to as the **c-inverse** of A, if and only if

$$AA_cA = A.$$

Remark 3.1. Note that if A is a nonsingular matrix then clearly

$$AA^{-1}A = A,$$

so that the c-inverse satisfies **only** this property of the proper inverse. Evidently, the latter satisfies other conditions as well. For example, it requires that

$$A^{-1}A, \qquad AA^{-1},$$

be symmetric matrices and, in fact, equal to the identity matrix. But this is not necessarily satisfied by the conditional inverse.

We now show that the c-inverse is not a vacuous concept.

Proposition 3.1. Let A be $m \times m$ and let B be a nonsingular, $m \times m$, matrix such that

$$BA = H,$$

where H is in (upper) Hermite form. Then, B is a c-inverse of A. Moreover, B' is the c-inverse of A', and we write $B = A_c$, $B' = (A')_c = A'_c$.

Proof: It is clear, by Proposition 2.9, that such a nonsingular matrix B exists, **because for square matrices an upper echelon form is in fact an upper Hermite form.** From Proposition 2.13, H is an idempotent matrix. Hence,

$$H = HH = BABA.$$

Pre-multiplying by B^{-1}, we find

$$ABA = B^{-1}H = A, \quad \text{or } \mathring{A}_cA = A.$$

We further note that $H'H' = (HH)' = H'$, due to the fact that H is idempotent. Consequently, $A'B'A' = H'B'^{-1} = A'$, thus completing the proof.

q.e.d.

For rectangular (non-square) matrices, we have

Proposition 3.2. Let A be $m \times n$ $(m \leq n)$ and

$$A_0 = \begin{pmatrix} A \\ 0 \end{pmatrix},$$

where A_0 is $n \times n$. Let B_0 be a nonsingular matrix of order n such that

$$B_0 A_0 = H_0,$$

and H_0 is in (upper) Hermite form. Partition

$$B_0 = (B, B_1),$$

where B is $n \times m$. Then, B is a c-inverse of A, denoted A_c.

Proof: By Proposition 3.1, B_0 is a c-inverse of A_0. Hence, we have

$$A_0 B_0 A_0 = A_0.$$

But

$$\begin{pmatrix} A \\ 0 \end{pmatrix} = A_0 = A_0 B_0 A_0 = \begin{bmatrix} AB & AB_1 \\ 0 & 0 \end{bmatrix} \begin{pmatrix} A \\ 0 \end{pmatrix} = \begin{pmatrix} ABA \\ 0 \end{pmatrix},$$

which shows that

$$A = ABA, \quad \text{or} \quad AA_c A = A.$$

q.e.d.

Remark 3.2. A similar result is obtained if $m \geq n$. One has only to deal with the transpose of A in Proposition 3.2 and note that if A_c is the c-inverse of A, A'_c is the c-inverse of A', which was proved in Proposition 3.1.

Evidently, c-inverses are not unique, since the matrix B reducing a given matrix A to Hermite form is not unique. This is perhaps best made clear by an example.

Example 3.1. Let

$$A = \begin{bmatrix} 2 & 3 & 1 \\ 4 & 5 & 1 \\ 1 & 1 & 0 \end{bmatrix}$$

and observe that

$$B_1 = \begin{bmatrix} -2 & 1 & 1 \\ 0 & 1 & -4 \\ -1 & 1 & -2 \end{bmatrix}, \quad B_2 = \begin{bmatrix} -3 & 2 & -1 \\ -1 & 2 & -6 \\ -2 & 2 & -4 \end{bmatrix},$$

have the property of reducing A to Hermite form, i.e.

$$B_1 A = B_2 A = \begin{bmatrix} 1 & 0 & -1 \\ 0 & 1 & 1 \\ 0 & 0 & 0 \end{bmatrix}.$$

A somewhat more stringent set of requirements defines the so-called least squares inverse.

Definition 3.2. Let A be $m \times n$ $(m \leq n)$. The $n \times m$ matrix A_s is said to be a **least squares inverse**, generally referred to as the **s-inverse** of A, if and only if

i. $AA_sA = A$,

ii. AA_s is symmetric.

Remark 3.3. Evidently, if A_s is an s-inverse, it is also a c-inverse. The converse, however, is not true.

That the class of s-inverses is not a vacuous one is shown by

Proposition 3.3. Let A be $m \times n$ $(m \leq n)$. Then,

i.

$$A_s = (A'A)_c A'$$

is an s-inverse of A.

ii. If A_s is an s-inverse of A, A'_s is not necessarily an s-inverse of A'.

Proof: To prove part i, we need to show that

$$AA_sA = A(A'A)_c A'A = A, \quad \text{and that} \quad AA_s = A(A'A)_c A'$$

is symmetric. To this end, we note that since $(A'A)_c$ is a c-inverse it satisfies

$$(A'A)(A'A)_c A'A = A'A. \tag{3.4}$$

Let A have rank $r \leq m$. From Proposition 2.15, there exist matrices C_1, C_2 of rank r and dimensions $m \times r$, $r \times n$, respectively, such that (rank factorization)

$$A = C_1 C_2.$$

Thus we can write Eq. (3.4) as

$$C_2'C_1'C_1C_2(A'A)_cC_2'C_1'C_1C_2 = C_2'C_1'C_1C_2.$$

Pre-multiply by

$$C_1(C_1'C_1)^{-1}(C_2C_2')^{-1}C_2$$

to obtain

$$C_1C_2(A'A)_cC_2'C_1'C_1C_2 = C_1C_2. \tag{3.5}$$

Bearing in mind the definition of A_s and the rank factorization of A, we see that Eq. (3.5) can also be written as

$$AA_sA = A.$$

Moreover,

$$AA_s = A(A'A)_cA'$$

is evidently symmetric because $(A'A)_c$ is a symmetric matrix.

To prove part ii, we note that if A_s is an s-inverse of A, $A_s' = A(A'A)_c$. Thus,

$$A'A_s'A' = (AA_sA)' = A',$$

so that it satisfies the first component of the definition for an s-inverse.[1] However, it need not satisfy the second component because

$$A'A_s' = (A'A)(A'A)_c$$

is the product of two symmetric matrices. However, the product of two symmetric matrices is not necessarily symmetric **unless** they commute!

q.e.d.

Remark 3.4. Evidently, since the c-inverse is not unique, the s-inverse is also not unique.

A unique pseudoinverse is defined in the following manner.

Definition 3.3. Let A be $m \times n$; the $n \times m$ matrix A_g is said to be a **generalized inverse**, generally referred to as the **g-inverse** of A, if and only if it satisfies

i. $AA_gA = A$,

ii. AA_g is symmetric,

---

[1]This is to be expected since an s-inverse is also a c-inverse, and the latter has this property.

iii. $A_g A$ is symmetric,

iv. $A_g A A_g = A_g$.

Remark 3.5. Note that the g-inverse mimics the corresponding conditions satisfied by a proper inverse, for if A is nonsingular and A^{-1} is its inverse,

$$AA^{-1}A = A,$$
$$AA^{-1} = I \text{ is symmetric,}$$
$$A^{-1}A = I \text{ is symmetric,}$$
$$A^{-1}AA^{-1} = A^{-1}.$$

Finally, we note that c-inverses, s-inverses and g-inverses are more generally referred to as **pseudoinverses**.

3.3   Properties of the Generalized Inverse

In this section, we examine a number of useful properties of the g-inverse. We begin with the existence and uniqueness properties.

Proposition 3.4. Let A be any $m \times n$ matrix. Then, the following statements are true:

i. There exists a unique matrix A_g satisfying the conditions of Definition 3.3;

ii. $A_g A$, $A A_g$ are idempotent matrices;

iii. The g-inverse of A' is A'_g.

Proof: It is clear that if A is the null matrix, the $n \times m$ null matrix is the g-inverse of A. Thus, suppose $\text{rank}(A) = r > 0$. By Proposition 2.15, there exist two matrices, namely C_1, which is $m \times r$ of rank r, and C_2, which is $r \times n$ of rank r, such that $A = C_1 C_2$. Define

$$A_g = C_2'(C_2 C_2')^{-1}(C_1' C_1)^{-1} C_1'$$

and observe that

$$AA_g = C_1 C_2 C_2'(C_2 C_2')^{-1}(C_1' C_1)^{-1} C_1' = C_1 (C_1' C_1)^{-1} C_1',$$
$$A_g A = C_2'(C_2 C_2')^{-1}(C_1' C_1)^{-1} C_1' C_1 C_2 = C_2'(C_2 C_2')^{-1} C_2.$$

This shows AA_g **and** $A_g A$ to be symmetric idempotent matrices, thus satisfying ii and iii of Definition 3.3, and proving part ii of the proposition.

Moreover,

$$AA_gA = C_1C_2C_2'(C_2C_2')^{-1}C_2 = C_1C_2 = A,$$
$$A_gAA_g = C_2'(C_2C_2')^{-1}(C_1'C_1)^{-1}C_1'C_1(C_1'C_1)^{-1}C_1' = A_g,$$

which shows the existence of the g-inverse. To show uniqueness, suppose B_g is another g-inverse of A. We show $A_g = B_g$, thus completing the proof of part i of the proposition.

Now, $AA_gA = A$. Postmultiplying by B_g, we have $AA_gAB_g = AB_g$. Because AB_g and AA_g are both symmetric, we obtain

$$AB_g = (AB_g)'(AA_g)' = AB_gAA_g = AA_g.$$

Similarly, $B_gA = B_gAA_gA = A_gAB_gA = A_gA$. Premultiplying the relation $AB_g = AA_g$ by B_g, we have

$$B_g = B_gAB_g = B_gAA_g = A_gAA_g = A_g.$$

To show the validity of part iii, we simply note that if A_g is a g-inverse of A, then transposing the four conditions of Definition 3.3 yields the conclusion that A_g' is the g-inverse of A', which completes the proof of the proposition.

q.e.d.

Let us now establish some other useful properties of the g-inverse.

Proposition 3.5. Let A be an $m \times m$ symmetric matrix of rank r $(r \leq m)$. Let D_r be the diagonal matrix containing its nonzero characteristic roots (in decreasing order of magnitude), and let P_r be the $m \times r$ matrix whose columns are the (orthonormal) characteristic vectors corresponding to the nonzero roots of A. Then

$$A_g = P_rD_r^{-1}P_r.$$

Proof: By the definition of characteristic roots and vectors,

$$AP = PD,$$

where P is the orthogonal matrix of characteristic vectors of A, and D is the diagonal matrix of the latter's characteristic roots arranged in decreasing order of magnitude. Because A is of rank r, D can be written as

$$D = \begin{bmatrix} D_r & 0 \\ 0 & 0 \end{bmatrix}.$$

Partition P by

$$P = (P_r, P_*),$$

where P_r is $m \times r$, and note

$$A = PDP' = (P_r, P_*) \begin{bmatrix} D_r & 0 \\ 0 & 0 \end{bmatrix} \begin{pmatrix} P'_r \\ P'_* \end{pmatrix} = P_r D_r P'_r.$$

We verify

i. $AA_g A = (P_r D_r P'_r)(P_r D_r^{-1} P'_r)(P_r D_r P'_r) = P_r D_r P'_r = A;$

ii. $AA_g = (P_r D_r P'_r)(P_r D_r^{-1} P'_r) = P_r P'_r,$ which is symmetric;

iii. $A_g A = (P_r D_r^{-1} P'_r)(P_r D_r P'_r) = P_r P'_r;$

iv. $A_g AA_g = P_r P'_r P_r D_r^{-1} P'_r = P_r D_r^{-1} P'_r = A_g.$

This shows that A_g, above, is the g-inverse of A.

q.e.d.

Corollary 3.1. If A is symmetric **and** idempotent, then

$$A_g = A.$$

Proof: If A is symmetric **and** idempotent, its characteristic roots are either zero or one. Hence, in the representation above $D_r = I_r$. Thus,

$$A = P_r P'_r, \qquad A_g = P_r P'_r.$$

q.e.d.

We have already seen that if A_g is the g-inverse of A, $(A_g)'$ is the g-inverse of A'. We now examine a number of other properties of the g-inverse that reveal it to be analogous to the ordinary (proper) inverse of a nonsingular matrix.

Proposition 3.6. Let A be $m \times n$. The following statements are true:

i. $(A_g)_g = A;$

ii. $r(A_g) = r(A);$

iii. $(A'A)_g = A_g A'_g;$

iv. $(AA_g)_g = AA_g;$

v. If $m = n$ and A is nonsingular, $A^{-1} = A_g.$

Proof: The proof of i is simple since A_g is the g-inverse of A, and thus it satisfies:

i. $A_g A A_g = A_g$;

ii. $A_g A$ is symmetric;

iii. $A A_g$ is symmetric;

iv. $A A_g A = A$.

But Definition 3.3 indicates that the above define A as the g-inverse of A_g.

To prove ii, we note that $A A_g A = A$ implies $r(A) \leq r(A_g)$, whereas $A_g A A_g = A_g$ implies $r(A_g) \leq r(A)$. Together, these relations imply

$$r(A) = r(A_g).$$

To prove iii, we verify that $A_g A'_g$ is the g-inverse of $A'A$. Since

i. $(A'A) A_g A'_g (A'A) = (A' A'_g A')(A A_g A) = A'A$;

ii. $(A'A) A_g A'_g = A' A'_g A' A'_g = (A_g A)'$ is symmetric;

iii. $A_g A'_g (A'A) = (A_g A)$ is symmetric;

iv. $A_g A'_g (A'A) A_g A'_g = (A_g A A_g)(A'_g A' A'_g) = A_g A'_g$,

$A_g A'_g$ is indeed the g-inverse of $A'A$.

Part iv may be proved by noting that $A A_g$ is a symmetric idempotent matrix, and that Corollary 3.1 states that $(A A_g)_g = A A_g$.

To prove v, we note that $A A^{-1} A = A$, $A A^{-1} = A^{-1} A = I$ is symmetric, and $A^{-1} A A^{-1} = A^{-1}$, which completes the proof.

q.e.d.

Corollary 3.2. Let A be an $m \times n$ matrix; let P be $m \times m$ and Q be $n \times n$, and let both be orthogonal. Then,

$$(PAQ)_g = Q' A_g P'.$$

Proof: We have

$$(PAQ)(PAQ)_g = PAQQ' A_g P' = P A A_g P',$$
$$(PAQ)_g (PAQ) = Q' A_g P' PAQ = Q' A_g AQ.$$

The symmetry of the matrices above follows from the symmetry of AA_g and $A_g A$, respectively. Moreover,

$$(PAQ)_g(PAQ)(PAQ)_g = Q'A_g P'PAQQ'A_g P' = Q'A_g AA_g P' = (PAQ)_g,$$
$$(PAQ)(PAQ)_g(PAQ) = PAQQ'A_g P'PAQ = PAA_g AQ = PAQ.$$

q.e.d.

Remark 3.6. It is worth entering a note of caution here. It is well known that if A, B are conformable nonsingular matrices, then $(AB)^{-1} = B^{-1}A^{-1}$. The results in iii and iv of the preceding proposition may suggest that the same is true of g-inverses. Unfortunately, it is not **generally** true that if A, B are $m \times n$ and $n \times q$, respectively, then $(AB)_g = B_g A_g$. This is true for the matrices in iii and iv of the preceding proposition, as well as for those in Corollary 3.2.

In the following discussion we consider a number of other instances where the relation above is valid.

Proposition 3.7. Let D be a diagonal matrix,

$$D = \begin{bmatrix} C & 0 \\ 0 & 0 \end{bmatrix}. \text{ Then } D_g = \begin{bmatrix} C^{-1} & 0 \\ 0 & 0 \end{bmatrix},$$

where the diagonal elements of C are nonzero.

Proof: Obvious.

Corollary 3.3. Let D, E be two diagonal matrices and put $F = DE$. Then, $F_g = E_g D_g$.

Proof: If either E or D is the zero matrix, the result holds; thus, let us assume that neither D nor E are null. Without loss of generality, put

$$D = \begin{bmatrix} C_1 & 0 & 0 \\ 0 & 0 & 0 \\ 0 & 0 & 0 \end{bmatrix}, \quad E = \begin{bmatrix} E_1 & 0 & 0 \\ 0 & E_2 & 0 \\ 0 & 0 & 0 \end{bmatrix},$$

where it is implicitly assumed that E contains more non-null elements than D, and note that

$$F = \begin{bmatrix} C_1 E_1 & 0 & 0 \\ 0 & 0 & 0 \\ 0 & 0 & 0 \end{bmatrix}.$$

By Proposition 3.7,

$$F_g = \begin{bmatrix} E_1^{-1}C_1^{-1} & 0 & 0 \\ 0 & 0 & 0 \\ 0 & 0 & 0 \end{bmatrix}$$

$$= \begin{bmatrix} E_1^{-1} & 0 & 0 \\ 0 & E_2^{-1} & 0 \\ 0 & 0 & 0 \end{bmatrix} \begin{bmatrix} C_1^{-1} & 0 & 0 \\ 0 & 0 & 0 \\ 0 & 0 & 0 \end{bmatrix} = E_g D_g.$$

q.e.d.

Proposition 3.8. Let A be $m \times n$ $(m \le n)$ of rank m. Then,

$$A_g = A'(AA')^{-1}, \qquad AA_g = I.$$

Proof: We verify that A_g is the g-inverse of A. To this end, we note that AA' is $m \times m$ of rank m; hence, the inverse exists. Moreover,

i. $AA_gA = AA'(AA')^{-1}A = A$;

ii. $AA_g = AA'(AA')^{-1} = I$ is symmetric (and idempotent);

iii. $A_gA = A'(AA')^{-1}A$ is symmetric (and idempotent);

iv. $A_gAA_g = A_g$,

which completes the proof.

q.e.d.

Corollary 3.4. Let A be $m \times n$ $(m \ge n)$ of rank n. Then,

$$A_g = (A'A)^{-1}A', \qquad A_gA = I.$$

Proof: Obvious.

Proposition 3.9. Let B be $m \times r$, C be $r \times n$, and let both be of rank r. Then,

$$(BC)_g = C_g B_g.$$

Proof: By Proposition 3.8,

$$B_g = (B'B)^{-1}B', \qquad C_g = C'(CC')^{-1}.$$

Putting $A = BC$, we may verify that

$$A_{\mathrm{g}} = C_{\mathrm{g}} B_{\mathrm{g}}.$$

q.e.d.

A further useful result is

Proposition 3.10. Let A be $m \times n$. Then, the following statements are true:

i. $I - AA_{\mathrm{g}}$, $I - A_{\mathrm{g}}A$ are symmetric, idempotent;

ii. $(I - AA_{\mathrm{g}})A = 0$, $A_{\mathrm{g}}(I - AA_{\mathrm{g}}) = 0$;

iii. $(I - AA_{\mathrm{g}})AA_{\mathrm{g}} = AA_{\mathrm{g}}(I - AA_{\mathrm{g}}) = 0$;

iv. $(I - A_{\mathrm{g}}A)A_{\mathrm{g}}A = A_{\mathrm{g}}A(I - A_{\mathrm{g}}A) = 0$.

Proof: Proposition 3.4 states that AA_{g}, $A_{\mathrm{g}}A$ are both symmetric and idempotent. Hence,

$$(I - AA_{\mathrm{g}})(I - AA_{\mathrm{g}}) = I - AA_{\mathrm{g}} - AA_{\mathrm{g}} + AA_{\mathrm{g}}AA_{\mathrm{g}} = I - AA_{\mathrm{g}}.$$

Similarly,

$$(I - A_{\mathrm{g}}A)(I - A_{\mathrm{g}}A) = I - A_{\mathrm{g}}A - A_{\mathrm{g}}A + A_{\mathrm{g}}AA_{\mathrm{g}}A = I - A_{\mathrm{g}}A,$$

which proves i.
   Since

$$AA_{\mathrm{g}}A = A, \qquad A_{\mathrm{g}}AA_{\mathrm{g}} = A_{\mathrm{g}},$$

the proof of ii is obvious.
   The proofs of iii and iv follow easily from that of ii.

q.e.d.

To conclude this section, we give some additional results for certain special types of matrices.

Proposition 3.11. Let B, C be, respectively, $m \times s$, $n \times s$, such that $BC' = 0$ and

$$A = \begin{pmatrix} B \\ C \end{pmatrix}.$$

Then,

$$A_{\mathrm{g}} = (B_{\mathrm{g}}, C_{\mathrm{g}}).$$

Proof: We verify that A_g is the g-inverse of A. We have to show that

$$A_g A = B_g B + C_g C$$

is symmetric. Now, B_g, C_g are, respectively, the g-inverses of B, C. Thus $B_g B$, $C_g C$ are both symmetric matrices, and consequently so is $A_g A$. Also

$$AA_g = \begin{bmatrix} BB_g & BC_g \\ CB_g & CC_g \end{bmatrix}.$$

From Proposition 3.6, we note that

$$BC_g = BC_g CC_g = BC'C'_g C_g = 0;$$

the last equality is valid because of the condition $BC' = 0$. Similarly,

$$CB_g = CB_g BB_g = CB'B'_g B = 0,$$

which shows that

$$AA_g = \begin{bmatrix} BB_g & 0 \\ 0 & CC_g \end{bmatrix},$$

which is, clearly, a symmetric matrix. Moreover,

$$AA_g A = \begin{bmatrix} BB_g & 0 \\ 0 & CC_g \end{bmatrix} \begin{pmatrix} B \\ C \end{pmatrix} = \begin{pmatrix} BB_g B \\ CC_g C \end{pmatrix} = \begin{pmatrix} B \\ C \end{pmatrix} = A,$$

$$A_g AA_g = (B_g, C_g) \begin{bmatrix} BB_g & 0 \\ 0 & CC_g \end{bmatrix}$$
$$= (B_g BB_g, C_g CC_g) = (B_g, C_g) = A_g,$$

thus completing the proof.

q.e.d.

Proposition 3.12. If B, C are any matrices, and

$$A = \begin{bmatrix} B & 0 \\ 0 & C \end{bmatrix},$$

then

$$A_g = \begin{bmatrix} B_g & 0 \\ 0 & C_g \end{bmatrix}.$$

Alternatively, if we put

$$A = B \otimes C,$$

then

$$A_g = B_g \otimes C_g.$$

Proof: Obvious by direct verification.

3.4   Linear Equations and Pseudoinverses

What motivated our exploration of the theory of pseudoinverses was the desire to characterize the class of solutions to the linear system

$$Ax = b,$$

where A is $m \times n$ and, generally, $m \neq n$. When A is not a square matrix, the question that naturally arises is whether the system is consistent, i.e. whether there exists at least one vector, x_*, that satisfies the system, and, if consistent, how many solutions there are and how they may be characterized.

The first question is answered by

Proposition 3.13. Let A be $m \times n$; then, the system of equations

$$Ax = b$$

is consistent if and only if for some c-inverse of A,

$$AA_c b = b.$$

Proof: Necessity: Suppose the system is consistent, and x_0 is a solution, i.e.

$$b = Ax_0.$$

Premultiply by AA_c to obtain

$$AA_c b = AA_c Ax_0 = Ax_0 = b,$$

which establishes necessity.

Sufficiency: Assume that for **some** c-inverse,

$$AA_c b = b.$$

Take

$$x = A_c b,$$

and observe that this is a **solution**, thus completing the proof.

q.e.d.

The question now arises as to how many solutions there are, given that there is at least one solution (i.e. the system is consistent). This is answered by

Proposition 3.14. Let A be $m \times n$, $(m \geq n)$ and suppose

$$Ax = b.$$

This is a **consistent** system if and only if there exists an arbitrary vector d, such that

$$x = A_c b + (I_n - A_c A)d$$

is a solution.

Proof: Necessity: Suppose the system is consistent; by Proposition 3.13, there exists a c-inverse such that

$$A A_c b = b.$$

Let

$$x = A_c b + (I_n - A_c A)d,$$

where d is **arbitrary**, and observe

$$Ax = A A_c b = b,$$

which shows x to be a solution, thus establishing necessity.

Sufficiency: Suppose x is **any** solution, i.e. it satisfies $b - Ax = 0$, so that the system is consistent. Pre-multiply by A_c to obtain

$$A_c b - A_c A x = 0.$$

Adding x to both sides of the equation, we have

$$x = A_c b + (I_n - A_c A)x,$$

which shows that the solution is of the desired form with $d = x$, thus completing the proof.

q.e.d.

Corollary 3.5. The statements of the proposition are true if A_c is replaced by A_g.

Proof: Clearly, for a consistent system,

$$x = A_g b + (I_n - A_g A)d$$

is a solution, where d is arbitrary.

Conversely, if x is any solution, so is $x = A_g b + (I_n - A_g A)x$, which completes the proof.

q.e.d.

Corollary 3.6. The solution to the system $Ax = b$, as above, is unique if and only if

$$A_g A = I_n.$$

Proof: If

$$A_g A = I_n,$$

then the general solution of the preceding corollary

$$x = A_g b + (I_n - A_g A)d,$$

becomes

$$x = A_g b,$$

which is unique, because the generalized inverse is unique.

Conversely, if the general solution above is unique **for every vector** d, then we must have $A_g A = I_n$.

q.e.d.

Corollary 3.7. The solution of the **consistent** system above is unique if and only if

$$r(A) = n.$$

Proof: From Corollary 3.4, if $\text{rank}(A) = n$, then $A_g A = I_n$. Corollary 3.5 then shows that the solution is unique.

Conversely, suppose the solution is unique. Then,

$$A_g A = I_n,$$

which shows that

$$n \leq r(A).$$

But the rank of A cannot possibly exceed n. Thus,

$$r(A) = n.$$

q.e.d.

It is clear from the preceding discussion that there are, in general, infinitely many solutions to the system considered above. Thus, for example, if $x_{\cdot i}$ is a solution, $i = 1, 2, \ldots, k$, then

$$x = \sum_{i=1}^{k} \gamma_i x_{\cdot i}$$

is also a solution, provided $\sum_{i=1}^{k} \gamma_i = 1$.

This prompts us to inquire how many linearly independent solutions there are; if we determine this, then all solutions can be expressed in terms of this linearly independent set. We have

Proposition 3.15. Let the (consistent) system

$$Ax = b$$

be such that A is $m \times n$ $(m \geq n)$ of rank $0 < r \leq n$, $b \neq 0$. Then, there are $n - r + 1$ linearly independent solutions.

Proof: Recall that since

$$A_g A A_g = A_g, \qquad A A_g A = A,$$

we have

$$r(A_g A) = r(A) = r.$$

Now, the general solution of the system can be written as

$$x = A_g b + (I_n - A_g A)d$$

for arbitrary d.

Consider, in particular, the vectors

$$x_{\cdot i} = A_g b + (I_n - A_g A)d_{\cdot i}, \qquad i = 0, 1, 2, \ldots, n,$$

where the $d_{\cdot i}$ are n-element vectors such that for $i = 0$, $d_{\cdot i} = 0$, while for $i \neq 0$ all the elements of $d_{\cdot i}$ are zero save the ith, which is unity. Write

$$X = (x_{\cdot 0}, x_{\cdot 1}, \ldots, x_{\cdot n}) = (A_g b, I_n - A_g A) \begin{bmatrix} 1 & e' \\ 0 & I \end{bmatrix},$$

where e is an n-element column vector all of whose elements are unity. Since the upper triangular matrix in the right member above is nonsingular, we conclude

$$r(X) = r(A_g b, I_n - A_g A) = 1 + n - r.$$

The last equality follows since $A_g b$ is orthogonal to $I_n - A_g A$, and thus the two are linearly independent. In addition,

$$r(I_n - A_g A) = n - r(A_g A) = n - r.$$

Thus, we see that the number of linearly independent solutions cannot exceed n (since we deal with the case $r > 0$) – and at any rate it is exactly $n - r + 1$.

q.e.d.

Remark 3.7. It should be pointed out that the $n-r+1$ linearly independent solutions above do not constitute a vector space since the vector 0 is not a solution in view of the condition $b \neq 0$.

Because there are many solutions to the typical system considered here, the question arises whether there are (linear) functions of solutions that are invariant to the particular choice of solution. This is answered by

Proposition 3.16. Let A be $m \times n$; the linear transformation

$$Gx,$$

where x is a solution of the **consistent** system

$$Ax = b,$$

is unique if and only if G lies in the space spanned by the rows of A.

Proof: The general form of the solution is

$$x = A_g b + (I_n - A_g A)d$$

for arbitrary d. Thus,

$$Gx = GA_g b + G(I_n - A_g A)d$$

is unique if and only if

$$G = GA_g A.$$

But, if the equation above is satisfied, G lies in the row space of A. Conversely, suppose G lies in the row space of A. Then, there exists a matrix C such that

$$G = CA.$$

Consequently,

$$GA_g A = CAA_g A = CA = G.$$

q.e.d.

3.5   Approximate Solutions

In the previous section, we examined systems of linear equations and gave necessary and sufficient conditions for their consistency, i.e. for the existence of solutions. Moreover, we gave a characterization of such solutions. Here, we shall examine **inconsistent systems**. Thus, a system

$$Ax = b$$

may have no solution, and thus it may be better expressed as

$$r(x) = Ax - b.$$

Nonetheless, we may wish to determine a vector x_* that is an approximate solution in the sense that

$$r(x_*)$$

is "small." The precise meaning of this terminology will be made clear below.

Definition 3.4. Let A be $m \times n$, and consider the system

$$r(x) = Ax - b.$$

A solution x_* is said to be a **least squares (LS) approximate solution** if and only if for all n-element vectors x

$$r(x)'r(x) \geq r(x_*)'r(x_*).$$

Remark 3.8. If the system $r(x) = Ax - b$ of Definition 3.4 is **consistent**, then any LS approximate solution corresponds to a solution in the usual sense of the previous section.

The question of when an LS approximate solution exists, and how it may be arrived at, is answered by

Proposition 3.17. Consider the system

$$r(x) = Ax - b.$$

The vector

$$x_* = Bb$$

is an LS solution to the system above if B is an s-inverse of A, i.e. it obeys

   i. $ABA = A$,

  ii. AB is symmetric.

Proof: We observe that for any n-element vector x,

$$
\begin{aligned}
(b &- Ax)'(b - Ax) \\
&= [(b - ABb) + (ABb - Ax)]'[(b - ABb) + (ABb - Ax)] \\
&= b'(I - AB)'(I - AB)b + (Bb - x)'A'A(Bb - x) \\
&= b'(I - AB)b + (Bb - x)'A'A(Bb - x).
\end{aligned}
$$

This is so because B obeys i and ii and the cross terms vanish, i.e.

$$(b - ABb)'(ABb - Ax) = b'(A - B'A'A)(Bb - x) = b'(A - ABA)(Bb - x) = 0.$$

Because $A'A$ is (at least) positive semidefinite, the quantity

$$(b - Ax)'(b - Ax) \text{ is minimized only if we take } x = Bb.$$

q.e.d.

Corollary 3.8. The quantity

$$b'(I - AB)b$$

is a lower bound for

$$(b - Ax)'(b - Ax).$$

Proof: Obvious.

Corollary 3.9. If B is a matrix that defines an LS solution,

$$AB = AA_\mathrm{g}.$$

Proof: We have

$$AB = AA_\mathrm{g}AB = A_\mathrm{g}'A'B'A' = A_\mathrm{g}'A' = AA_\mathrm{g}.$$

q.e.d.

We may now ask: what is the connection between g-inverses, LS solutions to inconsistent systems, and our discussion in the previous section? In part, this is answered by

Proposition 3.18. An n-element (column) vector x_* is an **LS solution** to an **inconsistent** system

$$r(x) = Ax - b,$$

where A is $m \times n$, if and only if x_* is a solution to the consistent system

$$Ax = AA_\mathrm{g}b.$$

Proof: First, we note that since A_g is also a c-inverse of A, Proposition 3.13 shows that

$$Ax = AA_\mathrm{g}b$$

is, indeed, a consistent system. Because it is a consistent system, Corollary 3.5 shows that the general form of the solution is

$$x_* = A_\mathrm{g}(AA_\mathrm{g}b) + (I_n - A_\mathrm{g}A)d = A_\mathrm{g}b + (I_n - A_\mathrm{g}A)d$$

for any arbitrary vector d.

With x_* as just defined, we have $b - Ax_* = b - AA_g b$ and, consequently,

$$(b - Ax_*)'(b - Ax_*) = b'(I_n - AA_g)b.$$

Proposition 3.17 shows that x_*, as above, is an LS solution to the (inconsistent) system $r(x) = Ax - b$. Conversely, suppose that x_* is **any** LS solution to the system above. Then it must satisfy the condition $(Ax_* - b)'(Ax_* - b) = b'(I_n - AA_g)b$, because A_g is also an s-inverse. Put

$$q = x_* - A_g b \quad \text{or} \quad x_* = q + A_g b.$$

Substitute in the equation above to obtain

$$\begin{aligned}
b'(I - AA_g)b &= (Ax_* - b)'(Ax_* - b) \\
&= (Aq + AA_g b - b)'(Aq + AA_g b - b) \\
&= b'(I - AA_g)b + q'A'Aq,
\end{aligned}$$

which immediately implies

$$Aq = 0.$$

Thus,

$$Ax_* = Aq + AA_g b = AA_g b,$$

which completes the proof.

q.e.d.

Remark 3.9. The import of Proposition 3.18 is that an LS solution to a (possibly) inconsistent system

$$r(x) = Ax - b$$

can be found by solving the associated (consistent) system

$$Ax = AA_g b.$$

The general class of solutions to this system was determined in Proposition 3.15. Thus, we see that there may be multiple (or infinitely many) LS solutions. If uniqueness is desired, it is clear that the solution must be made to satisfy additional conditions. This leads to

Definition 3.5. Consider the (possibly) inconsistent system

$$r(x) = Ax - b,$$

where A is $m \times n$. An n-element vector x_* is said to be a **minimum norm least squares** (MNLS) approximate solution if and only if

i. For all n-element vectors x

$$(b - Ax)'(b - Ax) \geq (b - Ax_*)'(b - Ax_*),$$

ii. For those x for which

$$(b - Ax)'(b - Ax) = (b - Ax_*)'(b - Ax_*)$$

we have

$$x'x > x'_* x_*.$$

This leads to the important

Proposition 3.19. Let

$$r(x) = Ax - b$$

be a (possibly) inconsistent system, where A is $m \times n$. The MNLS (approximate) solution is given by

$$x_* = A_g b,$$

and it is unique.

Proof: First, we note that

$$(b - AA_g b)'(b - AA_g b) = b'(I - AA_g)b,$$

which shows x_* to be an LS solution–because it attains the lower bound of

$$(b - Ax)'(b - Ax).$$

We must now show that this solution has minimum norm and that it is unique. Now, if x is **any** LS solution, it must satisfy

$$Ax = AA_g b.$$

Premultiply by A_g to obtain

$$A_g Ax = A_g AA_g b = A_g b.$$

Thus, any LS solution x also satisfies

$$x = A_g b - A_g Ax + x = A_g b + (I_n - A_g A)x.$$

Consequently, for **any** LS solution we have

$$x'x = b'A'_g A_g b + (x - A_g Ax)'(x - A_g Ax).$$

But if x is any LS solution,

$$A_g A x = A_g b.$$

Consequently,

$$x'x = b'A_g'A_g b + (x - A_g b)'(x - A_g b) = x_*'x_* + (x - x_*)'(x - x_*)$$

which shows that if

$$x \neq x_* = A_g b$$

then

$$x'x > x_*'x_*.$$

Uniqueness is an immediate consequence of the argument above. Thus, let x_0 be another MNLS solution and suppose $x_0 \neq x_*$. But x_0 must satisfy

$$x_0'x_0 = b'A_g'A_g b + (x_0 - A_g b)'(x_0 - A_g b).$$

Since we assume $x_0 \neq x_*$ we have $x_0'x_0 > x_*'x_*$, which is a contradiction. Moreover, A_g is unique, which thus completes the proof of the proposition.

$$\text{q.e.d.}$$

Remark 3.10. It is now possible to give a summary description of the role of the various pseudoinverses. Thus, the c-inverse is useful in broadly describing the class of solutions to the (consistent) system

$$Ax = b,$$

where A is $m \times n$.

The s-inverse is useful in describing the class of LS solutions to the possibly inconsistent system

$$r(x) = Ax - b,$$

i.e. in the case where no vector x may exist such that $r(x) = 0$. Neither the c-inverse nor the s-inverse of a matrix A is necessarily unique.

The g-inverse serves to characterize the solutions to both types of problems. Particularly, however, it serves to define the MNLS (approximate) solution to the inconsistent system

$$r(x) = Ax - b.$$

This means that of all possible least squares solutions to the inconsistent system above the g-inverse **chooses a unique vector by imposing the additional requirement that the solution vector exhibit minimal norm**.

This aspect should always be borne in mind in dealing with econometric applications of the g-inverse, since there is no particular **economic** reason to believe that the estimator of a vector exhibiting minimal norm is of any extraordinary significance.

Chapter 4

Matrix Vectorization

4.1 Introduction

It is frequently more convenient to write a matrix in **vector** form. For lack of a suitable term, we have coined for this operation the phrase "vectorization of a matrix." For example, if A is a matrix of parameters and $\tilde{A}$ the corresponding matrix of estimators, it is often necessary to consider the distribution of

$$\tilde{A} - A.$$

We have a convention to handle what we wish to mean by the expectation of a random matrix, but there is no convention regarding the "covariance matrix" of a matrix. Similarly, there is a literature regarding aspects of the (limiting) distribution of sequences of vectors, but not for matrices.

In differentiation with a view to obtaining the conditions that define a wide variety of estimators, vectorization of matrices offers great convenience as well.

To give a different example, when dealing with special matrices such as symmetric, triangular, diagonal, or Toeplitz, the corresponding elements are not all free but obey a number of restrictions. It would thus be convenient to have a means of displaying the "free" or unrestricted elements in a vector to facilitate discussion and avoid redundant differentiation.

In this chapter, we establish the proper conventions for vectorization, give a number of results that enable us to vectorize products of matrices, and use these tools to establish more convenient representations, for example, of the trace of a product of matrices; moreover, such conventions enable us to examine more precisely and efficiently the vector subspace(s) containing matrices whose elements are (linearly) restricted and various other ancillary topics.

P.J. Dhrymes, *Mathematics for Econometrics*,
DOI 10.1007/978-1-4614-8145-4_4, © The Author 2013

4.2   Vectorization of Matrices

We begin with

Convention 4.1. Let A be an $n \times m$ matrix; the notation $\text{vec}(A)$ will mean the nm-element **column** vector whose first n elements are of those of the first **column** of A, $a_{\cdot 1}$, the second n elements, those of the second column of A, $a_{\cdot 2}$, and so on. Thus,

$$\text{vec}(A) = (a'_{\cdot 1}, a'_{\cdot 2}, \ldots, a'_{\cdot m})'.$$

The notation $\text{rvec}(A)$ will mean the mn-element **row** vector, whose first m elements are those of the first row of A, $a_{1\cdot}$, the second m elements are those of the second row, $a_{2\cdot}$, and so on. Thus,

$$\text{rvec}(A) = (a_{1\cdot}, a_{2\cdot}, \ldots, a_{n\cdot}).$$

Evidently, $\text{vec}(A)$ and $\text{rvec}(A)$ contain **precisely** the same elements, but they are arranged in different order. An immediate consequence of Convention 4.1 is

Proposition 4.1. Let A, B be $n \times m$, and $m \times q$, respectively. Then,

$$\text{vec}(AB) = (I_q \otimes A)\,\text{vec}(B), \quad \text{or}$$
$$= (B' \otimes I_n)\,\text{vec}(A).$$

Proof: For the first representation, we note that the jth column AB is simply

$$Ab_{\cdot j}.$$

When AB is vectorized we find

$$\text{vec}(AB) = (I_q \otimes A)\,\text{vec}(B),$$

whose jth sub-vector is $Ab_{\cdot j}$. To show the validity of the second representation, we note that the jth column of AB can also be written as

$$\sum_{i=1}^{m} a_{\cdot i} b_{ij} = \sum_{i=1}^{m} b'_{ji} a_{\cdot i} = (b'_{\cdot j} \otimes I_n)\text{vec}(A),$$

where $b'_{\cdot j}$ is the transpose of the jth column of B (or the jth row of B'); consequently

$$\text{vec}(AB) = (B' \otimes I_n)\,\text{vec}(A).$$

q.e.d.

Vectorization of products involving more than two matrices is easily obtained by repeated application of Proposition 4.1. We give a few such results explicitly. Thus,

Corollary 4.1. Let A_1, A_2, A_3 be suitably dimensioned matrices. Then

$$\begin{aligned}
\text{vec}(A_1 A_2 A_3) &= (I \otimes A_1 A_2)\,\text{vec}(A_3) \\
&= (A_3' \otimes A_1)\,\text{vec}(A_2) \\
&= (A_3' A_2' \otimes I)\,\text{vec}(A_1).
\end{aligned}$$

Proof: By Proposition 4.1, taking

$$A = A_1 A_2, \qquad B = A_3,$$

we have

$$\text{vec}(A_1 A_2 A_3) = (I \otimes A_1 A_2)\,\text{vec}(A_3).$$

Taking $A_1 = A$, $\quad A_2 A_3 = B$, we have

$$\text{vec}(A_1 A_2 A_3) = (A_3' A_2' \otimes I)\,\text{vec}(A_1),$$

as well as

$$\text{vec}(A_1 A_2 A_3) = (I \otimes A_1)\,\text{vec}(A_2 A_3).$$

Applying Proposition 4.1 again, we find

$$\text{vec}(A_2 A_3) = (A_3' \otimes I)\,\text{vec}(A_2),$$

and hence $\text{vec}(A_1 A_2 A_3) = (A_3' \otimes A_1)\text{vec}(A_2)$.

q.e.d.

Corollary 4.2. Let A_1, A_2, A_3, A_4 be suitably dimensioned matrices. Then,

$$\begin{aligned}
\text{vec}(A_1 A_2 A_3 A_4) &= (I \otimes A_1 A_2 A_3)\,\text{vec}(A_4) \\
&= (A_4' \otimes A_1 A_2)\,\text{vec}(A_3) \\
&= (A_4' A_3' \otimes A_1)\,\text{vec}(A_2) \\
&= (A_4' A_3' A_2' \otimes I)\,\text{vec}(A_1).
\end{aligned}$$

Proof: The first representation follows if we apply Proposition 4.1, with $A = A_1 A_2 A_3$, $\quad B = A_4$. The second and third representations are obtained by taking $A = A_1 A_2$, $\quad B = A_3 A_4$ and then applying Proposition 4.1.

The fourth is obtained by taking $A = A_1$, $\quad B = A_2 A_3 A_4$ and then applying Proposition 4.1.

q.e.d.

Remark 4.1. The reader should note the pattern involved in these relations; thus, if we wish to vectorize the product of the n conformable matrices

$$A_1 A_2 A_3 \cdots A_n$$

by vectorizing A_i, we obtain

$$(A'_n A'_{n-1} \cdots A'_{i+1} \otimes A_1 A_2 \cdots A_{i-1}) \operatorname{vec}(A_i),$$

so that the matrices appearing to the right of A_i appear on the left of the Kronecker product sign $(\otimes)$ in transposed form and order, whereas those appearing on the left of A_i appear on the right of the Kronecker product sign in the original form and order.

We further have

Proposition 4.2. Let A, B be $m \times n$. Then, $\operatorname{vec}(A+B) = \operatorname{vec}(A) + \operatorname{vec}(B)$.

Proof: Obvious from Convention 4.1.

Corollary 4.3. Let A, B, C, D be suitably dimensioned matrices. Then,

$$
\begin{aligned}
\operatorname{vec}[(A+B)(C+D)] &= [(I \otimes A) + (I \otimes B)][\operatorname{vec}(C) + \operatorname{vec}(D)] \quad \text{or} \\
&= [(C' \otimes I) + (D' \otimes I)][\operatorname{vec}(A) + \operatorname{vec}(B)].
\end{aligned}
$$

Proof: By Proposition 4.1,

$$
\begin{aligned}
\operatorname{vec}[(A+B)(C+D)] & \\
&= [(I \otimes (A+B)] \operatorname{vec}(C+D) \\
&= [(C+D)' \otimes I] \operatorname{vec}(A+B) \\
&= [(C' \otimes I) + (D' \otimes I)][\operatorname{vec}(A) + \operatorname{vec}(B)].
\end{aligned}
$$

q.e.d.

We now turn our attention to the representation of the trace of products of matrices in terms of various functions of vectorized matrices. Thus,

Proposition 4.3. Let A, B be suitably dimensioned matrices. Then,

$$
\begin{aligned}
\operatorname{tr}(AB) &= \operatorname{vec}(A')' \operatorname{vec}(B) \\
&= \operatorname{vec}(B')' \operatorname{vec}(A).
\end{aligned}
$$

Proof: By definition, assuming A is $m \times n$ and B is $n \times q$,

$$\operatorname{tr}(AB) = \sum_{i=1}^{m} a_{i.} b_{.i}, \tag{4.1}$$

where $a_{i\cdot}$ is the ith row of A and $b_{\cdot i}$ is the ith column of B. But $a_{i\cdot}$ is simply the ith column of A' written in row form, and Eq. (4.1) then shows that $\operatorname{tr}(AB) = \operatorname{vec}(A')' \operatorname{vec}(B)$. Moreover, since

$$\operatorname{tr}(AB) = \operatorname{tr}(BA) = \sum_{j=1}^{q} b_{j\cdot} a_{\cdot j}, \tag{4.2}$$

we see that $\operatorname{tr}(AB) = \operatorname{vec}(B')' \operatorname{vec}(A)$.

q.e.d.

It is an easy consequence of Propositions 4.1 and 4.3 to establish a "vectorized representation" of the trace of the product of more than two matrices.

Proposition 4.4. Let A_1, A_2, A_3 be suitably dimensioned matrices. Then,

$$\begin{aligned}
\operatorname{tr}(A_1 A_2 A_3) &= \operatorname{vec}(A_1')'(A_3' \otimes I) \operatorname{vec}(A_2) \\
&= \operatorname{vec}(A_1')'(I \otimes A_2) \operatorname{vec}(A_3) \\
&= \operatorname{vec}(A_2')'(I \otimes A_3) \operatorname{vec}(A_1) \\
&= \operatorname{vec}(A_2')'(A_1' \otimes I) \operatorname{vec}(A_3) \\
&= \operatorname{vec}(A_3')'(A_2' \otimes I) \operatorname{vec}(A_1) \\
&= \operatorname{vec}(A_3')'(I \otimes A_1) \operatorname{vec}(A_2).
\end{aligned}$$

Proof: From Proposition 4.3, taking

$$A = A_1, \qquad B = A_2 A_3,$$

we have

$$\operatorname{tr}(A_1 A_2 A_3) = \operatorname{vec}(A_1')' \operatorname{vec}(A_2 A_3). \tag{4.3}$$

Using Proposition 4.1 we have

$$\begin{aligned}
\operatorname{vec}(A_2 A_3) &= (I \otimes A_2) \operatorname{vec}(A_3) \\
&= (A_3' \otimes I) \operatorname{vec}(A_2).
\end{aligned}$$

This together with Eq. (4.3) establishes

$$\begin{aligned}
\operatorname{tr}(A_1 A_2 A_3) &= \operatorname{vec}(A_1')' (A_3' \otimes I) \operatorname{vec}(A_2) \\
&= \operatorname{vec}(A_1')'(I \otimes A_2) \operatorname{vec}(A_3).
\end{aligned}$$

Noting that

$$\operatorname{tr}(A_1 A_2 A_3) = \operatorname{tr}(A_2 A_3 A_1)$$

and using exactly the same procedure as above shows

$$\operatorname{tr}(A_1 A_2 A_3) = \operatorname{vec}(A_2')'(I \otimes A_3) \operatorname{vec}(A_1)$$
$$= \operatorname{vec}(A_2')'(A_1' \otimes I) \operatorname{vec}(A_3).$$

Finally, since

$$\operatorname{tr}(A_1 A_2 A_3) = \operatorname{tr}(A_3 A_1 A_2),$$

we find by the same argument

$$\operatorname{tr}(A_1 A_2 A_3) = \operatorname{vec}(A_3')'(A_2' \otimes I) \operatorname{vec}(A_1)$$
$$= \operatorname{vec}(A_3')'(I \otimes A_1) \operatorname{vec}(A_2).$$

q.e.d.

Remark 4.2. The representation of the trace of the product of more than three matrices is easily established by using the methods employed in the proof of Proposition 4.4. For example,

$$\begin{aligned}
\operatorname{tr}(A_1 A_2 A_3 A_4) &= \operatorname{vec}(A_1')'(A_4' A_3' \otimes I) \operatorname{vec}(A_2) \\
&= \operatorname{vec}(A_1')'(A_4' \otimes A_2) \operatorname{vec}(A_3) \\
&= \operatorname{vec}(A_1')'(I \otimes A_2 A_3) \operatorname{vec}(A_4) \\
&= \operatorname{vec}(A_2')'(I \otimes A_3 A_4) \operatorname{vec}(A_1) \\
&= \operatorname{vec}(A_2')'(A_1' A_4' \otimes I) \operatorname{vec}(A_3) \\
&= \operatorname{vec}(A_2')'(A_1' \otimes A_3) \operatorname{vec}(A_4) \\
&= \operatorname{vec}(A_3')'(A_2' \otimes A_4) \operatorname{vec}(A_1) \\
&= \operatorname{vec}(A_3')'(I \otimes A_4 A_1) \operatorname{vec}(A_2) \\
&= \operatorname{vec}(A_3')'(A_2' A_1' \otimes I) \operatorname{vec}(A_4) \\
&= \operatorname{vec}(A_4')'(A_3' A_2' \otimes I) \operatorname{vec}(A_1) \\
&= \operatorname{vec}(A_4')'(A_3' \otimes A_1) \operatorname{vec}(A_2) \\
&= \operatorname{vec}(I_4')'(I \otimes A_1 A_2) \operatorname{vec}(A_3).
\end{aligned}$$

This example also shows why it is not possible to give all conceivable representations of the trace of the product of an arbitrary number of matrices.

4.3 Linearly Restricted Matrices

As we have seen above, if A is a real $n \times m$ matrix we define the vec operator as

$$\operatorname{vec}(A) = \begin{pmatrix} a_{.1} \\ a_{.2} \\ \vdots \\ a_{.m} \end{pmatrix},$$

where $a_{.j}$ is the (n-element) jth column of A. In this discussion we adopt the convention that $n \geq m$, which entails no loss of generality since if the condition is not satisfied for A, it is surely satisfied for A'.

Let V_{nm} be the space of real $n \times m$ matrices. If $n = m$ we may define[1] a basis for this space, say

$$\{e_{.i}e'_{.j} : i, j = 1, 2, \ldots n\},$$

where $e_{.j}$ is an n-element **column** vector all of whose elements are zero, except the jth, $j \leq n$, which is unity. Henceforth $e_{.j}$, will always have the meaning just indicated, and $e_{j.}$ will denote the corresponding **row** entity, unless otherwise indicated. Often, we may refer to such entities as **unit vectors**.

Any real $n \times n$ matrix may thus be written in terms of this basis as

$$A = \sum_{i=1}^{n} \sum_{j=1}^{n} a_{ij} e_{.i} e'_{.j},$$

where the a_{ij} are arbitrary elements in R.

For many purposes in econometrics, it is not convenient to think of matrices in terms of the space V_{nm}, but rather in terms of R^{nm}, the nm-dimensional real Euclidean space. To do so, we need to represent an $n \times m$ matrix A in the form of a (column) vector; this facility is made possible by the vec operator defined earlier in this chapter. From Chap. 1, we know that we may define a **basis** for this vector space; the basis in question may be chosen as the set of vectors

$$\{e_{.j} : j = 1, 2, \ldots, mn\}.$$

Thus, if $a = \text{vec}(A)$, the elements of a may be written in the form standard for vector spaces, namely

$$a = \sum_{j=1}^{mn} a_j e_{.j}, \quad a_j \in R, \quad \text{for all } j,$$

where a_j for $j = 1, 2, \ldots, n$ consists of the elements of the first column of A; for $j = n + 1, n + 2, \ldots, 2n$ consists of the elements in the second column of A; for $j = 2n + 1, 2n + 2, \ldots, 3n$ consists of the elements in the third column of A; $\ldots, \ldots$, and for $j = 2(m - 1) + 1, 2(m - 1) + 2, \ldots, mn$ consists of the mth (last) column of A.

---

[1]Even if $n > m$, the procedure given below will produce a basis as well, except that the unit vectors $e_{.j}$ will then be m-element column vectors.

What if the matrix A were to be symmetric, or strictly upper (lower) triangular, or just upper (lower) triangular, or diagonal? Would we be able to write such a matrix in terms of the basis above? The answer, in principle, is yes, but we would have to specify that the elements a_{ij} obey certain restrictions, namely

$$
\begin{array}{ll}
a_{ij} = a_{ji}, \text{ for } i \neq j & \text{if } A \text{ is a symmetric matrix} \\
a_{ij} = 0, \text{ for } j \geq i & \text{if } A \text{ is strictly lower triangular} \\
a_{ij} = 0, \text{ for } j > i & \text{if } A \text{ is lower triangular} \\
a_{ij} = 0, \text{ for } i \geq j & \text{if } A \text{ is strictly upper triangular} \\
a_{ij} = 0, \text{ for } i > j & \text{if } A \text{ is upper triangular} \\
a_{ij} = 0, \text{ for } i \neq j & \text{if } A \text{ is diagonal}
\end{array}
$$

Remark 4.3. What all these (square) matrices have in common is that their elements are subject to a set of linear restrictions. This means that the number of "free" elements is not n^2 but something less. Notice that among the matrices we examined earlier we had not considered **orthogonal** matrices, although their elements, too, are subject to restrictions. Specifically, if A is an orthogonal matrix,

$$
a'_{\cdot i} a_{\cdot j} = 1, \quad \text{if } i = j, \text{ and zero otherwise.}
$$

These restrictions, however, are **not linear!**

Finally, before we proceed with the main topic of this section, we introduce another type of matrix.

Definition 4.1. Let $A = (a_{ij})$ be an $n \times n$ matrix.

  i. It is said to be a Toeplitz matrix if $a_{i,j} = a_{i+s,j+s}$, for all i, j, s;

 ii. It is termed a symmetric Toeplitz matrix if, in addition to i, it satisfies the condition $a_{ij} = a_{ji}$, for $i \neq j$.

Remark 4.4. Notice that a symmetric Toeplitz matrix has, at most, n free elements, say α_i, $i = 1, 2, \ldots, n$, and, moreover, its diagonal contains only α_1, its jth supra-diagonal contains only the element α_j, $j > 1$, as does its jth sub-diagonal. In a non-symmetric Toeplitz matrix corresponding supra- and sub-diagonals need not contain the same element, so that for such matrices there are at most $2n - 1$ free elements.

A well known example of a symmetric Toeplitz matrix, from elementary econometrics, is the covariance matrix of the AR(1) process

$$u_t = \rho u_{t-1} + \epsilon_t, \quad t = 1, 2, \ldots, n,$$

for a sample of size n, where the ϵ_t are independent, identically distributed (i.i.d.) random variables with mean zero and variance σ^2.

Precisely, the covariance matrix of the vector $u = (u_1, u_2, \ldots, u_n)'$ is given by

$$\text{Cov}(u) = \Sigma = \frac{\sigma^2}{1 - \rho^2} V, \quad V = \begin{bmatrix} 1 & \rho & \rho^2 & \cdots & \cdots & \rho^{n-1} \\ \rho & 1 & \rho & \cdots & \cdots & \rho^{n-2} \\ \vdots & \vdots & \vdots & \cdots & \cdots & \vdots \\ \rho^{n-1} & \rho^{n-2} & \rho^{n-3} & \cdots & \cdots & 1 \end{bmatrix}.$$

For this particular example, it appears that we have only two distinct elements, namely σ^2 and ρ. Note, however, that in any given supra- or sub-diagonal all elements are the same, and moreover, corresponding supra- and sub-diagonals contain the same element.

Another example of a Toeplitz matrix, from a somewhat more advanced level of econometrics, is the covariance matrix of $\{u_t : t = 1, 2, \ldots, n\}$, which is a vector containing n observations on a covariance stationary process, i.e. one for which $\text{Cov}(u_{t+\tau}, u_t) = \kappa(|\tau|)$. This covariance matrix is easily determined to be

$$\text{Cov}(u) = K = \begin{bmatrix} \kappa(0) & \kappa(1) & \kappa(2) & \cdots & \cdots & \kappa(n-1) \\ \kappa(1) & \kappa(0) & \kappa(1) & \cdots & \cdots & \kappa(n-2) \\ \vdots & \vdots & \vdots & \cdots & \cdots & \vdots \\ \kappa(n-1) & \kappa(n-2) & \kappa(n-3) & \cdots & \cdots & \kappa(0) \end{bmatrix}.$$

It is clearly seen that K is a symmetric Toeplitz matrix and has at most n free elements, in this case the variance $\kappa(0)$, and the auto-covariances $\kappa(\tau)$, $\tau = 1, 2, \ldots, n-1$.

When the free elements of a symmetric Toeplitz matrix, say α_i, $i = 1, 2, \ldots n$, obey $a_i = 0$, for $i > k$, we are led to the special definition

Definition 4.2. Let $A = (a_{ij})$ be a symmetric Toeplitz matrix, with free elements α_i, $i = 1, 2, \ldots, n$; if $\alpha_i = 0$ for $i > k$, A is said to be a k-symmetric Toeplitz matrix.

Remark 4.5. An example of a k-symmetric Toeplitz matrix is given by

$$A = \begin{bmatrix} \alpha_1 & \alpha_2 & \alpha_3 & \cdots & \alpha_k & 0 & \cdots & \cdots & \cdots & \cdots & 0 \\ \alpha_2 & \alpha_1 & \alpha_2 & \cdots & \cdots & \alpha_k & 0 & \cdots & \cdots & \cdots & 0 \\ \vdots & \vdots & \vdots & \vdots & \vdots & \vdots & \vdots & \vdots & \vdots & \vdots & \vdots \\ \alpha_k & \vdots & \vdots & \vdots & \vdots & \vdots & \vdots & \vdots & \vdots & \vdots & \vdots \\ 0 & \vdots & \vdots & \vdots & \vdots & \vdots & \vdots & \vdots & \vdots & \vdots & 0 \\ \vdots & \vdots & \vdots & \vdots & \vdots & \vdots & \vdots & \vdots & \vdots & \vdots & \alpha_k \\ \vdots & \vdots & \vdots & \vdots & \vdots & \vdots & \vdots & \vdots & \vdots & \vdots & \vdots \\ \vdots & \vdots & \vdots & \vdots & \vdots & \vdots & \vdots & \vdots & \vdots & \vdots & \vdots \\ 0 & \cdots & 0 & \alpha_k & \alpha_{k-1} & \alpha_{k-2} & \cdots & \cdots & \cdots & \cdots & \alpha_1 \end{bmatrix}.$$

It may be verified that this is the covariance matrix of an $MA(k-1)$ (moving average of order $k-1$) process,

$$u_t = \sum_{i=0}^{k-1} \theta_i \epsilon_{t-i},$$

where $\{\epsilon_t : t = 0, \pm1, \pm2, \ldots\}$ is a white noise sequence[2] with mean zero, and variance σ^2, denoted by $WN(0, \sigma^2)$; for identification of parameters we must set $\theta_0 = 1$.

This becomes evident in the scalar case if we make the identification, in terms of the matrix K above, $\alpha_1 = \kappa(0)$, $\alpha_2 = \kappa(1)$, $\ldots$, $\alpha_k = \kappa(k-1)$ and note that $E(u_{t+s}u_t) = 0$ for $s \geq k$; moreover, we can express the κ's in terms of the *theta*'s.

4.3.1   Restricted Subspaces

In this subsection, we explore the implication of Remark 4.3 that the class of matrices whose elements are subject to linear restrictions lie in a particular subspace of R^{n^2} and determine its structure.

In examining any $n \times n$ matrix A, we shall adopt the convention that such a matrix is represented, for the purposes of our discussion, by vec(A), and is thus an element of R^{n^2}. When the discussion (of any problem involving such matrix) is completed, we shall employ the inverse operation[3] to **get back to** A **from** vec (A).

In this framework A becomes an n^2-element column vector, with **distinct** n-element sub-vectors. If the elements of A are subject to restrictions, the

---

[2]This term will be formally defined in a later chapter; for the moment the reader may think of white noise as a sequence of **zero mean uncorrelated random variables with finite variance.**

[3]For lack of established terminology, we term such operation **rematricizing.**

elements of vec(A) are subject to the same restrictions, and it is evident that vec(A) must lie in a **restricted subspace (RS)** of R^{n^2}.

We examine the nature of such subspaces and determine their bases in the discussion below.

The RS for Symmetric Matrices

Before we proceed with our discussion, it is convenient to introduce the following definition, which will greatly facilitate exposition.

Definition 4.3. Let P be an $m \times n$ matrix, and let $p = \text{vec}(P)$. The operator mat denotes the operation of rematricising p, i.e. it is defined by

$$P = \text{mat}(p).$$

Remark 4.6. Strictly speaking, we have defined the mat operator **only** in reference to a vectorized representation of a **given** matrix. Otherwise its meaning is ambiguous. For example, if we are given an arbirary q-element vector, say d, the operation $\text{mat}(d)$ could have several meanings, including no meaning at all. In particular, if q is a **prime number**, it **cannot** be represented as the **product of two integers** (other than itself and one), and in this case the operation $\text{mat}(d)$ has no useful meaning because it is sort of an identity, except that it may be a row or a column. If q has more than one factorization of the form $q = mn$, where m, n are integers, the operation $\text{mat}(d)$ could have several plausible meanings; even if the factorization is **unique**, the ambiguity still remains since $\text{mat}(d)$ may denote either an $m \times n$ matrix or an $n \times m$ matrix, for specic integers n and m. However, the context in which we have defined this operator gives us an unambiguous interpretation, namely if $a = \text{vec}(A)$, then $\text{mat}(a) = A$. This notational device offers considerable expository convenience.

The restrictions on a **symmetric** matrix are of the form $a_{ij} = a_{ji}$ for $i \neq j$. Thus, for an $n \times n$ matrix there are $n(n-1)/2$ such (linear) restrictions, and we conjecture that the restricted space is of dimension $n(n+1)/2$.

Assuming this is the case, we now establish a **basis** for this restricted space so that all its elements can be expressed in terms of the basis. More precisely, we need to determine a (minimal) set of linearly independent vectors in terms of which we may describe symmetric matrices. A little reflection will convince us that such matrices have, at most, only $s = n(n+1)/2$ **distinct** elements. What we need then is a matrix, say B_{sm}, such that given the **distinct** elements of a symmetric matrix, a linear transformation of these distinct elements will produce the desired symmetric matrix, according to the conventions established above.

Let the distinct elements be denoted by the vector

$$\alpha = (a_{11}, a_{21}, \ldots, a_{n1}, a_{22}, \ldots, a_{n2}, a_{33}, \ldots, a_{n3}, \ldots, a_{nn})', \qquad (4.4)$$

and note that

$$\text{vec}(A) = a = B_{sm}\alpha, \qquad (4.5)$$

so that when we rematricize, i.e. we mat $B_{sm}\alpha$, we obtain A. In Eq. (4.5)

$$B_{sm} = \begin{bmatrix}
I_n & 0 & 0 & 0 & \cdots & \cdots & 0 & 0 \\
e_{2 \cdot (n)} & 0 & 0 & 0 & \cdots & \cdots & 0 & 0 \\
0 & I_{n-1} & 0 & 0 & \cdots & \cdots & 0 & 0 \\
e_{3 \cdot (n)} & 0 & 0 & 0 & \cdots & \cdots & 0 & 0 \\
0 & e_{2 \cdot (n-1)} & 0 & 0 & \cdots & \cdots & 0 & 0 \\
0 & 0 & I_{n-2} & 0 & \cdots & \cdots & 0 & 0 \\
e_{4 \cdot (n)} & 0 & 0 & 0 & \cdots & \cdots & 0 & 0 \\
0 & e_{3 \cdot (n-1)} & 0 & 0 & \cdots & \cdots & 0 & 0 \\
0 & 0 & e_{2 \cdot (n-2)} & 0 & \cdots & \cdots & 0 & 0 \\
0 & 0 & 0 & I_{n-3} & \cdots & \cdots & 0 & 0 \\
\vdots & \vdots & \vdots & \vdots & \vdots & \vdots & \vdots & \vdots \\
\vdots & \vdots & \vdots & \vdots & \vdots & \vdots & \vdots & \vdots \\
e_{n \cdot (n)} & 0 & 0 & 0 & \cdots & \cdots & 0 & 0 \\
0 & e_{n-1 \cdot (n-1)} & 0 & 0 & \cdots & \cdots & 0 & 0 \\
0 & 0 & e_{n-2 \cdot (n-2)} & 0 & \cdots & \cdots & 0 & 0 \\
0 & 0 & 0 & e_{n-3 \cdot (n-3)} & \cdots & \cdots & 0 & 0 \\
\vdots & \vdots & \vdots & \vdots & \vdots & \vdots & \vdots & \vdots \\
0 & 0 & 0 & 0 & \cdots & \cdots & e_{2 \cdot (2)} & 0 \\
0 & 0 & 0 & 0 & \cdots & \cdots & 0 & I_1
\end{bmatrix}, \qquad (4.6)$$

and the notation $e_{i \cdot (r)}$ indicates an r-element **row** vector all of whose elements are zero except the ith, $i \leq r$, which is unity. It is easily verified that the matrix B_{sm} of Eq. (4.6) has $s = n(n+1)/2$ linearly independent columns, and it is of dimension $n^2 \times s$; hence, it is a basis for the restricted subspace of R^{n^2}. The dimension of this subspace is s, and it contains the class of symmetric matrices with real elements.

What the transformation in Eq. (4.5) does is to take any vector in R^s and, according to our convention, transform it into the symmetric matrix that has the elements of this vector as its **distinct** elements.

If we denote the restricted subspace of symmetric matrices by V_{sm}, then from a formal point of view the matrix B_{sm} effects the transformation:

$$B_{sm} : R^s \to V_{sm},$$

i.e. **any** element of R^s, through B_{sm}, creates a **unique** element of V_{sm}.

In applications, we often need the opposite operation, i.e. given any symmetric matrix A, we need to extract, in vector form, its **distinct** elements. The matrix that performs this operation for symmetric matrices is

$$C_{sm} = \text{diag}(I_n, I^*_{n-1}, I^*_{n-2}, \cdot, I^*_1), \quad I^*_{n-i} = (0, I_{n-i}), \ i = 1, 2, \ldots n - 1, \quad (4.7)$$

where the zero matrix is of dimension $n - i \times i$. Thus, C_{sm} is of dimension $s \times n^2$ and, evidently, of rank s. Precisely, given any symmetric matrix A we obtain

$$\alpha = C_{sm}a, \quad a = \text{vec}(A). \quad (4.8)$$

It follows, therefore, from Eqs. (4.5) and (4.8) that

$$\alpha = C_{sm}a = C_{sm}B_{sm}\alpha, \quad a = B_{sm}\alpha = B_{sm}C_{sm}a, \quad (4.9)$$

and Eq. (4.9) suggests that the matrices $C_{sm}B_{sm}$ and $B_{sm}C_{sm}$ behave more or less like **identity operators**. We examine this topic in the next section.

Remark 4.7. The matrices C_{sm}, B_{sm}, and analogous matrices we shall define below for other types of restricted matrices, are termed, respectively, the **selection** and **restoration** matrices. The selection matrix is also occasionally termed the **elimination** matrix. The terms are self explanatory and justified in that the elimination matrix eliminates from A the **redundant** elements (or alternatively selects the distinct elements) to give us the vector of **distinct or unrestricted** elements, α, whereas the restoration matrix takes the vector of distinct elements and **restores** the original vector a, such that mat $(a) = A$.

Notice that B_{sm}, the restoration matrix, is also a basis of the RS of R^{n^2} which contains the class of symmetric matrices. As a basis of this vector subspace B_{sm} is **not unique** since if H is **any nonsingular** ($s \times s$) matrix, $B^*_{sm} = B_{sm}H$ is **also a basis**, but it is **not a restoration** matrix for a.

Properties of the Matrices B_{sm}, C_{sm}

It is evident from Eq. (4.9) that C_{sm} is a left inverse of B_{sm} and that $C_{sm}B_{sm} = I_s$ because α contains **distinct** elements, or at least because no linear dependence is **known** to exist among its elements. The matrix $B_{sm}C_{sm}$, however, which is $n^2 \times n^2$, cannot be of full rank (i.e. the identity matrix) due to the fact that the rank of B_{sm}, C_{sm} is $s < n^2$ or, equivalently, because the elements of a exhibit linear dependencies.

Example 4.1. Let $n = 3$, and note that

$$B_{sm} = \begin{bmatrix} 1 & 0 & 0 & 0 & 0 & 0 \\ 0 & 1 & 0 & 0 & 0 & 0 \\ 0 & 0 & 1 & 0 & 0 & 0 \\ 0 & 1 & 0 & 0 & 0 & 0 \\ 0 & 0 & 0 & 1 & 0 & 0 \\ 0 & 0 & 0 & 0 & 1 & 0 \\ 0 & 0 & 1 & 0 & 0 & 0 \\ 0 & 0 & 0 & 0 & 1 & 0 \\ 0 & 0 & 0 & 0 & 0 & 1 \end{bmatrix}, \quad C_{sm} = \begin{bmatrix} 1 & 0 & 0 & 0 & 0 & 0 & 0 & 0 & 0 \\ 0 & 1 & 0 & 0 & 0 & 0 & 0 & 0 & 0 \\ 0 & 0 & 1 & 0 & 0 & 0 & 0 & 0 & 0 \\ 0 & 0 & 0 & 0 & 1 & 0 & 0 & 0 & 0 \\ 0 & 0 & 0 & 0 & 0 & 1 & 0 & 0 & 0 \\ 0 & 0 & 0 & 0 & 0 & 0 & 0 & 0 & 1 \end{bmatrix}.$$

Consequently,

$$C_{sm}B_{sm} = I_6, \quad B_{sm}C_{sm} = \begin{bmatrix} 1 & 0 & 0 & 0 & 0 & 0 & 0 & 0 & 0 \\ 0 & 1 & 0 & 0 & 0 & 0 & 0 & 0 & 0 \\ 0 & 0 & 1 & 0 & 0 & 0 & 0 & 0 & 0 \\ 0 & 1 & 0 & 0 & 0 & 0 & 0 & 0 & 0 \\ 0 & 0 & 0 & 0 & 1 & 0 & 0 & 0 & 0 \\ 0 & 0 & 0 & 0 & 0 & 1 & 0 & 0 & 0 \\ 0 & 0 & 1 & 0 & 0 & 0 & 0 & 0 & 0 \\ 0 & 0 & 0 & 0 & 0 & 1 & 0 & 0 & 0 \\ 0 & 0 & 0 & 0 & 0 & 0 & 0 & 0 & 1 \end{bmatrix}.$$

Notice that $B_{sm}C_{sm}$ is essentially the matrix B_{sm}, with three extra columns of zeros, and transforms the orginal vector a into itself **utilizing the restrictions** $a_{ij} = a_{ji}$ for $i \neq j$.

RS for Lower Triangular Matrices

In discussing lower triangular (or other types of restricted) matrices we shall be considerably more brief than in the previous discussion; the framework and considerations explained in the case of symmetric matrices broadly apply to the cases we consider below.

Thus, let $A = (a_{ij})$ be a lower triangular matrix; the restrictions on the elements a_{ij} are of the form

$$a_{ij} = 0, \quad \text{for } j > i.$$

The **nonzero** elements of A are given by

$$\alpha = C_{LT}a, \quad a = \text{vec}(A), \tag{4.10}$$

where C_{LT} is as in the right member of Eq. (4.7). However, in the restoration operation, i.e. in

$$a = B_{LT}\alpha,$$

the matrix B_{LT} is **not** as defined in the right member of Eq. (4.6); rather it is defined by $B_{LT} = C'_{LT}$, i.e.

$$a = B_{LT}\alpha = C'_{LT}C_{LT}a$$

so that $C'_{LT}C_{LT}$ is a **diagonal** matrix, and its diagonal contains $n(n+1)/2$ nonzero elements and $n(n-1)/2$ zero elements.

RS for Strictly Lower Triangular Matrices

Strictly lower diagonal matrices are defined by the restrictions

$$A = (a_{ij}), \quad a_{ij} = 0, \quad \text{for } j \geq i.$$

The selection matrix, C_{slt}, is given by

$$C_{slt} = \text{diag}(I^*_{n-1}, I^*_{n-2}, \ldots, I^*_1, 0), \quad I^*_{n-i} = (0, I_{n-i}), \ i = 1, 2, \ldots, n-1. \tag{4.11}$$

where the last element of the (block) diagonal matrix above, 0, is of dimension $1 \times n$. We note that C_{slt} is $s+1 \times n^2$ of rank $s = n(n-1)/2$, and its distinct elements are given by

$$\alpha = C_{slt}a, \tag{4.12}$$

where the last element of α is **zero**.

Example 4.2. Let A be a 3×3 strictly lower triangular matrix. Its distinct elements are $\alpha = (a_{21}, a_{31}, a_{32}, 0)'$, adopting the convention of putting the diagonal element last. The matrices C_{slt} and $C_{slt}C'_{slt}$ are given by

$$C_{slt} = \begin{bmatrix} 0 & 1 & 0 & 0 & 0 & 0 & 0 & 0 & 0 \\ 0 & 0 & 1 & 0 & 0 & 0 & 0 & 0 & 0 \\ 0 & 0 & 0 & 0 & 0 & 1 & 0 & 0 & 0 \\ 0 & 0 & 0 & 0 & 0 & 0 & 0 & 0 & 0 \end{bmatrix}, \quad C_{slt}C'_{slt} = \begin{bmatrix} 1 & 0 & 0 & 0 \\ 0 & 1 & 0 & 0 \\ 0 & 0 & 1 & 0 \\ 0 & 0 & 0 & 0 \end{bmatrix}, \tag{4.13}$$

and one may easily verify that

$$a = \text{vec}(A), \quad \alpha = C_{slt}a, \quad a = C'_{slt}\alpha = C'_{slt}C_{slt}a. \tag{4.14}$$

Remark 4.8. The case with upper triangular, and strictly upper triangular matrices may be handled in one of two ways: first, if A is (strictly) upper triangular then A' is (strictly) lower triangular, in which case the discussion above will suffice; second, the matrices C_{UT}, C_{sut} may be defined appropriately. The application of the first approach is self evident. As for the second, the matrix C_{UT} is a block matrix, with diagonal blocks I_i, $\quad i = 1, 2, 3, \ldots, n$, i.e.

$$C_{UT} = \mathrm{diag}(I_1, I_2, I_3, \cdots, I_n), \tag{4.15}$$

the remaining blocks being suitably dimensioned zero matrices; for the strictly upper triangular case we have

$$
C_{sut} =
\begin{bmatrix}
0 & e_1. & 0 & 0 & \cdots & 0 \\
0 & 0 & e_1. & 0 & \cdots & 0 \\
0 & 0 & e_2. & 0 & \cdots & 0 \\
0 & 0 & 0 & e_1. & \cdots & 0 \\
0 & 0 & 0 & e_2. & \cdots & 0 \\
0 & 0 & 0 & e_3. & \cdots & 0 \\
\vdots & \vdots & \vdots & \vdots & \vdots & \vdots \\
0 & 0 & 0 & 0 & \cdots & e_1. \\
0 & 0 & 0 & 0 & \cdots & e_2. \\
0 & 0 & 0 & 0 & \cdots & e_3. \\
0 & 0 & 0 & 0 & \cdots & e_4. \\
0 & 0 & 0 & 0 & \cdots & e_5. \\
\vdots & \vdots & \vdots & \vdots & \vdots & \vdots \\
0 & 0 & 0 & 0 & \cdots & e_{n-2}. \\
0 & 0 & 0 & 0 & \cdots & e_{n-1}.
\end{bmatrix}. \tag{4.16}
$$

Otherwise the situation remains as before; notice, in addition, that C_{sut} is very similar to C_{UT}. We chose this more extensive representation to minimize confusion that may arise because of the missing diagonal elements; the later would necessitate striking out columns from the diagonal blocks of C_{UT}. Evidently C_{UT} is a matrix of dimension $n(n+1)/2 \times n^2$ and C_{sut} is of dimension $n(n-1)/2 \times n^2$.

Remark 4.9. A strictly lower (or upper) triangular matrix is a special case of a class of matrices termed nilpotent.

Definition 4.4. An $n \times n$ matrix A is said to be **nilpotent** if and only if

$$A^r = 0, \quad \text{for some integer } r.$$

The **smallest** integer, say ν, such that $A^\nu = 0$, is said to be the **nilpotency index** of the matrix A; notice that if $A^\nu = 0$, then $A^r = 0$, for all $r \geq \nu$.

If A is a nilpotent matrix, with nilpotency index $r < n$, it has the canonical representation

$$A = PQP^{-1}, \quad Q = (q_{ij}), \quad \text{such that}$$

$$q_{ij} = 0, \quad \text{if } j \geq i - n + r + 1$$
$$= 1, \quad \text{otherwise.} \tag{4.17}$$

RS for Diagonal Matrices

Let C_d and C_d' be, respectively, the selection and restoration matrices for the diagonal matrix A. If $\alpha = (a_{11}, a_{22}, \ldots, a_{nn})'$ are the **diagonal** elements of A, then

$$a = \text{vec}(A), \quad \alpha = C_d a, \quad a = C_d' C_d a, \quad C_d = \text{diag}(e_{1\cdot}, e_{2\cdot}, \ldots, e_{n\cdot}). \tag{4.18}$$

It may thus be verified that C_d is $n \times n^2$, of rank n, and that $C_d' C_d$ is an $n^2 \times n^2$ **diagonal** matrix, n of whose diagonal elements are one and the rest, $n(n-1)$, are zero.

RS for Toepltiz Matrices

We recall that a Toeplitz matrix A is defined by the restrictions

$$a_{ij} = a_{i+r, j+r}, \quad \text{for all } i, j, r.$$

It is then easy to verify that A has at most $2n - 1$ unrestricted elements, namely those that are in its **first row and column**. The restoration matrix, B_T,

$$B_T : R^{2n-1} \to R^{n^2},$$

maps the unrestricted space R^{2n-1} into a subspace of R^{n^2}, such that if $\alpha \in R^{2n-1}$, then

$$a = B_T \alpha, \quad \text{mat}(a) = A, \tag{4.19}$$

and A is a Toeplitz matrix. The matrix B_T, which is $n^2 \times 2n - 1$, is given by

$$B_T' = \begin{bmatrix} I_n & 0 & I_{n-1}^{\circ\prime} & 0 & 0 & \cdots & \cdots & 0 & 0 & \cdots & 0 & I_1^{\circ} \\ 0 & e_{1\cdot}' & 0 & e_{2\cdot}' & e_{1\cdot}' & \cdots & \cdots & e_{n-1\cdot}' & e_{n-2\cdot}' & \cdots & e_{1\cdot}' & 0 \end{bmatrix},$$

$$I_{n-i}^{\circ} = (I_{n-i}, 0), \quad i = 1, 2, \ldots n - 1, \tag{4.20}$$

where $e_{i\cdot}$ is in this case an $n-1$-element **row** vector all of whose elements are zero except the ith, which is one, and the zero matrix in I_{n-i}° is $n - i \times i$.

The selection (or elimination) matrix C_T is evidently given by

$$C_T = \text{diag}(I_n, I_{n-1} \otimes e_{1\cdot}), \tag{4.21}$$

where $e_{1\cdot}$ is an n-element **row** vector all of whose elements are zero, except the first, which is unity and thus C_T is seen to be a $(2n-1) \times n^2$ matrix of rank $2n - 1$.

RS for Symmetric and k-Symmetric Toeplitz Matrices

We have seen in the discussion above that it is rather difficult and cumbersome to produce, from first principles, the selection and restoration matrices for Toeplitz matrices. It is even more cumbersome to do so for **symmetric** and k-**symmetric** Toeplitz matrices. Thus, we need to introduce more structure into our discussion. To this end, let

$$H_r = \sum_{i=1}^{n-r} e_{\cdot i} e_{\cdot i+r}', \quad r = 0, 1, 2, \ldots, n-1, \tag{4.22}$$

and note that

$$H_0 = I_n, \quad H_k H_s = H_{k+s}, \quad k + s \leq n. \tag{4.23}$$

Using the set $\{H_r : r = 0, 1, 2, \ldots, n-1\}$, it is easy to show that the symmetric Toeplitz matrix A, with distinct elements $\alpha = (a_1, a_2, \ldots, a_n)'$, is given by

$$A = a_1 H_0 + \sum_{r=1}^{n-1} a_{r+1} H_r^*, \quad H_r^* = H_r + H_r'. \tag{4.24}$$

Since

$$\mathrm{vec}(A) = \sum_{r=0}^{n-1} \mathrm{vec}(H_r^*) a_{r+1}, \quad H_0^* = I_n,$$

we see that the restoration matrix, B_{ST}, is given by

$$\mathrm{vec}(A) = a = B_{ST}\alpha, \quad B_{ST} = (\mathrm{vec}(H_0^*), \mathrm{vec}(H_1^*), \ldots, \mathrm{vec}(H_{n-1}^*)). \tag{4.25}$$

The elimination matrix is of course easily determined, as for any other symmetric matrix, as

$$C_{ST} = (I_n, 0, 0, \ldots, 0). \tag{4.26}$$

For k-symmetric Toeplitz matrices, the restoration matrix is obtained by redefining

$$H_r^* = 0, \quad r \geq k, \tag{4.27}$$

so that

$$a = B_{kST}\alpha, \quad B_{kST} = [\mathrm{vec}(H_0^*), \mathrm{vec}(H_1^*), \ldots, \mathrm{vec}(H_{k-1}^*)], \quad \alpha = (a_1, a_2, \ldots, a_k)'. \tag{4.28}$$

The elimination matrix is similarly obtained as

$$C_{kST} = (I_k^*, 0, \ldots, 0), \quad I_k^* = (I_k, 0), \tag{4.29}$$

where the zero matrix is of dimension $k \times n - k$.

4.3.2 The Structure of Restricted Subspaces

In the previous section, we examined a number of restricted subspaces and obtained explicitly the restoration (basis) and elimination matrices from first principles. In this section, we discuss more abstractly several aspects of such (linearly) restricted subspaces in terms of certain properties of the restoration matrix. We will discover that the restoration and elimination matrices are closely connected and that the latter may be derived from the former. Thus, most of the ensuing discussion will be in the context of restoration matrices.

If we examine the restoration (and elimination) matrices of the previous section, we easily determine that they all share the following properties:

i. Their elements are either zero or one;

ii. The restoration matrices are $q \times p$, $(q > p,)$ and of rank p;

iii. The elimination matrices are $q \times p$, $(q < p,)$ and of rank q.

We denote the restoration matrix, generically, by B and set $q = n^2$; we denote the elimination matrix, generically, by C and set $p = n^2$. Thus, B is $n^2 \times p$ of rank p, and C is $q \times n^2$ of rank q. From Chap. 3, we determine that their (unique) generalized inverse is given, respectively, by $B_g = (B'B)^{-1}B'$, $C_g = C'(CC')^{-1}$. To verify this assertion, we observe that

$$BB_gB = B(B'B)^{-1}B'B = B, \quad CC_gC = CC'(CC')^{-1}C = C$$
$$BB_g = B(B'B)^{-1}B', \quad CC_g = CC'(CC')^{-1} = I_q$$
$$B_gB = (B'B)^{-1}B'B = I_p, \quad C_gC = C'(C'C)^{-1}C$$
$$B_gBB_g = B_g, \quad C_gCC_g = C_g,$$

which confirms that B_g and C_g, as defined above, are indeed the generalized inverse of B and C, respectively.

As we pointed out in the earlier discussion, the RS of the special matrices examined above is described by the (dimension expanding) transformation

$$B : R^p \to R^q, \quad p < q,$$

so that it defines a subspace of R^q; since B may be taken to be a **basis** for this subspace, any element of the RS, say $a = \text{vec}(A)$, may be expressed **uniquely** as

$$a = B\alpha, \quad \alpha \in R^p, \quad \alpha = h(a), \quad a = \text{vec}(A), \quad (4.30)$$

for some (vector) function $h(\cdot)$. We should emphasize that uniqueness is to be understood in **terms of the basis matrix** B. Evidently, if another basis is employed, say $B_* = BQ$, for some $p \times p$ nonsingular matrix Q, then we can write

$$a = B_* \alpha_*, \quad \text{where} \quad B_* = BQ, \quad \alpha_* = Q^{-1}\alpha.$$

Thus, fixing the basis B, we have the following useful characterization of the (linearly) restricted subspace induced by B, which we shall denote by $RS(B)$.

Lemma 4.1. Let $RS(B)$ be the restricted subspace induced by the transformation

$$B : R^p \to R^q, \quad p < q.$$

Then, the following statements are equivalent:

   i. $A \in RS(B)$;

   ii. $\mathrm{vec}(A) = a = B\alpha$, $\alpha \in R^p$, with $\alpha = h(a)$, for some vector valued function h;

   iii. $(I_q - BB_g)a = 0$.

Proof: By definition, if $A \in RS(B)$, there exists a vector $\alpha \in R^p$ such that $a = B\alpha$. From the properties of the generalized inverse, we obtain

$$B_g a = B_g B\alpha = \alpha,$$

which defines the function $h(a) = B_g a$, thus completing the proof that i implies ii.

   If ii is given, then, by definition, mat $(a) = A$, so that $A \in RS(B)$, thus showing that ii implies i. Moreover, from ii we deduce that

$$\alpha = B_g a, \quad B\alpha = BB_g a, \quad \text{or} \quad (I_q - BB_g)a = 0,$$

thus showing that ii implies iii.

   Finally, given iii, we have

$$a = BB_g a;$$

putting $\alpha = B_g a$, we note that $\alpha \in R^p$, and we find

$$a = B\alpha, \quad \text{for} \quad \alpha \in R^p, \quad \text{such that} \quad \alpha = h(a),$$

thus showing that iii implies ii.

q.e.d.

Corollary 4.4. In the context of $RS(B)$, the elimination matrix is given by

$$C = B_g. \tag{4.31}$$

Proof: From Lemma 4.1, if $A \in RS(B)$ and $a = \text{vec}(A)$, then $a = B\alpha$, for $\alpha \in R^p$; moreover, $\alpha = B_g a$, so that B_g extracts from A its distinct elements and is thus the elimination matrix.

q.e.d.

We may summarize the conclusions above in the following proposition.

Proposition 4.5. Let $RS(B)$ be the restricted subspace of R^{n^2} induced by the transformation

$$B : R^p \to R^{n^2},$$

where B is an $n^2 \times p$ matrix of rank p, whose elements are either zero or one. The following statements are true:

i. The restoration matrices for symmetric, lower triangular, strictly lower triangular, diagonal, Toeplitz, symmetric Toeplitz, and k-symmetric Toeplitz matrices are of the type B, as above, with the correspondence

Lower Triangle (LT)	$p = n(n+1)/2$		
Strictly LT	$p = n(n-1)/2$	Toeplitz (T)	2n-1
Symmetric	$p = n(n+1)/2$	Symmetric T	$p = n$
Diagonal	$p = n$	k-Symmetric T	$p = k$;

ii. The corresponding selection, or elimination matrix, generically denoted by C, is given by

$$C = B_g$$

and, moreover,

$$B_g = (B'B)^{-1}B'$$
$$0 = (I_{n^2} - BB_g)a.$$

4.3.3 Permutation Matrices and the vec Operator

The discussion in this section may be motivated *inter alia* by the following consideration. Suppose the elements of a matrix, say M, of dimension $n \times m$, are a function of a vector γ, of considerably smaller dimension, and it is desired to differentiate the matrix $M'M$, whose dimension is $m \times m$. Whatever the convention may be for differentiating a matrix with respect to a vector, a subject we examine in the next chapter, we may begin by vectorizing

$$\text{vec}(M'M) = (I_m \otimes M')\text{vec}(M), \quad \text{or} \quad (M' \otimes I_m)\text{vec}(M').$$

The first representation is suitable for differentiating $M'M$ "with respect to its second factor", whereas the second is suitable for differentiating it with respect to the first factor. It is evident that $\text{vec}\,(M)$ and $\text{vec}\,(M')$ are both nm-element column vectors that contain the **same elements but in different order**; thus, it would be very convenient if we can determine the nature of a matrix, say P, such that $\text{vec}(M') = P\text{vec}(M)$. Evidently, P does nothing more than rearrange (permute) the elements of $\text{vec}\,(M)$ so that their order corresponds to that of the elements of $\text{vec}\,(M')$.

Definition 4.5. A square matrix of order n, denoted by P_n, is said to be a **permutation matrix** if and only if it can be obtained by permuting the columns (or rows) of the identity matrix I_n.

A partial list of the properties of the permutation matrix is given in the discussion below.

Lemma 4.2. Let P_n be a permutation matrix; the following statements are true:

   i. P_n is a product of elementary matrices (see Sect. 2.4);

  ii. P_n is an orthogonal matrix, i.e. $P_n'P_n = I_n$, or $P_n' = P_n^{-1}$.

Proof: The proof of i is evident from the discussion of Sect. 2.4.

The proof of ii is as follows. Since P_n consists of permutations of the identity matrix I_n, let E_i be the elementary matrices of type one, i.e. those that interchange two columns (or rows). By construction, P_n contains a number of such interchanges, say r in number. Thus $P_n = \prod_{i=1}^{r} E_i$; since the elementary matrices are orthogonal, it follows that P_n is also orthogonal; more precisely, by construction, the columns of P_n contain one unit element and all others are zero, whence it follows that $P_n' = P_n^{-1}$.

$$\text{q.e.d.}$$

Lemma 4.3. Let A be a real $n \times m$ matrix; there exists a **unique** permutation matrix P_{mn} such that

$$\text{vec}(A') = P_{mn}a, \quad a = \text{vec}(A).$$

Proof: Define the matrix

$$P_{mn} = (I_m \otimes e_{1.}', I_m \otimes e_{2.}', \dots, I_m \otimes e_{n.}')',$$

where $e_i.$ is an n-element row vector all of whose elements are zero except the ith, $i \leq n$, which is unity; notice also that the dimension of P_{mn} is $nm \times nm$ and, moreover, that the first block, viz. $I_m \otimes e_1.$ operating on a, extracts the first element, then the $(n+1)$, then the $(2n+1)$, and so on until the $(m-1)n+1$ element. Thus, the first m-element sub-vector of $P_{mn} a$ consists of $(a_{11}, a_{12}, a_{13}, \ldots, a_{1m})'$, which is simply the first m-element sub-vector of vec (A'). Similarly, for the second, third and so on until the nth block component of P_{mn}. Notice further that, by construction, the latter is indeed a permutation matrix.

As for uniqueness, suppose there exists another matrix, say P_{mn}^*, that produces the result

$$\text{vec}(A') = P_{mn}^* a, \quad a = \text{vec}(A).$$

Subtracting, we find $(P_{mn} - P_{mn}^*)a = 0$, and, since a is arbitrary, we conclude $P_{mn} = P_{mn}^*$, thus proving uniqueness.

q.e.d.

Remark 4.10. The order mn of the subscript of the permutation matrix above is not arbitrary. In fact, we have the following convention: if P_{ij} is the permutation matrix that transforms vec (A) into vec (A') and A is $n \times m$, P_{ij} consists of blocks of the form $I_m \otimes e_i.$. Notice that the dimension of the identity matrix (m) is equal to the number of columns of the matrix A, whereas the dimension of the unit vectors $e_i.$, (n,) corresponds to the number of rows, and there are as many unit vectors as their dimension, i.e. if the row vectors contain say 5 elements there are **five unit vectors each of dimension 5**. A consequence of that is that the number of the **identity matrices in this representation is equal to the number of unit vectors involved.** The convention then is to write P_{ij} as P_{mn}, so that the first subscript corresponds to the dimension of the identity matrix contained therein and the second subscript corresponds to the dimension of the unit vectors. In light of this convention, $P_{mn} \neq P_{nm}$; more precisely, P_{nm} transforms $b = \text{vec}(B)$, where B is $m \times n$, into vec (B') and, according to Lemma 4.3, has typical block element of the form $I_n \otimes e_i.$, where now the dimension of the unit vectors is m. We have the immediate corollary.

Corollary 4.5. If P_{mn} is a permutation matrix,

$$P_{mn} = P_{nm}'.$$

Moreover, when $n = m$, the permutation matrix P_{nn} is its own inverse, and it is symmetric.

Proof: Let A be $n \times m$ and P_{mn} be defined by the representation $a^* = P_{mn}a$, $a^* = \text{vec}(A') = \text{rvec}(A)'$. Since A' is $m \times n$, we also have the representation

$$a = P_{nm}a^*, \quad \text{which implies} \quad a^* = P_{mn}a = P_{mn}P_{nm}a^*.$$

Because a, and hence a^*, are arbitrary, we conclude that

$$P_{mn}P_{nm} = I_{nm}, \quad \text{or} \quad P_{mn} = P_{nm}^{-1} = P_{nm}',$$

Finally, if $n = m$, the last equation above implies

$$P_{nn} = P_{nn}^{-1}, \quad \text{as well as} \quad P_{nn} = P_{nn}',$$

which shows P_{nn} to be symmetric and its own inverse.

q.e.d.

Remark 4.11. Perhaps the reader will gain greater insight into the properties of permutation matrices by a more explicit representation of P_{mn}. A close look at the form of P_{mn} given in Lemma 4.3, and a little reflection, shows that the successive rows and columns of that matrix are given by

Successive rows					Successive columns				
$e_{1\cdot}$	$e_{m+1\cdot}$	$e_{2m+1\cdot}$	$\cdots$	$e_{(n-1)m+1\cdot}$	$e_{\cdot1}$	$e_{\cdot m+1}$	$e_{\cdot 2m+1}$	$\cdots$	$e_{\cdot(n-1)m+1}$
$e_{2\cdot}$	$e_{m+2\cdot}$	$e_{2m+2\cdot}$	$\cdots$	$e_{(n-1)m+2\cdot}$	$e_{\cdot2}$	$e_{\cdot m+2}$	$e_{\cdot 2m+2}$	$\cdots$	$e_{\cdot(n-1)m+2}$
$e_{3\cdot}$	$e_{m+3\cdot}$	$e_{2m+3\cdot}$	$\cdots$	$e_{(n-1)m+3\cdot}$	$e_{\cdot3}$	$e_{\cdot m+3}$	$e_{\cdot 2m+3}$	$\cdots$	$e_{\cdot(n-1)m+3}$
$\vdots$	$\vdots$	$\vdots$	$\vdots$	$\vdots$	$\vdots$	$\vdots$	$\vdots$	$\vdots$	$\vdots$
$e_{m\cdot}$	$e_{2m\cdot}$	$e_{3m\cdot}$	$\cdots$	$e_{nm\cdot}$	$e_{\cdot m}$	$e_{\cdot 2m}$	$e_{\cdot 3m}$	$\cdots$	$e_{\cdot nm}$,

where all unit vectors (rows as well as columns) have nm elements.

Viewed in light of the representation above, the result that P_{mn} is orthogonal becomes quite transparent, as does the fact that when $n = m$, $P_{nn} = P_{nn}' = P_{nn}^{-1}$.

The fact that $P_{mn}' = P_{nm}$ when $n \neq m$ is best illustrated with an example.

Example 4.3. Let A be 3×2; the corresponding P_{23} permutation matrix is given by

$$P_{23} = = \begin{bmatrix} I_2 \otimes e_{1\cdot} \\ I_2 \otimes e_{2\cdot} \\ I_2 \otimes e_{3\cdot} \end{bmatrix} = \begin{bmatrix} e_{1\cdot} & 0 \\ 0 & e_{1\cdot} \\ e_{2\cdot} & 0 \\ 0 & e_{2\cdot} \\ e_{3\cdot} & 0 \\ 0 & e_{3\cdot} \end{bmatrix}, \quad P_{23}' = \begin{bmatrix} 1 & 0 & 0 & 0 & 0 & 0 \\ 0 & 0 & 1 & 0 & 0 & 0 \\ 0 & 0 & 0 & 0 & 1 & 0 \\ 0 & 1 & 0 & 0 & 0 & 0 \\ 0 & 0 & 0 & 1 & 0 & 0 \\ 0 & 0 & 0 & 0 & 0 & 1 \end{bmatrix}$$

$$
= \begin{bmatrix} e_{1.} & 0 & 0 \\ 0 & e_{1.} & 0 \\ 0 & 0 & e_{1.} \\ e_{2.} & 0 & 0 \\ 0 & e_{2.} & 0 \\ 0 & 0 & e_{2.} \end{bmatrix} = \begin{pmatrix} I_3 \otimes e_{1.} \\ I_3 \otimes e_{2.} \end{pmatrix} = P_{32},
$$

where in the representation of P_{23} the $e_{i.}$ for $i = 1, 2, 3$ denote 3-element unit (row) vectors, and in the representation of P_{32}, for $i = 1, 2$ they denote 2-element unit (row) vectors.

Before we leave this topic we give three additional results. We produce an "analytic" representation of P_{mn}, and determine its trace and characteristic roots.

To produce an analytic expression we proceed through the Lemma below.

Lemma 4.4. The permutation matrix P_{mn} has an analytic expression as

$$
P_{mn} = \sum_{i=1}^{m} \sum_{j=1}^{n} (S'_{ij} \otimes S_{ij}), \tag{4.32}
$$

where $S_{ij} = e^{*}_{.i} e'_{.j}$, $e^{*}_{.i}$ is an m-element unit (column) vector and $e_{.j}$ is, according to our notational convention, an n-element (column) unit vector.

Proof: Let A be an arbitrary real $n \times m$ matrix; then,

$$
A' = I_m A' I_n = \left(\sum_{i=1}^{m} e^{*}_{.i} e^{*'}_{.i} \right) A' \left(\sum_{j=1}^{n} e_{.j} e'_{.j} \right)
$$

$$
= \sum_{i=1}^{m} \sum_{j=1}^{n} e^{*}_{.i} (e^{*'}_{.i} A' e_{.j}) e'_{.j} = \sum_{i=1}^{m} \sum_{j=1}^{n} e^{*}_{.i} (e'_{.j} A e^{*}_{.i}) e'_{.j}
$$

$$
= \sum_{i=1}^{m} \sum_{j=1}^{n} S_{ij} A S_{ij}.
$$

It follows from Corollary 4.1 that

$$
\operatorname{vec}(A') = \sum_{i=1}^{m} \sum_{j=1}^{n} (S'_{ij} \otimes S_{ij}) \operatorname{vec}(A).
$$

Because P_{mn} is the (permutation) matrix that transforms $\text{vec}(A)$ into $\text{vec}(A')$, we conclude that

$$P_{mn} = \sum_{i=1}^{m}\sum_{j=1}^{n}(S'_{ij} \otimes S_{ij}).$$

q.e.d.

The next lemma establishes the trace of the permutation matrix P_{mn}.

Lemma 4.5. Let P_{mn} be the permutation matrix of Lemma 4.4. Then,

$$\text{tr}P_{mn} = 1 + d,$$

where d is the greatest common divisor of $m - 1$ and $n - 1$.

Proof: We may write

$$S_{ij} = e^{*}_{\cdot i} e'_{\cdot j}, \quad \text{so that} \tag{4.33}$$

$$\sum_{i,j}(S'_{ij} \otimes S_{ij}) = \sum_{j=1}^{n}\left(\sum_{i=1}^{m} e_{\cdot j} \otimes (e^{*'}_{\cdot i} \otimes e^{*}_{\cdot i}) \otimes e'_{\cdot j}\right)$$

$$= \sum_{j=1}^{n}\left[e_{\cdot j} \otimes \sum_{i=1}^{m}\left(e^{*'}_{\cdot i} \otimes e^{*}_{\cdot i}\right) \otimes e'_{\cdot j}\right]$$

$$= \sum_{j=1}^{n}\left(e_{\cdot j} \otimes I_{m} \otimes e'_{\cdot j}\right).$$

It may be verified that the rightmost member of the last equation in the equation system Eq. (4.33) is a matrix all of whose elements are zero, except for the elements in positions

$$\{((j-1)m + s, (s-1)n + j) : s = 1, 2, \ldots, m, \quad j = 1, 2, \ldots n\},$$

which are unity. The question is: how many of these are diagonal elements? For such elements, we must have

$$(j-1)m + s = (s-1)n + j,$$

and the summation over s and j will produce the trace of P_{mn}. We note that the choice $j = s = 1$ gives us a diagonal element, so that the trace is at least one. To determine how many other terms we have, rewrite the equation above as

$$(j-1)(m-1) = (s-1)(n-1),$$

and introduce the Kronecker delta, such that $\delta(a, b) = 1$ if $a = b$, and zero otherwise.

In this notation, we can write

$$\operatorname{tr} P_{mn} = 1 + \sum_{s=2}^{m} \sum_{j=2}^{n} \delta[(j-1)(m-1), (s-1)(n-1)]$$

$$= 1 + \sum_{s=1}^{m-1} \sum_{j=1}^{n-1} \delta[j(m-1), s(n-1)].$$

Let d be the greatest common divisor of $m-1$ and $n-1$, so that

$$m - 1 = c_1 d, \quad n - 1 = c_2 d,$$

where c_1 and c_2 are **integers** and d is an **integer equal to or greater than one**. The condition

$$j(m-1) = s(n-1), \quad \text{or} \quad jc_1 = sc_2,$$

is evidently satisfied by the pairs $j = kc_2$, $s = kc_1$, $k = 1, 2, 3, \ldots d$. To verify this, note that $j = kc_2$ has range $\{c_2, 2c_2, 3c_2, \ldots, dc_2 = n-1\}$, which is contained in the range of summation over j. Similarly $s = kc_1$ has range $\{c_1, 2c_1, 3c_1, \ldots, dc_1 = m-1\}$, which is contained in the range of summation over s. Consequently,

$$\operatorname{tr} P_{mn} = 1 + \sum_{s=1}^{m-1} \sum_{j=1}^{n-1} \delta[j(m-1), s(n-1)] = 1 + d.$$

q.e.d.

The following inference is immediately available.

Corollary 4.6. Let P_{mn} be the permutation matrix of Lemma 4.5, and suppose that $n = m$. Then,

$$\operatorname{tr} P_{nn} = n, \quad |P_{nn}| = (-1)^{n(n-1)/2}. \tag{4.34}$$

Proof: The proof of the first assertion is immediate from Lemma 4.5 since the greatest common divisor of $n-1$ and $n-1$ is, evidently, $d = n - 1$.

For the second part, we note that since P_{nn} is a symmetric orthogonal matrix, its characteristic roots are real; hence, they are either plus or minus 1.[4] Because the total number of roots is n^2, let k be the number of positive unit roots. Then, using the first part, we have

$$n = k - (n^2 - k), \quad \text{or} \quad k = \frac{n(n+1)}{2},$$

---

[4]For an explicit derivation, see Sect. 2.7.

so that the number of **negative** unit roots is

$$n^2 - \frac{n(n+1)}{2} = \frac{n(n-1)}{2}, \quad \text{and thus} \quad |P_{nn}| = (-1)^{n(n-1)/2}.$$

q.e.d.

Remark 4.12. If we are dealing with P_{mn}, $n > m$, the determinant is not easily found, because P_{mn} **is not symmetric**. This means that some of its roots may be **complex**. However, for all orthogonal matrices, whether symmetric or not, their roots obey

$$\lambda^2 = 1, \quad \text{which is satisfied by} \quad \lambda = \pm 1, \quad \text{or} \quad \lambda = \pm i,$$

where i is the **imaginary unit**, obeying $i^2 = -1$ and remembering that complex roots appear as pairs of complex conjugates, the absolute value of i being $(i)(-i) = 1$.

We summarize the preceding discussion in

Proposition 4.6. Let P_{mn} be a permutation matrix, i.e. a matrix resulting from permuting the columns (or rows) of the identity matrix I_{mn}, and let A be a real $n \times m$ matrix, $n \geq m$.
The following statements are true:

i. There exists a unique permutation matrix, say P_{mn}, such that

$$\text{vec}(A') = P_{mn}\text{vec}(A);$$

ii. The matrix P_{mn} is orthogonal, i.e.

$$P'_{mn} = P^{-1}_{mn};$$

iii. If P_{nm} is the permutation matrix that, for **any** real $m \times n$ matrix B, produces vec $(B') = P_{nm}\text{vec}(B)$,

$$P'_{nm} = P_{mn};$$

iv. The permutation matrix P_{mn} has the analytic expression

$$P_{nm} = \sum_{i=1}^{m}\sum_{j=1}^{n}(S'_{ij} \otimes S_{ij}), \quad S_{ij} = e^*_{\cdot i}e'_{\cdot j};$$

v. tr $P_{nm} = 1 + $ d, where d is the greatest common divisor of $n - 1$ and $m - 1$;

vi. If A is a square matrix, i.e. $n = m$, then, $P'_{nn} = P_{nn}^{-1} = P_{nn}$.

vii. Since P_{nn} is a symmetric orthogonal matrix, its characteristic roots are ± 1, and it has $n(n+1)/2$ positive unit roots, and $n(n-1)/2$ negative unit roots;

viii. The determinant of P_{nn} is given by

$$|P_{nn}| = (-1)^{n(n-1)/2}.$$

4.3.4 Permutation and the vec Operator

We begin with the following proposition.

Proposition 4.7. Let A be $n \times m$, B be $r \times s$, and P_{rn} be a permutation matrix. The following statement is true:

$$P_{rn}(A \otimes B) = (B \otimes A)P_{sm}. \tag{4.35}$$

Proof: Let X be an arbitrary $(s \times m)$ matrix such that vec (X) is an sm-element vector, and note that

BXA' is a matrix of dimension $r \times n$, so that $P_{rn}\mathrm{vec}(BXA') = \mathrm{vec}(AX'B')$.

Expanding the two sides above by the methods developed above, we find

$$P_{rn}(A \otimes B)\mathrm{vec}(X) = P_{rn}\mathrm{vec}(BXA') = \mathrm{vec}(AX'B')$$

$$= (B \otimes A)\mathrm{vec}(X') = (B \otimes A)P_{sm}\mathrm{vec}(X).$$

Since $\mathrm{vec}(X)$ is arbitrary, the conclusion follows.

q.e.d.

Corollary 4.7. Let A, B, P_{rn}, P_{sm} be as in Proposition 4.7, and further suppose that rank$(A) = r \leq m$. The following statements are true:

i. $P_{rn}(A \otimes B)P'_{sm} = (B \otimes A)$.

ii. The $mn \times mn$ (square) matrix $S = P_{nm}(A' \otimes A)$ has the properties:

    1. It is symmetric;

    2. rank$(S) = r^2$, rank$(A) = r$;

    3. tr$(S) =$ tr$(A'A)$;

4. $S^2 = AA' \otimes A'A$.

Proof: To prove i, we note from Proposition 4.7 that

$$P_{rn}(A \otimes B) = (B \otimes A)P_{sm}.$$

Post-multiplying by P'_{sm}, we obtain the desired result.

To prove part ii.1, observe that

$$S' = \{[P_{nm}(A' \otimes A)P_{mn}]P'_{mn}\}' = [(A \otimes A')P_{mn}]' = P_{nm}(A' \otimes A).$$

To prove ii.2, we note that P_{nm} is nonsingular and hence that

$$\text{rank}(S) = \text{rank}(A' \otimes A).$$

From Proposition 2.68 (singular value decomposition), we deduce that rank(A) = rank $(A'A)$. From the properties of Kronecker matrices we know that if (λ, x) is a pair of characteristic root and vector for a (square) matrix A_1, and if (μ, y) is a similar entity with respect to another (square) matrix A_2, then $(\lambda \otimes \mu, x \otimes y)$ is a pair of characteristic root and vector for the Kronecker product $A_1 \otimes A_2$. Hence, putting $C = (A' \otimes A)$, we have, for CC',

$$(A'A \otimes AA')(x \otimes y) = \lambda x \otimes \mu y.$$

By Corollary 2.8, the nonzero characteristic roots of $A'A$ and AA' are **identical**; since the characteristic roots of $A'A \otimes AA'$ are

$$\{(\lambda_i \mu_j) : i = 1, 2, \ldots m, \quad j = 1, 2, \ldots, n\},$$

and rank(A) = rank(A $'$) = r, we conclude that the number of nonzero roots of $A'A \otimes AA'$ is r^2, and hence that the rank of CC' is r^2. But this implies rank $(S) = r^2$, which completes the proof of ii.2.

To prove ii.3, we note, from the construction of the matrix P_{nm} in the proof of Lemma 4.3, that the diagonal blocks of the matrix S are given by $a'_{i\cdot} \otimes a_{i\cdot} = a'_{i\cdot}a_{i\cdot}$, so that

$$\text{tr}S = \sum_{i=1}^{n}\left(\text{tr}a'_{i\cdot}a_{i\cdot}\right) = \left(\sum_{i=1}^{n}\sum_{j=1}^{m}a_{ij}^2\right) = \text{tr}A'A,$$

which completes the proof of ii.3.

The proof of ii.4 is straightforward in view of Proposition 4.7; this is so because

$$S^2 = P_{nm}(A' \otimes A)P_{nm}(A' \otimes A) = (A \otimes A')(A' \otimes A) = (AA' \otimes A'A).$$

q.e.d.

Chapter 5

Vector and Matrix Differentiation

5.1 Introduction

Frequently, we need to differentiate quantities like $\operatorname{tr}(AX)$ with respect to the elements of X, or quantities like Ax, $z'Ax$ with respect to the elements of (the vectors) x and/or z.

Although no new concept is involved in carrying out such operations, they involve cumbersome manipulations, and thus it is desirable to derive such results in vector and/or matrix notation and have them easily available for reference.

Throughout this chapter, and in applications, we employ the following convention.

Convention 5.1. Let

$$y = \psi(x),$$

where y, x are, respectively, m- and n-element column vectors. The symbol

$$\frac{\partial y}{\partial x} = \left[\frac{\partial y_i}{\partial x_j}\right], \qquad i = 1, 2, \ldots, m, \quad j = 1, 2, \ldots, n,$$

will denote the matrix of first-order partial derivatives (Jacobian matrix) of the transformation from x to y such that the ith row contains the (parial) derivatives of the ith element of y with respect to the elements of x, namely

$$\frac{\partial y_i}{\partial x_1}, \frac{\partial y_i}{\partial x_2}, \ldots, \frac{\partial y_i}{\partial x_n}.$$

Remark 5.1. Notice that if y, above, is a scalar, then Convention 5.1 implies that $\partial y/\partial x$ is a **row vector**. If we wish to represent it as a column vector we do so by writing $\partial y/\partial x'$, or $(\partial y/\partial x)'$. Many authors prefer the first alternative; in this volume, however, we always write the partial derivative of a **scalar** with respect to a **vector** as a **row**, and if we need to represent it as a column we write $(\partial y/\partial x)'$.

5.2   Derivatives of Functions of the Form $y = Ax$

We begin with the simple proposition.

Proposition 5.1. If

$$y = Ax,$$

where A is $m \times n$ that does not depend on x, and the latter is an n-element column vector

$$\frac{\partial y}{\partial x} = A.$$

Proof: Since the ith element of y is given by

$$y_i = \sum_{k=1}^{n} a_{ik} x_k,$$

it follows that

$$\frac{\partial y_i}{\partial x_j} = a_{ij},$$

and hence that

$$\frac{\partial y}{\partial x} = A.$$

q.e.d.

If the vector x above is a function of another set of variables, say those contained in the r-element column vector α, then we have

Proposition 5.2. Let

$$y = Ax$$

be as in Proposition 5.1, but suppose that x is a function of the r-element vector α, while A is independent of α and x. Then

$$\frac{\partial y}{\partial \alpha} = \frac{\partial y}{\partial x}\frac{\partial x}{\partial \alpha} = A\frac{\partial x}{\partial \alpha}.$$

Proof: Since $y_i = \sum_{k=1}^{n} a_{ik} x_k$,

$$\frac{\partial y_i}{\partial \alpha_j} = \sum_{k=1}^{n} a_{ik} \frac{\partial x_k}{\partial \alpha_j}.$$

But the right member of the equation above is simply the (i, j) element of $A(\partial x / \partial \alpha)$. Hence,

$$\frac{\partial y}{\partial \alpha} = A \frac{\partial x}{\partial \alpha}.$$

q.e.d.

Remark 5.2. Convention 5.1 enables us to define routinely the first-order derivative of one vector with respect to another, but it is not sufficient to enable us to obtain second-order derivatives. This is so because it is not clear what is meant by the derivative of a **matrix** with respect to a vector. In particular the entity $\partial y / \partial \alpha$ is a **matrix** and Convention 5.1 gives no guidance on this issue. To help us derive an appropriate convention, consider the structure of second order derivative matrices (Hessians) and note that for a scalar function $z = \phi(\gamma)$, where γ is an r-element vector, we obtain

$$\frac{\partial^2 z}{\partial \gamma \partial \gamma} = \begin{bmatrix} \frac{\partial^2 z}{\partial \gamma_1 \partial \gamma_1} & \frac{\partial^2 z}{\partial \gamma_1 \partial \gamma_2} & \frac{\partial^2 z}{\partial \gamma_1 \partial \gamma_3} & \cdots & \cdots & \frac{\partial^2 z}{\partial \gamma_1 \partial \gamma_r} \\ \frac{\partial^2 z}{\partial \gamma_2 \partial \gamma_1} & \frac{\partial^2 z}{\partial \gamma_2 \partial \gamma_2} & \frac{\partial^2 z}{\partial \gamma_2 \partial \gamma_3} & \cdots & \cdots & \frac{\partial^2 z}{\partial \gamma_2 \partial \gamma_r} \\ \vdots & \vdots & \vdots & \vdots & \vdots & \vdots \\ \frac{\partial^2 z}{\partial \gamma_r \partial \gamma_1} & \frac{\partial^2 z}{\partial \gamma_r \partial \gamma_2} & \frac{\partial^2 z}{\partial \gamma_r \partial \gamma_3} & \cdots & \cdots & \frac{\partial^2 z}{\partial \gamma_r \partial \gamma_r} \end{bmatrix}.$$

It would seem then appropriate, in the case where z is an m-element column vector (i.e. ϕ is an m-element vector function), to define

$$\frac{\partial^2 z}{\partial \gamma \partial \gamma} = \begin{pmatrix} \frac{\partial^2 z_1}{\partial \gamma \partial \gamma} \\ \frac{\partial^2 z_2}{\partial \gamma \partial \gamma} \\ \vdots \\ \frac{\partial^2 z_m}{\partial \gamma \partial \gamma} \end{pmatrix}.$$

Each of the sub-matrices above are of dimension $r \times r$, so the matrix $\frac{\partial^2 z}{\partial \gamma \partial \gamma}$ is of dimension $mr \times r$. From Convention 5.1 we see that the **rows** of $\partial z / \partial \gamma$ are of the form $\partial z_i / \partial \gamma$. Hence, the **columns** of $(\partial z / \partial \gamma)'$ can be written as $(\partial z_i / \partial \gamma)'$ which is what we need to deal with. This suggests

Convention 5.2. Let

$$y = \psi(x)$$

be as in Convention 5.1. The symbol

$$\frac{\partial^2 y}{\partial x \partial x}$$

means

$$\frac{\partial^2 y}{\partial x \partial x} = \frac{\partial}{\partial x} \operatorname{vec} \left[\left(\frac{\partial y}{\partial x} \right)' \right],$$

so that it is a matrix of dimension $(mn) \times n$.

It is also convenient, and as part of this convention, to introduce the notation

$$\frac{\partial Y}{\partial \gamma},$$

where Y is a matrix of dimension $m \times n$ and γ an r-element column vector, to be defined by

$$\frac{\partial Y}{\partial \gamma} = \frac{\partial}{\partial \gamma} \operatorname{vec}(Y),$$

where the right member above is a matrix of dimension $mn \times r$.

With this in mind Convention 5.2 is also useful in handling the case where A depends on the vector α.

An easy consequence of Convention 5.2 are the two propositions below.

Proposition 5.3. Let

$$y = Ax$$

be as in Proposition 5.2. Then,

$$\frac{\partial^2 y}{\partial \alpha \partial \alpha} = \frac{\partial}{\partial \alpha} \operatorname{vec} \left[\left(\frac{\partial y}{\partial \alpha} \right)' \right] = (A \otimes I_r) \frac{\partial^2 x}{\partial \alpha \partial \alpha}.$$

Proof: By Proposition 5.2,

$$\frac{\partial y}{\partial \alpha} = A \frac{\partial x}{\partial \alpha}.$$

By Convention 5.2 and Proposition 4.1,

$$\frac{\partial^2 y}{\partial \alpha \partial \alpha} = \frac{\partial}{\partial \alpha} \operatorname{vec} \left[\left(A \frac{\partial x}{\partial \alpha} \right)' \right]$$

$$= \frac{\partial}{\partial \alpha} \operatorname{vec} \left[\left(\frac{\partial x}{\partial \alpha} \right)' A' \right]$$

$$= \frac{\partial}{\partial \alpha}(A \otimes I_r)\, \text{vec}\left[\left(\frac{\partial x}{\partial \alpha}\right)'\right]$$

$$= (A \otimes I_r)\frac{\partial}{\partial \alpha}\, \text{vec}\left[\left(\frac{\partial x}{\partial \alpha}\right)'\right] = (A \otimes I_r)\frac{\partial^2 x}{\partial \alpha \partial \alpha}.$$

q.e.d.

Proposition 5.4. Let

$$y = Ax,$$

where y is $m \times 1$, A is $m \times n$, x is $n \times 1$, and both A and x depend on the r-element vector α. Then,

$$\frac{\partial y}{\partial \alpha} = (x' \otimes I_m)\frac{\partial A}{\partial \alpha} + A\frac{\partial x}{\partial \alpha}.$$

Proof: We may write

$$y = \sum_{i=1}^{n} a_{\cdot i} x_i,$$

where $a_{\cdot i}$ is the ith column of A. Hence,

$$\frac{\partial y}{\partial \alpha} = \sum_{i=1}^{n}\frac{\partial a_{\cdot i}}{\partial \alpha}x_i + \sum_{i=1}^{n} a_{\cdot i}\frac{\partial x_i}{\partial \alpha}$$

$$= (x' \otimes I_m)\frac{\partial A}{\partial \alpha} + A\frac{\partial x}{\partial \alpha}.$$

q.e.d.

5.3 Derivatives of Functions of the Form $y = z'Ax$

In this section, we consider the differentiation of bilinear and quadratic forms.

Proposition 5.5. Let

$$y = z'Ax,$$

where z is $m \times 1$, A is $m \times n$, x is $n \times 1$, and A is independent of z and x. Then,

$$\frac{\partial y}{\partial z} = x'A', \qquad \frac{\partial y}{\partial x} = z'A.$$

Proof: Define

$$z'A = c'$$

and note that $y = c'x$. Hence, by Proposition 5.1, we have

$$\frac{\partial y}{\partial x} = c' = z'A.$$

Similarly, we can write

$$y = x'A'z,$$

and employing the same device, we obtain

$$\frac{\partial y}{\partial z} = x'A'.$$

q.e.d.

For the special case where y is given by the quadratic form

$$y = x'Ax,$$

we have

Proposition 5.6. Let

$$y = x'Ax,$$

where x is $n \times 1$, and A is $n \times n$ and independent of x. Then,

$$\frac{\partial y}{\partial x} = x'(A + A').$$

Proof: By definition,

$$y = \sum_{j=1}^{n} \sum_{i=1}^{n} a_{ij}x_i x_j.$$

Differentiating with respect to the kth element of x, we have

$$\frac{\partial y}{\partial x_k} = \sum_{j=1}^{n} a_{kj}x_j + \sum_{i=1}^{n} a_{ik}x_i, \qquad k = 1, 2, \ldots, n,$$

and consequently

$$\frac{\partial y}{\partial x} = x'A' + x'A = x'(A' + A).$$

q.e.d.

Corollary 5.1. For the special case where A is a symmetric matrix and

$$y = x'Ax,$$

we have

$$\frac{\partial y}{\partial x} = 2x'A.$$

Proof: Obvious from Proposition 5.6.

Corollary 5.2. Let A, y, and x be as in Proposition 5.6; then,

$$\frac{\partial^2 y}{\partial x \partial x'} = A' + A,$$

and, for the special case where A is symmetric,

$$\frac{\partial^2 y}{\partial x \partial x} = 2A.$$

Proof: Obvious, if we use Convention 5.2 and note that

$$\frac{\partial y}{\partial x} = x'(A' + A).$$

q.e.d.

For the case where z and/or x are functions of another set of variables, we have

Proposition 5.7. Let

$$y = z'Ax,$$

where z is $m \times 1$, A is $m \times n$, x is $n \times 1$, and both z and x are functions of the r-element vector α, whereas A is independent of α, x and z. Then,

$$\frac{\partial y}{\partial \alpha} = x'A'\frac{\partial z}{\partial \alpha} + z'A\frac{\partial x}{\partial \alpha},$$

$$\frac{\partial^2 y}{\partial \alpha \partial \alpha} = \left(\frac{\partial z}{\partial \alpha}\right)'A\left(\frac{\partial x}{\partial \alpha}\right) + \left(\frac{\partial x}{\partial \alpha}\right)'A'\left(\frac{\partial z}{\partial \alpha}\right) + (x'A' \otimes I_r)\frac{\partial^2 z}{\partial \alpha \partial \alpha}$$

$$+ (z'A \otimes I_r)\frac{\partial^2 x}{\partial \alpha \partial \alpha}.$$

Proof: We have

$$\frac{\partial y}{\partial \alpha} = \frac{\partial y}{\partial z}\frac{\partial z}{\partial \alpha} + \frac{\partial y}{\partial x}\frac{\partial x}{\partial \alpha} = x'A'\frac{\partial z}{\partial \alpha} + z'A\frac{\partial x}{\partial \alpha},$$

which proves the first part. For the second, note that

$$\frac{\partial^2 y}{\partial \alpha \partial \alpha} = \frac{\partial}{\partial \alpha}\left(\frac{\partial y}{\partial \alpha}\right)' = \frac{\partial}{\partial \alpha}\left(\frac{\partial y}{\partial \alpha}\right)'.$$

But

$$\frac{\partial y}{\partial \alpha} = \left(\frac{\partial z}{\partial \alpha}\right)' Ax + \left(\frac{\partial x}{\partial \alpha}\right)' A'z,$$

and, by the results of Proposition 5.4,

$$\frac{\partial}{\partial \alpha}\left(\frac{\partial z}{\partial \alpha}\right)' Ax = (x'A' \otimes I_r)\frac{\partial^2 z}{\partial \alpha \partial \alpha'} + \left(\frac{\partial z}{\partial \alpha}\right)' A \left(\frac{\partial x}{\partial \alpha}\right)$$

$$\frac{\partial}{\partial \alpha}\left(\frac{\partial x}{\partial \alpha}\right)' A'z = (z'A \otimes I_r)\frac{\partial^2 x}{\partial \alpha \partial \alpha} + \left(\frac{\partial x}{\partial \alpha}\right)' A' \left(\frac{\partial z}{\partial \alpha}\right),$$

which proves the validity of the proposition.

q.e.d.

Remark 5.3. Note that, despite appearances, the matrix $\partial^2 y/\partial \alpha \partial \alpha$ is symmetric, as required. This is so because $(x'A' \otimes I_r)\partial^2 z/\partial \alpha \partial \alpha$ is of the form

$$\sum_{i=1}^{m} \frac{\partial^2 z_i}{\partial \alpha \partial \alpha} c_i,$$

where c_i is the ith element of $x'A'$; evidently, the matrices

$$\frac{\partial^2 z_i}{\partial \alpha \partial \alpha}, \qquad i = 1, 2, \dots, m,$$

are all symmetric.

Corollary 5.3. Consider the quadratic form

$$y = x'Ax,$$

where x is $n \times 1$, A is $n \times n$, and x is a function of the r-element vector α, whereas A is independent of α. Then

$$\frac{\partial y}{\partial \alpha} = x'(A' + A)\frac{\partial x}{\partial \alpha},$$

$$\frac{\partial^2 y}{\partial \alpha \partial \alpha} = \left(\frac{\partial x}{\partial \alpha}\right)' (A' + A) \left(\frac{\partial x}{\partial \alpha}\right) + (x'(A' + A) \otimes I_r)\frac{\partial^2 x}{\partial \alpha \partial \alpha}.$$

Proof: Since

$$\frac{\partial y}{\partial \alpha} = \frac{\partial y}{\partial x}\frac{\partial x}{\partial \alpha},$$

Proposition 5.6 guarantees the validity of the first part. For the second part, applying the arguments of Proposition 5.7, we see that

$$\frac{\partial^2 y}{\partial \alpha \partial \alpha'} = \frac{\partial}{\partial \alpha}\left(\frac{\partial y}{\partial \alpha}\right) = \frac{\partial}{\partial \alpha}\left(\frac{\partial y}{\partial \alpha}\right)'.$$

$$\left(\frac{\partial y}{\partial \alpha}\right)' = \left(\frac{\partial x}{\partial \alpha}\right)'(A' + A)x.$$

$$\frac{\partial}{\partial \alpha}\left(\frac{\partial y}{\partial \alpha}\right)' = \left(\frac{\partial x}{\partial \alpha}\right)'(A' + A)\left(\frac{\partial x}{\partial \alpha}\right) + (x'(A' + A) \otimes I_r)\frac{\partial^2 x}{\partial \alpha \partial \alpha'}.$$

q.e.d.

Corollary 5.4. Consider the same situation as in Corollary 5.2, but suppose in addition that A is symmetric. Then,

$$\frac{\partial y}{\partial \alpha} = 2x'A\frac{\partial x}{\partial \alpha},$$

$$\frac{\partial^2 y}{\partial \alpha \partial \alpha} = 2\left(\frac{\partial x}{\partial \alpha}\right)'A\left(\frac{\partial x}{\partial \alpha}\right) + (2x'A \otimes I)\frac{\partial^2 x}{\partial \alpha \partial \alpha}.$$

Proof: Obvious from Corollary 5.1.

5.4 Differentiation of the Trace

Let us now turn our attention to the differentiation of the trace of matrices. In fact, the preceding discussion has anticipated most of the results to be derived below. We begin with another convention.

Convention 5.3. If it is desired to differentiate, say, $\mathrm{tr}(AB)$ with respect to the elements of A, the operation involved will be interpreted as the "rematricization" of the vector

$$\frac{\partial \, \mathrm{tr}(AB)}{\partial \, \mathrm{vec}(A)},$$

i.e. we first obtain the vector

$$\phi = \frac{\partial \, \mathrm{tr}(AB)}{\partial \, \mathrm{vec}(A)}$$

and then put the resulting (column) vector in matrix form, using the operator mat defined in Chap. 4. Doing so yields

$$\text{mat}(\phi') = \frac{\partial \, \text{tr}(AB)}{\partial A}.$$

With this in mind, we establish

Proposition 5.8. Let A be a square matrix of order m. Then,

$$\frac{\partial \, \text{tr}(A)}{\partial A} = I_m.$$

If the elements of A are functions of the r-element vector α, then

$$\frac{\partial \, \text{tr}(A)}{\partial \alpha} = \frac{\partial \, \text{tr}(A)}{\partial \, \text{vec}(A)} \frac{\partial \, \text{vec}(A)}{\partial \alpha} = \text{vec}(I)' \frac{\partial \, \text{vec}(A)}{\partial \alpha}.$$

Proof: We note that $\text{tr}(A) = \text{tr}(A \cdot I_m)$. From Proposition 4.3, we have $\text{tr}(A) = \text{vec}(I_m)' \, \text{vec}(A)$. Thus,

$$\frac{\partial \, \text{tr}(A)}{\partial \, \text{vec}(A)} = \text{vec}(I_m)'.$$

Rematricizing this vector we obtain

$$\frac{\partial \, \text{tr}(A)}{\partial A} = \text{mat}[\text{vec}(I_m)] = I_m,$$

which proves the first part. For the second part, we note that Proposition 5.2 implies

$$\frac{\partial \, \text{tr}(A)}{\partial \alpha} = \frac{\partial \, \text{tr}(A)}{\partial \, \text{vec}(A)} \frac{\partial \, \text{vec}(A)}{\partial \, \text{vec}(\alpha)} = \text{vec}(I)' \frac{\partial \, \text{vec}(A)}{\partial \alpha}.$$

q.e.d.

We shall now establish results regarding differentiation of the trace of products of a number of matrices. We have

Proposition 5.9. Let A be $m \times n$, and X be $n \times m$; then

$$\frac{\partial \, \text{tr}(AX)}{\partial X} = A'.$$

If X is a function of the elements of the vector α, then

$$\frac{\partial \, \text{tr}(AX)}{\partial \alpha} = \frac{\partial \, \text{tr}(AX)}{\partial \, \text{vec}(X)} \frac{\partial \, \text{vec}(X)}{\partial \alpha} = \text{vec}(A')' \frac{\partial \, \text{vec}(X)}{\partial \alpha}.$$

Proof: By Proposition 4.3,

$$\operatorname{tr}(AX) = \operatorname{vec}(A')' \operatorname{vec}(X).$$

Thus, by Proposition 5.1,

$$\frac{\partial \operatorname{tr}(AX)}{\partial \operatorname{vec}(X)} = \operatorname{vec}(A')'.$$

Rematricizing this result, we have

$$\frac{\partial \operatorname{tr}(AX)}{\partial X} = A',$$

which proves the first part. For the second part, we have, by Proposition 5.2,

$$\frac{\partial \operatorname{tr}(AX)}{\partial \alpha} = \operatorname{vec}(A')' \frac{\partial \operatorname{vec}(X)}{\partial \alpha}.$$

q.e.d.

Proposition 5.10. Let A be $m \times n$, X be $n \times m$, and B be $m \times m$; then

$$\frac{\partial \operatorname{tr}(AXB)}{\partial X} = A'B'.$$

If X is a function of the r-element vector α, then

$$\frac{\partial \operatorname{tr}(AXB)}{\partial \alpha} = \operatorname{vec}(A'B')' \frac{\partial \operatorname{vec}(X)}{\partial \alpha}.$$

Proof: We note that

$$\operatorname{tr}(AXB) = \operatorname{tr}(BAX).$$

But Proposition 5.9 implies

$$\frac{\partial \operatorname{tr}(AXB)}{\partial \operatorname{vec}(X)} = \operatorname{vec}(A'B')',$$

and thus

$$\frac{\partial \operatorname{tr}(AXB)}{\partial X} = A'B'.$$

For the second part, it easily follows that

$$\frac{\partial \operatorname{tr}(AXB)}{\partial \alpha} = \frac{\partial \operatorname{tr}(AXB)}{\partial \operatorname{vec}(X)} \frac{\partial \operatorname{vec}(X)}{\partial \alpha} = \operatorname{vec}(A'B')' \frac{\partial \operatorname{vec}(X)}{\partial \alpha}.$$

q.e.d.

Proposition 5.11. Let A be $m \times n$, X be $n \times q$, B be $q \times r$, and Z be $r \times m$; then

$$\frac{\partial \operatorname{tr}(AXBZ)}{\partial X} = A'Z'B',$$

$$\frac{\partial \operatorname{tr}(AXBZ)}{\partial Z} = B'X'A'.$$

If X and Z are functions of the r-element vector α, then

$$\frac{\partial \operatorname{tr}(AXBZ)}{\partial \alpha} = \operatorname{vec}(A'Z'B')' \frac{\partial \operatorname{vec}(X)}{\partial \alpha} + \operatorname{vec}(B'X'A')' \frac{\partial \operatorname{vec}(Z)}{\partial \alpha}.$$

Proof: Since

$$\operatorname{tr}(AXBZ) = \operatorname{tr}(BZAX),$$

Proposition 5.9 implies

$$\frac{\partial \operatorname{tr}(AXBZ)}{\partial X} = A'Z'B',$$
$$\frac{\partial \operatorname{tr}(AXBZ)}{\partial Z} = B'X'A'.$$

For the second part, we note

$$\frac{\partial \operatorname{tr}(AXBZ)}{\partial \alpha} = \frac{\partial \operatorname{tr}(AXBZ)}{\partial \operatorname{vec}(X)} \frac{\partial \operatorname{vec}(X)}{\partial \alpha} + \frac{\partial \operatorname{tr}(AXBZ)}{\partial \operatorname{vec}(Z)} \frac{\partial \operatorname{vec}(Z)}{\partial \alpha},$$

and it further implies that

$$\frac{\partial \operatorname{tr}(AXBZ)}{\partial \alpha} = \operatorname{vec}(A'Z'B')' \frac{\partial \operatorname{vec}(X)}{\partial \alpha} + \operatorname{vec}(B'X'A')' \frac{\partial \operatorname{vec}(Z)}{\partial \alpha}.$$

q.e.d.

Finally, we have

Proposition 5.12. Let A be $m \times m$, X be $q \times m$, and B be $q \times q$; then

$$\frac{\partial \operatorname{tr}(AX'BX)}{\partial X} = B'XA' + BXA.$$

If X is a function of the r-element vector α, then

$$\frac{\partial \operatorname{tr}(AX'BX)}{\partial \alpha} = \operatorname{vec}(X)'[(A' \otimes B) + (A \otimes B')] \frac{\partial \operatorname{vec}(X)}{\partial \alpha}.$$

Proof: From Remark 4.2, we see that $\operatorname{tr}(AX'BX) = \operatorname{vec}(X)'(A' \otimes B)\operatorname{vec}(X)$, and from Proposition 5.6 we conclude

$$\frac{\partial \operatorname{tr}(AX'BX)}{\partial \operatorname{vec}(X)} = \operatorname{vec}(X)'[(A' \otimes B) + (A \otimes B')].$$

Matricizing this vector we have, from Corollary 4.1 and Proposition 4.2,

$$\frac{\partial \operatorname{tr}(AX'BX)}{\partial X} = B'XA' + BXA.$$

The second part of the proposition follows immediately from Corollary 5.3 and the result above.

q.e.d.

Remark 5.4. The preceding results indicate that differentiating the trace of products of matrices with respect to the elements of one of the matrix factors is a special case of differentiation of linear, bilinear, and quadratic forms. For this reason, it is not necessary to derive second-order derivatives, because the latter are easily derivable from the corresponding results regarding linear, bilinear, and quadratic forms, i.e. quantities of the form

$$Ax, \qquad z'Ax, \qquad x'Ax,$$

where A is a matrix, and z, x are appropriately dimensioned vectors.

5.5 Differentiation of Determinants

We now consider certain other aspects of differentiation of functions of matrices that are also important in econometrics.

Proposition 5.13. Let A be a square matrix of order m; then

$$\frac{\partial |A|}{\partial A} = A^*,$$

where A^* is the matrix of cofactors (of the elements of A). If the elements of A are functions of the r elements of the vector α, then

$$\frac{\partial |A|}{\partial \alpha} = \operatorname{vec}(A^*)' \frac{\partial \operatorname{vec}(A)}{\partial \alpha}.$$

Proof: To prove the first part of the proposition, it is sufficient to obtain the typical (i, j) element of the matrix $\partial |A|/\partial A$. The latter is given by

$$\frac{\partial |A|}{\partial a_{ij}}.$$

Expand the determinant by the elements of the ith row and find, by Proposition 2.23, $|A| = \sum_{k=1}^{m} a_{ik} A_{ik}$, where A_{ik} is the cofactor of a_{ik}. Evidently, A_{ik} does **not** contain a_{ik}. Consequently,

$$\frac{\partial |A|}{\partial a_{ij}} = A_{ij},$$

and thus

$$\frac{\partial |A|}{\partial A} = A^*,$$

as was to be proved. For the second part, we note that

$$\frac{\partial |A|}{\partial \alpha} = \frac{\partial |A|}{\partial \operatorname{vec}(A)} \frac{\partial \operatorname{vec}(A)}{\partial \alpha}.$$

But it is easy to see, from Convention 5.2, that

$$\frac{\partial |A|}{\partial \operatorname{vec}(A)} = \operatorname{vec} \left(\frac{\partial |A|}{\partial A} \right)' = \operatorname{vec}(A^*)'.$$

Hence,

$$\frac{\partial |A|}{\partial \alpha} = \operatorname{vec}(A^*)' \frac{\partial \operatorname{vec}(A)}{\partial \alpha}.$$

q.e.d.

Corollary 5.5. Assume, in addition to the conditions of Proposition 5.13, that A is nonsingular, and let $B = A^{-1} = A^*/|A|$. Then,

$$\frac{\partial |A|}{\partial A} = |A| B', \qquad \frac{\partial \ln |A|}{\partial A} = B',$$

$$\frac{\partial |A|}{\partial \alpha} = |A| \operatorname{vec}(B')' \frac{\partial \operatorname{vec}(A)}{\partial \alpha}.$$

Proof: In the proof of Proposition 5.13, note that $A_{ik} = |A| b_{ki}$, where b_{ki} is the (k, i) element of B, and that

$$\frac{1}{|A|} = \frac{\partial \ln |A|}{\partial |A|}.$$

q.e.d.

Corollary 5.6. If in Proposition 5.13 α is assumed to be a scalar, then

$$\frac{\partial |A|}{\partial \alpha} = \operatorname{tr} \left(A^{*\prime} \frac{\partial A}{\partial \alpha} \right),$$

and if A is nonsingular, then

$$\frac{\partial |A|}{\partial \alpha} = |A| \operatorname{tr} \left(B \frac{\partial A}{\partial \alpha} \right),$$

$$\frac{\partial \ln |A|}{\partial \alpha} = \operatorname{tr} \left(B \frac{\partial A}{\partial \alpha} \right).$$

Proof: If α is a scalar, then

$$\frac{\partial \operatorname{vec}(A)}{\partial \alpha} = \operatorname{vec} \left(\frac{\partial A}{\partial \alpha} \right),$$

where, obviously,

$$\frac{\partial A}{\partial \alpha} = \left[\frac{\partial a_{ij}}{\partial \alpha} \right].$$

Using Propositions 4.3 and 5.13, we see that

$$\frac{\partial |A|}{\partial \alpha} = \operatorname{vec}(A^*)' \operatorname{vec} \left(\frac{\partial A}{\partial \alpha} \right) = \operatorname{tr} \left(A^{*'} \frac{\partial A}{\partial \alpha} \right).$$

If A is nonsingular, then

$$A^{*'} = |A| B$$

so that, in this case,

$$\frac{\partial |A|}{\partial \alpha} = |A| \operatorname{tr} \left(B \frac{\partial A}{\partial \alpha} \right), \qquad \frac{\partial \ln |A|}{\partial \alpha} = \operatorname{tr} \left(B \frac{\partial A}{\partial \alpha} \right).$$

q.e.d.

Corollary 5.7. Let A of Proposition 5.13 be a symmetric matrix, and define[1]

$$\alpha = \left(a'_{.1}, a^{*'}_{.2}, a^{*'}_{.3} \cdots a^{*'}_{.m} \right)',$$

where

$$a^{*'}_{.j} = (a_{jj}, a_{j+1,j}, \ldots, a_{mj}), \qquad j = 2, \ldots, m.$$

---

[1]In fact we dealt with such issues in Chap. 4; the elimination or selection matrix discussed in Remark 4.7, say, S, produces the distinct elements of the symmetric matrix A in the column vector α, by the operation $\alpha = S\operatorname{vec}(A)$, while the restoration matrix, also discussed therein, operates on α to produce (restore) $\operatorname{vec}(A)$. This is the matrix H defined below, so that H $\alpha = \operatorname{vec}(A)$.

Further, define the restoration matrix discussed in Remark 4.7 (of Chap. 4)

$$
H =
\begin{bmatrix}
I_m & 0 & 0 & \cdot & \cdot & 0 \\
e'_{\cdot 2m} & 0 & 0 & \cdot & \cdot & \cdot \\
0 & I_{m-1} & 0 & \cdot & \cdot & \cdot \\
e'_{\cdot 3m} & 0 & 0 & \cdot & \cdot & \cdot \\
0 & e'_{\cdot 2m-1} & 0 & \cdot & \cdot & \cdot \\
0 & 0 & I_{m-2} & \cdot & \cdot & \cdot \\
\vdots & & & & & \\
e'_{\cdot mm} & 0 & 0 & \cdot & \cdot & \cdot \\
0 & e'_{\cdot m-1,m-1} & 0 & \cdot & \cdot & \cdot \\
\cdot & \cdot & & \cdot & \cdot & \cdot \\
\cdot & \cdot & & \cdot & e'_{\cdot 22} & 0 \\
0 & & \cdot & \cdot & 0 & 1
\end{bmatrix}
$$

such that H is $m^2 \times m(m+1)/2$ and $e_{\cdot js}$ is an s-element column vector $(s = 2, 3, \ldots, m)$ all of whose elements are zero except the jth, which is unity. Then,

$$
\frac{\partial |A|}{\partial \alpha} = \mathrm{vec}(A^*)' H
$$

and if A is nonsingular

$$
\frac{\partial |A|}{\partial \alpha} = |A|\, \mathrm{vec}(B')' H, \qquad \frac{\partial \ln |A|}{\partial \alpha} = \mathrm{vec}(B')' H.
$$

Proof: We note that α is $m(m+1)/2 \times 1$ and contains the **distinct elements** of A. Moreover, $\mathrm{vec}(A) = H\alpha$. It is then immediate that

$$
\frac{\partial |A|}{\partial \alpha} = \frac{\partial |A|}{\partial\, \mathrm{vec}(A)} \frac{\partial\, \mathrm{vec}(A)}{\partial \alpha} = \mathrm{vec}(A^*)' H.
$$

If A is nonsingular, then

$$
A^{*'} = |A| B;
$$

since

$$
\frac{\partial \ln |A|}{\partial |A|} = \frac{1}{|A|},
$$

we thus have

$$
\frac{\partial \ln |A|}{\partial \alpha} = \mathrm{vec}(B')' H, \qquad \frac{\partial |A|}{\partial \alpha} = |A|\, \mathrm{vec}(B')' H.
$$

q.e.d.

Remark 5.5. Note that the operations

$$\text{vec}(A^*)'H \quad \text{or} \quad \text{vec}(B')'H$$

simply rearrange the elements of $\text{vec}(A^*)$ or $\text{vec}(B)$. In particular,

$$\text{vec}(A^*)'H = \left(A_{\cdot 1}^{*'} + A_{12}e_{\cdot 2m}' + \cdots + A_{1m}e_{\cdot mm}', A_{\cdot 2}^{*'} + A_{23}e_{\cdot 2m-1}' + \cdot \right.$$

$$\left. + A_{2m}e_{\cdot m-1}', m-1, \ldots, A_{\cdot m-1}^{*'} + A_{m-1,m}e_{\cdot 22}', A_{mm}\right),$$

where

$$A^* = (A_{ij}), \qquad A_{\cdot j}^* = (A_{jj}, A_{j+1,j} \cdots A_{mj}), \qquad j = 1, 2, \ldots, m.$$

The question then is what we should mean by

$$\frac{\partial |A|}{\partial A}.$$

We adopt the following convention.

Convention 5.4.

$$\frac{\partial |A|}{\partial a_{ij}} = \frac{\partial |A|}{\partial \alpha_{ij}} \quad i \geq j \quad i, j = 1, 2, \ldots, m$$

$$\frac{\partial |A|}{\partial a_{ij}} = \frac{\partial |A|}{\partial \alpha_{ji}} \quad i < j.$$

This enables us to write

$$\frac{\partial |A|}{\partial A} = A^*,$$

thus preserving the generality of Proposition 5.13 without being inconsistent with the results of the differentiation as stated in the corollary.

Another useful result is

Proposition 5.14. Let X be $n \times m$ and B be $n \times n$, and put

$$A = X'BX.$$

Then,

$$\frac{\partial |A|}{\partial X} = \left[\text{tr}\left(A^{*'}\frac{\partial A}{\partial x_{ik}}\right)\right] = BXA^{*'} + B'XA^*,$$

where A^* is the matrix of cofactors of A.

Proof: We shall prove this result by simply deriving the (i, k) element of the matrix $\partial |A| / \partial X$. By the usual chain rule for differentiation, we have

$$\frac{\partial |A|}{\partial x_{ik}} = \sum_{r,s=1}^{m} \frac{\partial |A|}{\partial a_{rs}} \frac{\partial a_{rs}}{\partial x_{ik}}.$$

But

$$\frac{\partial |A|}{\partial a_{rs}} = A_{rs}$$

so that

$$\frac{\partial |A|}{\partial x_{ik}} = \sum_{r,s=1}^{m} A_{rs} \frac{\partial a_{rs}}{\partial x_{ik}} = \text{tr}\left(A^* \, ' \frac{\partial A}{\partial x_{ik}}\right),$$

which proves the first part of the representation. Next, we note that, formally, we can put

$$\frac{\partial A}{\partial x_{ik}} = \frac{\partial X'}{\partial x_{ik}} BX + X'B \frac{\partial X}{\partial x_{ik}}.$$

But

$$\frac{\partial X'}{\partial x_{ik}} = e_{\cdot k} e'_{\cdot i}, \qquad \frac{\partial X}{\partial x_{ik}} = e_{\cdot i} e'_{\cdot k},$$

where $e_{\cdot k}$ is an m-element (column) vector all of whose elements are zero except the kth, which is unity, and $e_{\cdot i}$ is an n-element (column) vector all of whose elements are zero except the ith, which is unity. For simplicity of notation only, put

$$A^{*'} = |A|A^{-1}$$

and note that

$$\text{tr}\left(A^{*'} \frac{\partial A}{\partial x_{ik}}\right) = |A| \, \text{tr} \, A^{-1}(e_{\cdot k} e'_{\cdot i} BX + X'Be_{\cdot i} e'_{\cdot k})$$

$$= |A|[\text{tr}(a^{\cdot k} b_{i \cdot} X) + \text{tr}(X'b_{\cdot i} a^{k \cdot})] = |A|(b_{i \cdot} X a^{\cdot k} + b'_{\cdot i} X a^{k \cdot '}),$$

where $b_{\cdot i}$ is the ith column of B, $b_{i \cdot}$ is the ith row of B, $a^{\cdot k}$ is the kth column of A^{-1} and $a^{k \cdot}$ its kth row. Thus, the (i, k) element of

$$\partial |A| / \partial X$$

is given by

$$|A|(b_{i \cdot} X a^{\cdot k} + b'_{\cdot i} X a^{k \cdot '}).$$

But this is, of course, the (i, k) element of

$$|A|(BXA^{-1} + B'XA'^{-1}) = BXA^* + B'XA^*.$$

Consequently,

$$\frac{\partial |A|}{\partial X} = BXA^{*'} + B'XA^*.$$

q.e.d.

Corollary 5.8. If in Proposition 5.14 A is nonsingular, then

$$\frac{\partial |A|}{\partial X} = |A|(BXA^{-1} + B'XA'^{-1}).$$

Proof: Evident from Proposition 5.14.

Corollary 5.9. If in Proposition 5.14 A is nonsingular, then

$$\frac{\partial \ln |A|}{\partial X} = BXA^{-1} + B'XA'^{-1}.$$

Proof: We have

$$\frac{\partial \ln |A|}{\partial X} = \frac{\partial \ln |A|}{\partial |A|} \frac{\partial |A|}{\partial X} = \frac{1}{|A|} \frac{\partial |A|}{\partial X},$$

and the conclusion follows from Corollary 5.7.

q.e.d.

Corollary 5.10. If in Proposition 5.14 B is symmetric, then

$$\frac{\partial |A|}{\partial X} = 2BXA^*.$$

Proof: Obvious since if B is symmetric so is A, and thus

$$A^{*'} = A^*.$$

q.e.d.

Proposition 5.15. Let X be $m \times n$ and B be $m \times m$, and suppose that the elements of X are functions of the elements of the vector α. Put

$$A = X'BX.$$

Then

$$\frac{\partial |A|}{\partial \alpha} = \text{vec}(X)'[(A^{*'} \otimes B') + (A^* \otimes B)] \frac{\partial \, \text{vec}(X)}{\partial \alpha}.$$

Proof: By the usual chain rule of differentiation, we have

$$\frac{\partial |A|}{\partial \alpha} = \frac{\partial |A|}{\partial \, \text{vec}(X)} \frac{\partial \, \text{vec}(X)}{\partial \alpha}.$$

But

$$\frac{\partial |A|}{\partial \text{ vec}(X)} = \left[\text{vec}\left(\frac{\partial |A|}{\partial X}\right)\right]'.$$

From Corollary 4.1 and Proposition 5.14, we obtain

$$\text{vec}\left(\frac{\partial |A|}{\partial X}\right) = [(A^* \otimes B) + (A^{*'} \otimes B')]\,\text{vec}(X).$$

Thus, we conclude

$$\frac{\partial |A|}{\partial \alpha} = \text{vec}(X)'[(A^* \otimes B) + (A^{*'} \otimes B')]\frac{\partial \text{ vec}(X)}{\partial \alpha}.$$

q.e.d.

Corollary 5.11. If in Proposition 5.15 A is nonsingular, then

$$\frac{\partial |A|}{\partial \alpha} = |A|\,\text{vec}(X)'[(A^{'-1} \otimes B) + (A^{-1} \otimes B')]\frac{\partial \text{ vec}(X)}{\partial \alpha}.$$

Proof: Obvious if we note that

$$|A|A^{-1} = A^{*'}.$$

Corollary 5.12. If in Proposition 5.15 A is nonsingular, then

$$\frac{\partial \ln |A|}{\partial \alpha} = \text{vec}(X)'[(A^{'-1} \otimes B) + (A^{-1} \otimes B')]\frac{\partial \text{ vec}(X)}{\partial \alpha}.$$

Proof: Obvious.

Corollary 5.13. If in Proposition 5.15 B is symmetric, then

$$\frac{\partial |A|}{\partial \alpha} = 2\,\text{vec}(X')'[A^* \otimes B]\frac{\partial \text{ vec}(X)}{\partial \alpha}.$$

Proof: Obvious since if B is symmetric so is A, and thus

$$A^{*'} = A^*.$$

q.e.d.

The results above exhaust those aspects of differentiation of determinants that are commonly found useful in econometrics.

5.6 Differentiation of Inverse of a Matrix

We begin with

Proposition 5.16. Let A be $m \times m$ and nonsingular. Then, the "derivative of the inverse" $\partial A^{-1}/\partial A$ is given by

$$\frac{\partial \operatorname{vec}(A^{-1})}{\partial \operatorname{vec}(A)} = -(A'^{-1} \otimes A^{-1}).$$

If the elements of A are functions of the elements of the vector α, then

$$\frac{\partial \operatorname{vec}(A^{-1})}{\partial \alpha} = -(A'^{-1} \otimes A^{-1}) \frac{\partial \operatorname{vec}(A)}{\partial \alpha}.$$

Proof: We begin by taking the derivative of A^{-1} with respect to an element of A. From the relation

$$A^{-1}A = I$$

we easily see that

$$0 = \frac{\partial A^{-1}}{\partial a_{rs}} A + A^{-1} \frac{\partial A}{\partial a_{rs}}, \tag{5.1}$$

from which we obtain

$$\frac{\partial A^{-1}}{\partial a_{rs}} = -A^{-1} \frac{\partial A}{\partial a_{rs}} A^{-1}. \tag{5.2}$$

But $\partial A/\partial a_{rs}$ is a matrix all of whose elements are zero except the (r, s) element, which is unity. Consequently,

$$\frac{\partial A}{\partial a_{rs}} = e_{\cdot r} e'_{\cdot s}, \tag{5.3}$$

where $e_{\cdot j}$ is an m-element vector all of whose elements are zero except the jth, which is unity. Using Eq. (5.3) in Eq. (5.2), we find

$$\frac{\partial A^{-1}}{\partial a_{rs}} = -a^{\cdot r} a^{s \cdot}, \qquad r, s = 1, 2, \ldots, m, \tag{5.4}$$

where $a^{\cdot r}$ is the rth column and $a^{s \cdot}$ is the sth row of A^{-1}. Vectorizing Eq. (5.4) yields, by Proposition 4.1,

$$\frac{\partial \operatorname{vec}(A^{-1})}{\partial a_{rs}} = -(a^{s \cdot'} \otimes a^{\cdot r}), \qquad r, s = 1, 2, \ldots, m. \tag{5.5}$$

From Eq. (5.5) we see, for example, that

$$\frac{\partial \operatorname{vec}(A^{-1})}{\partial a_{11}} = -(a^{1 \cdot'} \otimes a^{\cdot 1}), \qquad \frac{\partial \operatorname{vec}(A^{-1})}{\partial a_{21}} = -(a^{1 \cdot'} \otimes a^{\cdot 2})$$

and so on; thus,

$$\frac{\partial \text{ vec}(A^{-1})}{\partial a_{.1}} = -(a^{1.\prime} \otimes A^{-1})$$

or, in general,

$$\frac{\partial \text{ vec}(A^{-1})}{\partial a_{.s}} = -(a^{s.\prime} \otimes A^{-1}), \qquad s = 1, 2, \ldots, m. \tag{5.6}$$

But Eq. (5.6) implies

$$\frac{\partial \text{ vec}(A^{-1})}{\partial \text{ vec}(A)} = -(A^{\prime -1} \otimes A^{-1}), \tag{5.7}$$

which proves the first part of the proposition. For the second part, we note that by the chain rule of differentiation

$$\frac{\partial \text{ vec}(A^{-1})}{\partial \alpha} = \frac{\partial \text{ vec}(A^{-1})}{\partial \text{ vec}(A)} \frac{\partial \text{ vec}(A)}{\partial \alpha},$$

and the desired result follows immediately from Eq. (5.7).

q.e.d.

Since the result of the proposition above may not be easily digestible, let us at least verify that it holds for a simple case. To this end we have the following corollary.

Corollary 5.14. Suppose in Proposition 5.16 α is a **scalar**; then

$$\frac{\partial A^{-1}}{\partial \alpha} = -A^{-1} \frac{\partial A}{\partial \alpha} A^{-1}.$$

Proof: From Proposition 5.16 we have, formally,

$$\frac{\partial \text{ vec}(A^{-1})}{\partial \alpha} = -(A^{\prime -1} \otimes A^{-1}) \frac{\partial \text{ vec}(A)}{\partial \alpha}.$$

Matricizing the vector above, using Corollary 4.1, yields

$$\frac{\partial A^{-1}}{\partial \alpha} = -A^{-1} \frac{\partial A}{\partial \alpha} A^{-1}.$$

q.e.d.

Chapter 6

DE Lag Operators GLSEM and Time Series

6.1 The Scalar Second-Order Equation

In this chapter we deal with econometric applications of (vector) difference equations **with constant coefficients**, as well as with aspects of the statistical theory of time series and their application in econometrics.

Definition 6.1. The second order non-stochastic difference equation (DE) is given by

$$a_0 y_t + a_1 y_{t-1} + a_2 y_{t-2} = g(t), \tag{6.1}$$

where y_t is the **scalar** dependent variable, the a_i, $i = 0, 1, 2$, are the (constant) coefficients, and $g(t)$ is the (non-random) real-valued "forcing function"; in such models usually we normalize by setting $a_0 = 1$.

If the function g is a **random variable**, then y_t is also a random variable and the equation above is said to be a **stochastic** DE.

A solution to the **non-stochastic** DE of Eq. (6.1) is a real valued function $y^*(t)$, $t = 1, 2, 3, \ldots$, which, given $a_i, i = 0, 1, 2$, and g(t) satisfies that equation.

The relation

$$a_0 y_t + a_1 y_{t-1} + a_2 y_{t-2} = 0$$

is said to be the **homogeneous part** of the DE in Eq. (6.1).

In elementary discussions, the solution to a DE as in Eq. (6.1) is obtained in two steps. First, we consider the homogeneous part

$$a_0 y_t + a_1 y_{t-1} + a_2 y_{t-2} = 0, \tag{6.2}$$

P.J. Dhrymes, *Mathematics for Econometrics*,
DOI 10.1007/978-1-4614-8145-4_6, © The Author 2013

and find the most general form of its solution, called the **general solution to the homogeneous part**. Then, we find just one solution to Eq. (6.1), called the **particular** solution. The sum of the general solution to the homogeneous part and the particular solution is said to be the **general solution** to the equation. What is meant by the "general solution", denoted, say, by y_t^*, is that y_t^* satisfies Eq. (6.1) and it can be made to satisfy any pre-specified set of "initial conditions". If $g(t)$ and the coefficients are specified and if, further, we are given the values assumed by y_t for $t = 0$, $t = -1$, (initial conditions) we can compute y_1 from Eq. (6.1); then, given y_1, y_0, y_{-1} we can compute y_2 and so on. Thus, given the coefficients and the function $g(\cdot)$, the behavior (of the homogeneous part) of y_t depends solely on the "initial conditions", and if these are also specified then the behavior of y_t is completely determined. Thus, for a solution of Eq. (6.1) to be a "general solution" it must be capable of accommodating any pre-specified set of initial conditions.

From the results above it is evident that the general solution to the homogeneous part simply carries forward the influence of initial conditions. Generally, in economics, we would not want to say that initial conditions are very crucial to the development of a system, but rather that it is the external forces impinging on the system that, in the long run, are ultimately responsible for its development. This introduces the concept of the **stability** of a (non-stochastic) difference equation. A DE as in Eq. (6.1) is said to be **stable** if and only if the general solution to its homogeneous part obeys[1]

$$\lim_{t \to \infty} y_t^{\mathrm{H}} = 0. \tag{6.3}$$

As we mentioned earlier, we also need to find a particular solution in order to obtain the general solution to Eq. (6.1). But a routine way of doing so is not easily obtained in this elementary context. In view of the fact that most, if not all, of the discussion in this chapter will involve **stochastic** DE and we are not particularly interested in their short term behavior, this elementary line of inquiry will not be pursued further.

It will facilitate matters if we introduce the notion of the lag operator L.[2]

---

[1] The superscript H in Eq. (5.3) indicates that it is the the solution to the **homogeneous part**.

[2] In the statistical and, more generally, the mathematical literature this operator is termed the **backward operator** and is denoted by B. But lag operator is a more convenient term and has a long tradition in the econometrics literature, so we retain it and shall use it exclusively wherever appropriate.

6.2 The Lag Operator L and Its Algebra

If $x(t)$ is a function of "time", the **lag operator** L is defined by

$$Lx(t) = x(t-1). \tag{6.4}$$

Powers of the operator are defined as successive applications, i.e.

$$L^2 x(t) = L[Lx(t)] = Lx(t-1) = x(t-2),$$

and in general

$$L^k x(t) = x(t-k), \qquad k > 0. \tag{6.5}$$

For $k = 0$, we have the **identity operator**

$$L^0 \equiv I, \qquad L^0 x(t) = x(t). \tag{6.6}$$

It is apparent that

$$L^k L^s = L^s L^k = L^{s+k}. \tag{6.7}$$

Moreover,

$$(c_1 L^{s_1} + c_2 L^{s_2})x(t) = c_1 L^{s_1} x(t) + c_2 L^{s_2} x(t) \tag{6.8}$$
$$= c_1 x(t-s_1) + c_2 x(t-s_2).$$

One can further show that the set

$$\{I, L, L^2, \ldots\}$$

over the field of real (or complex) numbers, together with the operations above, induces a **vector space**. But what is of importance to us is that the set of **polynomial operators**, whose typical element is

$$\sum_{i=0}^{n} c_i L^i,$$

induces an algebra that is **isomorphic to the algebra of polynomials** in a real or complex indeterminate. This means that to determine the outcome of a set of operations on polynomials in the lag operator one need only carry out such operations with respect to an ordinary polynomial in the real or complex indeterminate, ψ, and then substitute for ψ and its powers L and its powers. Perhaps a few examples will make this clear.

Example 6.1. Let

$$P_1(L) = c_{01} I + c_{11} L + c_{21} L^2,$$
$$P_2(L) = c_{02} I + c_{12} L + c_{22} L^2 + c_{32} L^3,$$

and suppose we desire the product $P_1(L) \cdot P_2(L)$. We consider, by the usual rules of multiplying polynomials,

$$
\begin{aligned}
P_1(\psi)P_2(\psi) = {} & (c_{01}c_{02}) + (c_{11}c_{02} + c_{12}c_{01})\psi \\
& + (c_{11}c_{12} + c_{22}c_{01} + c_{21}c_{02})\psi^2 \\
& + (c_{01}c_{32} + c_{11}c_{22} + c_{12}c_{21})\psi^3 \\
& + (c_{11}c_{32} + c_{22}c_{21})\psi^4 + c_{32}c_{21}\psi^5,
\end{aligned}
$$

and consequently

$$
\begin{aligned}
P_1(L)P_2(L) = {} & c_{01}c_{02}I + (c_{11}c_{02} + c_{12}c_{01})L + (c_{11}c_{12} + c_{22}c_{01} + c_{21}c_{02})L^2 \\
& + (c_{01}c_{32} + c_{11}c_{22} + c_{12}c_{21})L^3 + (c_{11}c_{32} + c_{22}c_{21})L^4 + c_{32}c_{21}L^5.
\end{aligned}
$$

Example 6.2. Let

$$
P_1(L) = I - \lambda L,
$$

and suppose we wish to find its inverse $I/P_1(L)$. To do so, we consider the inverse of $1 - \lambda\psi$; if $|\psi| \leq 1$ and $|\lambda| < 1$, then we know that

$$
\frac{1}{1 - \lambda\psi} = \sum_{i=0}^{\infty} \lambda^i \psi^i.
$$

Hence, under the condition $|\lambda| < 1$,

$$
\frac{1}{P_1(L)} = \sum_{i=0}^{\infty} \lambda^i L^i.
$$

Although the discussion above is heuristic, and rather sketchy at that, it is sufficient for the purposes we have in mind. The reader interested in more detail is referred to Dhrymes (1971, 1982).

If we use the apparatus of polynomial lag operators, we see that we can write Eq. (6.1) as

$$
[a_0 L^0 + a_1 L + a_2 L^2]y_t = g(t) \tag{6.9}
$$

Consequently, we may write the solution formally as

$$
y_t = [A(L)]^{-1}g(t), \tag{6.10}
$$

where

$$
A(L) = a_0 I + a_1 L + a_2 L^2. \tag{6.11}
$$

The question then arises as to the meaning and definition of the inverse of this polynomial operator. In view of the isomorphism referred to earlier, we consider the polynomial equation

$$
a_0 + a_1\psi + a_2\psi^2 = 0 \tag{6.12}
$$

and its roots ψ_1, ψ_2. By the fundamental theorem of algebra, we can write

$$A(\psi) = a_2(\psi - \psi_1)(\psi - \psi_2) = a_2(\psi_1\psi_2)\left(1 - \frac{\psi}{\psi_1}\right)\left(1 - \frac{\psi}{\psi_2}\right), \qquad (6.13)$$

where $\psi_i, i = 1, 2$, are the roots of the polynomial equation $\psi^2 + (a_1/a_2)\psi + (a_0/a_2) = 0$.

Putting

$$\lambda_1 = \frac{1}{\psi_1}, \qquad \lambda_2 = \frac{1}{\psi_2}, \qquad (6.14)$$

we thus write

$$A(L) = (I - \lambda_1 L)(I - \lambda_2 L), \qquad (6.15)$$

and for A(L) to be **invertible** we require that $|\lambda_i| < 1, i = 1, 2$. If these conditions hold then the solution exhibited in Eq. (6.10) can be given a concrete meaning and, thus, can be written as

$$y_t = \sum_{i=-\infty}^{t} \sum_{j=-\infty}^{t} \lambda_1^i \lambda_2^j L^{i+j} g(t) = \sum_{k=-\infty}^{t} \left(\sum_{j=-\infty}^{k} \lambda_1^{k-j} \lambda_2^j \right) g(t - k). \qquad (6.16)$$

Remark 6.1. Connecting the development above with the discussion of this non-stochastic DE in Eq. (6.1), the following observations are valid:

i. It may be shown that the roots $\lambda_i, i = 1, 2$, (which are assumed to be real) can form the basis of the (general) solution to the homogeneous part

$$y_t^H = c_1 \lambda_1^t + c_2 \lambda_2^t, \text{ if distinct, or } y_t^H = c_1 \lambda_1^t + c_2 t \lambda_2^t, \qquad (6.17)$$

if not, i.e. they are repeated.

ii. The solution exhibited in Eq. (6.16) is the **particular solution**, referred to in the discussion of the previous section.

The undetermined coefficients, c_1, c_2, can be used to impose any desired initial conditions for y_0, y_{-1}. Notice also that because $|\lambda_i| < 1, i = 1, 2$, the (general) solution to the homogeneous part converges to zero with t, i.e. it vanishes. This explains why initial conditions do not appear in the (particular) solution exhibited in Eq. (6.16) because, in that context, the process (DE) has "been in existence since the indefinite past".

Notice also that this interpretation makes sense, i.e. if $g(\cdot)$ is a bounded function, then y_t is also a bounded function, which is guaranteed by the stability condition one would normally impose on Eq. (6.1).

6.3 Vector Difference Equations

In this section, we will be considerably more formal than earlier because the results of this section are directly relevant to the analysis of time series and simultaneous equations models.

Definition 6.2. The equation

$$A_0 y_t + A_1 y_{t-1} + A_2 y_{t-2} + \cdots + \ldots, + A_r y_{t-r} = g(t), \qquad (6.18)$$

where A_i, $i = 0, 1, \ldots, r$, are $m \times m$ matrices of constants and y_t and $g(t)$ are m-element (column) vectors, is said to be an rth-**order vector difference equation with constant coefficients** (VDECC), provided the matrix A_0 is nonsingular.

If $g(t)$ is **not random**, the VDECC is said to be non-stochastic, while if **it is random** the VDECC is said to be **stochastic**.

Convention 6.1. Since the matrix A_0 in the VDECC is nonsingular, no loss of generality is entailed by taking $A_0 = I_m$, which we shall do in the discussion to follow. This convention is indeed required because of **identification restrictions** when we are dealing with **stochastic VDECC** in the context of time series analysis, or even in the non-stochastic case unless $g(t)$ is more precisely specified. In particular, note that if all we say about it is that it is even a continuous bounded function, then $Hg(t)$, where H is a non-singular matrix, is **also a continuous bounded function**. By imposing the normalization above we preclude this eventuality.

In discussing the characterization of the solution of Eq. (6.18), it is useful to note that **we can always transform it to an equivalent system that is a first-order** VDECC. This is done as follows. Define

$$\zeta_t = (y_t', y_{t-1}', \ldots, y_{t-r+1}')',$$

$$A^* = \begin{bmatrix} -A_1 & -A_2 & \cdots & & -A_r \\ I_m & 0 & \cdots & & 0 \\ 0 & I_m & \cdots & & 0 \\ 0 & 0 & \cdots & I_m & 0 \end{bmatrix},$$

and notice that the homogeneous part can be written as[3]

$$\zeta_t = A^* \zeta_{t-1}.$$

---

[3]The dimension of A^* is $mr \times mr$, and that of ζ_t is $mr \times 1$.

Indeed, Eq. (6.18) can be written as

$$\zeta_t = A^* \zeta_{t-1} + e_{.1} \otimes g(t), \tag{6.19}$$

where $e_{.1}$ is an r-element (column) vector all of whose elements are zero except the first, which is unity. Thus, in the following discussions we only deal with the first-order VDECC

$$y_t = Ay_{t-1} + g(t), \tag{6.20}$$

where y_t, $g(t)$ are m-element column vectors and A is an $m \times m$ matrix of constants. We shall further impose

Condition 6.1. *The matrix A in Eq. (6.20) is diagonalizable, i.e. we can write*

$$A = S\Lambda S^{-1},$$

which suggests that its characteristic roots are **distinct***, where S, Λ are, respectively, the matrices of characteristic vectors and roots of A.*

Remark 6.2. The need for Condition 6.1 will be fully appreciated below. For the moment let us note that in view of Propositions 2.39, 2.40 and 2.44, Condition 6.1 is only mildly restrictive.

Definition 6.3. A **solution** to the VDECC in Eq. (6.20) is a vector y_t^* such that y_t^* satisfies Eq. (6.20) together with a properly specified set of initial conditions, say

$$y_0 = \bar{y}_0.$$

6.3.1 Factorization of High Order Polynomials

To examine the nature of solutions to systems like those of Eq. (6.18) or Eq. (6.20) the discussions of previous sections are inadequate. We recall from a previous section that

$$I, L, L^2, \ldots L^n, \ldots, \text{ where } I = L^0,$$

is a basis for a vector space defined over the field of real or complex indeterminates. Moreover, the space of polynomials in the lag operator L (of the form $P(L) = \sum_0^n p_i L^i$) **is isomorphic** to the space of real (or complex) polynomials in the generic real variable t. The practical implication of this is that if we wish to perform any operations, such as additions, subtractions, multiplications and inversions on a polynomial in the lag operator, we first replace L and its powers by t and its powers, perform the operations in question, replace t and its power by L and its powers and this yields the desired result.

In the previous section we discussed such issues in the context of Eqs. (6.9) and (6.10) which involved **scalar** DE. To find the solution to the VDECC in Eqs. (6.18) and/or (6.20) we need to deal with

$$(I - AL)y_t = g(t), \text{ whose formal solution is } y_t = [I - AL]^{-1}g(t),$$

and it involves **not scalar polynomials, but matrix polynomials in the lag operator L.**

Naturally, this requires further discussion. It will turn out, however, that there is no fundamental difference between the two issues.

In this section we generalize the result above to the case of matrix polynomials in the lag operator L, say

$$P(L) = I - \sum_{i=1}^{n} p_i L^i, \text{ where the } p_i \text{ are } m \times m \text{ matrices.}$$

and establish its relevance in the representation of the solution to the VDECC of Eq. (6.20), as well as to the representation of autoregressions of order n, $AR(n)$, and autoregressive-moving average processes of order (m, n), $ARMA(m, n)$.

Since P(L) is a **matrix** whose elements are polynomials of degree n, by definition, its inverse is the transpose of the matrix of cofactors divided by the determinant $|P(L)|$. To this end, consider the (matrix) polynomial isomorphic to the one above, i.e.

$$P(t) = 1 - \sum_{i=1}^{n} p_i t^i, \tag{6.21}$$

whose (s,k) element is

$$P(t)_{(s,k)} = \delta_{(s,k)} - \sum_{i=1}^{n} p_{i(s,k)} t^i,$$

$\delta_{(s,k)}$ being the Kronecker δ, i.e. $\delta_{ij} = 1$, if i=j and zero otherwise.

Moreover each such element is a polynomial of degree n in the real indeterminate t. By definition the cofactor of $P(t)_{(s,k)}$ is the determinant of the matrix obtained from $P(t)$ by striking out its s th row and k th column. Hence its determinant is a polynomial of degree $(m-1)n$, generically denoted by $c(t)$. Finally, the determinant of $P(t)$ is a polynomial of degree mn, denoted by $d(t)$. Consequently, the inverse of P(t), if it exists, has elements which are ratios of polynomials of the form $(c(t)/d(t))$. Since c(t) is evidently well defined, the existence of $[P(t)]^{-1}$ is solely determined by whether or not $1/d(t)$ is well defined. But, since $d(t)$ is a polynomial of degree mn, we have

$$d(t) = 1 + \sum_{j=1}^{mn} d_j t^j = d_{mn} \left(\frac{1}{d_{mn}} + \sum_{j=1}^{mn} \frac{d_j}{d_{mn}} t^j \right).$$

By the fundamental theorem of algebra, if $t_s, s = 1, 2, \ldots, mn$ are the **roots of the polynomial equation**

$$\left(\frac{1}{d_{mn}} + \sum_{j=1}^{mn} \frac{d_j}{d_{mn}} t^j \right) = 0,$$

we have

$$d(t) = \prod_{s=1}^{mn} \left(1 - \frac{1}{t_s} t \right), \quad \text{because} \quad \frac{1}{d_{mn}} = \prod_{s=1}^{mn} (-t_s). \tag{6.22}$$

Putting

$$\lambda_s = \frac{1}{t_s}, \quad s = 1, 2, \ldots, mr \tag{6.23}$$

we have

$$[P(L)]^{-1} = \left(\prod_{s=1}^{mr} \frac{I}{I - \lambda_s L} \right) C(L), \quad C(L) = [c_{ji}(L)].$$

Let us now apply these results to the solution of Eq. (6.20). What corresponds to $P(L)$ in this case is $(I - AL)$. Thus, $d(t) = 0$ corresponds to

$$a(t) = |I - At| = t^m \left| \frac{1}{t} I - A \right| = 0$$

whose roots are exactly those of

$$|\lambda I - A| = 0, \quad \text{because t=0 \textbf{is not a root}, and where } \lambda = \frac{1}{t}. \tag{6.24}$$

But Eq. (6.24) gives **the characteristic equation of the matrix** A **and thus the roots of** $a(t) = 0$ **are simply the characteristic roots of that matrix!**

It follows then that

$$[I - AL]^{-1} = \left(\frac{I}{\prod_{s=1}^{m} (I - \lambda_s L)} \right) C(L), \quad C(L) = [c_{ij}(L)], \tag{6.25}$$

where $c_{ij}(L)$ is a (scalar) polynomial of degree $m - 1$ and is the cofactor of the (j, i) elements of $I - AL$.

Remark 6.3. The need for Condition 6.1 now becomes evident, since if we had repeated roots we would not have been able to write in Eq. (6.25) the determinant of $I - AL$ as the product of the polynomials $(I - \lambda_s L), s = 1, 2, \ldots, m$.

It may be shown that, as in the scalar case, the general solution to the homogeneous part of the VDECC in Eq. (6.20) is given by

$$y_t^H = \sum_{s=1}^{m} c_s \lambda_s^t, \tag{6.26}$$

where the $c_s, s = 1, 2, \ldots, m$ are the undetermined constants, that can accommodate any desired initial condition, say $y_0 = \bar{y}_0$. The condition for stability, as in the scalar case, is that $|\lambda_s| < 1, s = 1, 2, \ldots, m$ (or, equivalently, that the roots of the polynomial $|I_m - At| = 0$ obey $|t_s| > 1$.

Since A is diagonalizable as required by Condition 6.1, and its characteristic roots are less than one in absolute value

$$\lim_{t \to \infty} A^t = S(\lim_{t \to \infty} \Lambda^t)S^{-1} = 0,$$

which implies

$$\lim_{t \to \infty} y_t^H = 0. \tag{6.27}$$

Finally, the particular solution to Eq. (6.20) is given by

$$y_t = \left(\prod_{s=1}^{m} \frac{I}{(I - \lambda_s L)} \right) C(L)g(t), \tag{6.28}$$

and the inverse of each of the polynomials $I - \lambda_s L$ has the valid expansion

$$\frac{I}{I - \lambda_s L} = \sum_{k=0}^{\infty} \lambda_s^k L^k.$$

6.4   Applications

6.4.1   Preliminaries

In dealing with applications below we will effect a slight change in notation; the dependent variables, the $y's$ of the previous discussions, will no longer be written as columns but as **rows** and will be denoted by y_t as would other random vectors. For example, in the new notation, we would write the stochastic version of Eq. (6.18) as

$$y_t. A_0 + y_{t-1}. A_1 + y_{t-2}. A_2 + \cdots + \ldots, + y_{t-r}. A_r = u_t.,$$

where y_t. is an m-element **row vector**, the $A_j, j = 0, 2, \ldots, r$ are $m \times m$ matrices and u_t. an m-element **row vector of random variables**. This has a great advantage over the "usual" notation in that it allows us to use exactly the same notation both when we discuss theoretical issues and when we are engaged in estimation of unknown parameters. Thus, if we have a set of observations on a random vector of interest, say $u_{t\cdot}$, we can write the matrix of, say, T observations on that vector as

$$U = (u_{t\cdot}), \quad \text{which is of dimension } T \times m;$$

if we specify that $u'_{t\cdot} \sim N(0, \Sigma)$, we can estimate, in the usual fashion, by maximum likelihood methods or otherwise,

$$\hat{\Sigma} = \frac{U'U}{T}.$$

6.4.2 GLSEM

The GLSEM (General Linear Structural Econometric Model) is a vector difference equation of the general form

$$y_{t\cdot}B_0^* = -\sum_{i=1}^{r} y_{t-i\cdot}B_i + \sum_{j=1}^{s} p_{t-j\cdot}C_j + p_{t\cdot}C_0 + u_{t\cdot}, \tag{6.29}$$

where $y_{t\cdot}$ is an m-element row vector containing the model's **jointly dependent** or **endogenous** variables, i.e. those variables whose values are determined by the economic system modeled, at time t, and $u_{t\cdot}$ is an m-element row vector of random variables. Typically, it is assumed that

$$\{u'_{t\cdot}: t = 0, \pm 1, \pm 2, \ldots\}$$

is a sequence of independent identically distributed (i.i.d.) random vectors with mean zero and positive definite covariance matrix Σ.

The variables in the k-element row vector $p_{t\cdot}$, and their lags, $p_{t-j\cdot}$, $j = 1, 2, \ldots, s$, are the **exogenous variables** of the system, i.e. those variables whose behavior is determined **outside** the system modeled. In particular, they are taken to be **independent** (in a probabilistic sense) of the error process $\{u'_{t\cdot}: t = 0, \pm 1, \pm 2, \ldots\}$. The basic exogenous variables (and their lags), together with the lagged endogenous variables, are termed the **predetermined** variables of the system.

The basic specification of the GLSEM requires, in addition, that the matrix B_0^* be **nonsingular** and that certain elements of $\{B_0^*, B_i: i = 1, 2, \ldots, r\}$, $\{C_j: j = 0, 1, 2, \ldots, s\}$ be known a priori to be zero or normalized to unity, or otherwise **known**. There are certain other requirements that are irrelevant to this discussion and, thus, need not be considered.

The so called final form of the system in Eq. (6.29), is simply the particular solution to this VDECC and is given formally (in column vector form) by

$$y'_{t\cdot} = \frac{A(L)}{b(L)} p'_{t\cdot} + \frac{G(L)}{b(L)} u'_{t\cdot}, \tag{6.30}$$

where

$$\frac{A(L)}{b(L)} = [B^{*\prime}(L)]^{-1}C^{\prime}(L), \quad [B^{*\prime}(L)]^{-1} = \frac{G(L)}{b(L)}, \quad b(L) = |B^{*}(L)|$$

$$B^{*}(L) = B_{0}^{*} + LB_{1} + L^{2}B_{2} + \ldots + L^{r}B_{r}, \quad C(L) = C_{0} + LC_{1} + L^{2}C_{2} + \ldots$$
$$+ L^{s}C_{s}$$

$$A(L) = G^{\prime}(L)C^{\prime}(L).$$

We arrive at Eq. (6.30) as follows: first we write Eq. (6.29) in the more compact form

$$y_{t}.B^{*}(L) = p_{t}.C(L) + u_{t}.;$$

then we invert $B^{*}(L)$ to obtain

$$y_{t}. = p_{t}.C(L)[B^{*}(L)]^{-1} + u.[B^{*}(L)]^{-1},$$

and finally we transpose $y_{t}.$ so it becomes a **column** vector, to obtain the representation in Eq. (6.30), bearing in mind that $[B^{*\prime}(L)]^{-1} = \frac{G(L)}{b(L)}$.

In this context, stability conditions are customarily expressed by the requirement that the roots of

$$b(\xi) = 0, \quad b(\xi) = |B_{0}^{*} + B_{1}\xi + B_{2}\xi^{2} + \ldots + B_{r}\xi^{r}|, \qquad (6.31)$$

be greater than unity in absolute value.

Remark 6.4. This is an opportune time to clarify the requirements for stability. In the discussion of Sect. 6.3, particularly the equation appearing just above Eq. (6.22), we required for stability that the roots of that polynomial equation should be **greater than one in absolute value**. However for the expansion of the inverse of $(I - \lambda L)$ we required that λ be **less than one in absolute value**. While the discussion in that context provided sufficient explanation so these two statements were neither contradictory nor ambiguous, it would be desirable to have a more formal relationship between the two requirements. This is provided in the discussion of the rth order VDECC, Eq. (6.18), and its transformation to a first order VDECC system in Eq. (6.19). The polynomial equation corresponding to the discussion in Sect. 6.3.1, involving the equation just before Eq. (6.22) through Eq. (6.24), leads to the following considerations:

$$|I - At| = 0; \text{ since } t = 0 \text{ is not a root, this is equivalent to} \qquad (6.32)$$

$$|\lambda I - A| = 0, \text{ where } \lambda = \frac{1}{t}. \qquad (6.33)$$

Hence the **characteristic roots of the matrix** A **are the inverses of the roots of the polynomial equation** $|I - At| = 0$. In point of fact this relationship is an instance of the following result from the theory of polynomial equations.

Assertion 1. If x_i is a root of the polynomial equation $\sum_{j=0}^{n} a_j x^j = 0$, then $y_j = 1/x_j$ is a root of the polynomial equation $\sum_{j=0}^{n} a_j y^{n-j} = 0$.

This is exactly the situation with the polynomial equations $|I - At| = 0$ and $|\lambda I - A| = 0$, **the latter being also the characteristic equation of the matrix** A.

We conclude this section by addressing another issue that is often confusing, namely whether the presence of identities makes the routine application of the stability conditions developed above inappropriate. While identities introduce **singularities** in GLSEM, they do not essentially complicate the determination of issues regarding the stability of the GLSEM. Identities, which are exact, non-stochastic relationships, are frequently encountered in GLSEM modeling. Thus, suppose that the last m^* of the m equations of the GLSEM represent identities. Specifically, the identities are of the form

$$y_{t.}^{(1)} H_{12} + y_{t.}^{(2)} = \sum_{j=1}^{k} p'_{t-j}. D_j + p_{t.} D_0, \qquad (6.34)$$

thus expressing the m^* elements of $y_{t.}^{(2)}$, i.e. the "last" m^* endogenous variables as exact functions of the remaining $m - m^*$ endogenous and all the exogenous variables of the system, contained in $p_{t.}$. Note that, in this model, the covariance matrix of $u_{t.}^{*'}$ is

$$\mathrm{Cov}(u_{t.}^{*'}) = \begin{bmatrix} \Sigma_{11} & 0 \\ 0 & 0 \end{bmatrix}, \qquad \Sigma_{11} > 0.$$

The evident singularity of the covariance matrix of the system creates no difficulties in the present context. In general, exogenous variables are not involved in identities, which most often simply refer to renamed variables. For example if we have one equation determining **consumer durables**, another determining **consumer non-durables** and in another part of the system we use **total consumption** as an explanatory variable, it is evident that **that total consumption is simply that sum of the first two, i.e. the sum of consumer durables and consumer non-durables.** So substituting out the identities means that where the third variable occurs, whether in current or lagged form, we substitute (in the appropriate form) the sum of the first two variables. In the discussion below it is **assumed that no exogenous variables are involved in the identities** so that we have in, Eq. (6.34), $D_j = 0$ for all j. Partitioning the matrices of coefficients conformably, we can rewrite the system with the identities in Eq. (6.34) as

$$(y_{t\cdot}^{(1)}, y_{t\cdot}^{(2)}) \begin{bmatrix} B_{0(11)}^* & H_{12} \\ B_{0(21)}^* & I_{m^*} \end{bmatrix} = -\sum_{i=1}^{r} (y_{t-i\cdot}^{(1)}, y_{t-i\cdot}^{(2)}) \begin{bmatrix} B_{i(11)} & H_{12} \\ B_{i(21)} & I_{m^*} \end{bmatrix}$$

$$+ \sum_{j=1}^{k} p_{t-j\cdot} \begin{bmatrix} C_{j(11)} & 0 \\ C_{j(21)} & 0 \end{bmatrix} + +p_{t\cdot} \begin{bmatrix} C_{0(11)} & 0 \\ C_{0(21)} & 0 \end{bmatrix} + (u_{t\cdot}^{(1)}, 0), \qquad (6.35)$$

where $B_{0(11)}^*, B_{i(11)}$ are $m-m^* \times m-m^*$, H_{12} is $m-m^* \times m^*$, $B_{0(21)}^*, B_{i(21)}$ are $m^* \times m - m^*$, $i = 1, 2, \ldots, r$ and $C_{j(11)}, C_{j(21)}, j = 0, 1, 2, \ldots, k$ are $m - m^* \times m - m^*$, $m^* \times m - m^*$, respectively.

Making the substitutions $y_{t-i\cdot}^{(2)} = -y_{t-i\cdot}^{(1)}. H_{12}, i = 0, 1, \ldots, r$, and clearing of redundant equations we can write the GLSEM **without the identities** as

$$y_{t\cdot}^{(1)} [B_{0(11)}^* - H_{12} B_{0(21)}^*] = -\sum_{i=1}^{r} y_{t-i\cdot}^{(1)} . [B_{i(11)} - H_{12} B_{i(21)}]$$

$$+ \sum_{j=0}^{k} p_{t-j\cdot} \begin{bmatrix} C_{j(11)} \\ C_{j(21)} \end{bmatrix} + u_{t\cdot}^{(1)}. \qquad (6.36)$$

The question we have raised has the more specific formulation: would we get the same results whether we proceed with the system in Eq. (6.35) or in Eq. (6.36)? The answer is yes, except that the system in Eq. (6.35) exhibits a number of **zero roots**, due to the singularities induced by the identities. But its non-zero roots are precisely those obtained from the system in Eq. (6.36).

In dealing with this issue it is convenient to introduce the matrix

$$M = \begin{bmatrix} I_{m-m^*} & -H_{12} \\ 0 & I_{m^*} \end{bmatrix}, \qquad (6.37)$$

so that, with x a (scalar) real or complex indeterminate, we obtain

$$M \left(B_0^* + \sum_{i=1}^{r} B_i x^i \right)$$

$$= \begin{bmatrix} B_{0(11)}^* - H_{12} B_{0(21)}^* + \sum_{i=1}^{r} (B_{i(11)} - H_{21} B_{i(21)}) x^i & 0 \\ B_{0(21}^* + B_{0(21)}^* + \sum_{i=1}^{r} B_{i(21)} x^i & (\sum_{i=0}^{r} x^i) I_{m^*} \end{bmatrix}.$$

$$(6.38)$$

Consequently, since the determinant of M **is unity**

$$\left| B_0^* + \sum_{i=1}^{r} B_i x^i \right|$$

$$= \left| \left[\begin{array}{cc} B_{0(11)}^* - H12 B_{0(21)}^* + \sum_{i=1}^{r}(B_{i(11} - H_{21} B_{i(21)})x^i & 0 \\ B_{0(21}^* + \sum_{i=1}^{r} B_{i(21)} x^i & \left(\sum_{i=0}^{r} x^i\right) I_{m^*} \end{array} \right] \right| = 0.$$

$$(6.39)$$

But Eq. (6.39) **shows that the non-zero roots of**

$$\left| B_0^* + \sum_{i=1}^{r} B_i x^i \right| = 0, \qquad (6.40)$$

are precisely the roots of

$$\left| (B_{0(11}^* - H_{21} B_{0(21)}^*) + \sum_{i=1}^{r}(B_{i(11} - H_{21} B_{i(21)})x^i \right| = 0, \qquad (6.41)$$

which, for stability of the system, are required to be **greater than one in absolute value.** Consequently, one can obtain or verify the stability condition either from Eq. (6.35) or Eq. (6.36). When using Eq. (6.35) one should bear in mind that, due to singularities induced by the identities, this polynomial equation has $m^* r$ **zero roots.**

6.4.3 Applications to Time Series Models

Our objective here is not to undertake a general review of time series, but rather to examine those aspects of time series that have found reasonably extensive applications in econometrics. In the process we shall provide the basic definitions and concepts that underlie these applications,

Definition 6.4. A time series is a collection of random variables indexed on a linear set T, i.e. $\{x_t: \ t \in T\}$. The set in question is ordered in the sense there is a "before" and "after". If the set T is **discrete** we shall refer to such time series as **stochastic sequences,** while if it is **continuous** we shall refer to them as **stochastic processes.** An example of the latter is Brownian motion which refers to the position of a particle suspended in a liquid.

Example 6.3. The series giving consumption of durables by quarter in the US is a time series (stochastic sequence) in the sense above in that consumption is recorded discretely (quarterly) and consumption of durables in the third quarter of 1999 **precedes** consumption of durables in the fourth quarter of

2001 and conversely consumption of durables in the second quarter of 2012 succeeds (follows) consumption of durables in the first quarter of 2010. Some entities in economics, such as those dealing with activities of individuals **at a specific time** cannot be so ordered, as e.g. households' adjusted gross income as defined by the internal revenue service. Samples (observations) on such entities are referred to as **cross section** samples, and **we shall not deal with them in this discussion**.

The typical set we shall use in our discussion below is given by

$$\{T: \text{ if } t \in T, \quad t = 0, \pm1, \pm2, \pm3, \ldots\} \tag{6.42}$$

Convention 6.2. The probability distribution of a stochastic sequence is considered to have been specified if for **arbitrary** n we can specify the joint distribution of the collection $\{x_{t_i}: \ t_i \in T, \ i = 1, 2, \ldots, n.\}$

Definition 6.5. A stochastic sequence $\{x_t: \ t \in T\}$ is said to be **strictly (or strongly) stationary** if and only if for arbitrary $n, k \in T$ the joint distributions of

$$\{x_{t_i}: \quad i = 1, 2, \ldots, n\} \text{ and } \{x_{t_i+k} : i = 1, 2, \ldots, n\}$$

are identical.

Definition 6.6. A stochastic sequence $\{x_t: \ t \in T\}$ is said to be **covariance (or weakly) stationary** if and only if, for all $t, h \in T$

$$Ex_t = \mu, \quad \text{Cov}(x_{t+h}, x_t) = c(|h|), \tag{6.43}$$

provided the two moments above exist.[4]

Evidently, **if a strictly stationary sequence possesses at least a second moment, it is also covariance stationary**.

The function

$$\{c(t + h, t) = c(|h|): \quad h = 0, \pm1, \pm2, \ldots, , \ldots\}$$

is said to be the **auto-covariance** function, and the function

$$\{\rho(h) = c(|h|)/c(0): \quad h = 0, \pm1, \pm2, \ldots, , \ldots\}$$

is said to be the **auto-correlation function**. Evidently, $c(|h|) \le c(0)$, $c(0)$ being the variance. We also have

---

[4]By convention $\text{Cov}(x_{t+h}, x_t) = E(x_{t+h} - \mu)(x_t - \mu) = c([t + h] - t) = c(h)$. But since $E(x_{t+h} - \mu)(x_t - \mu) = E(x_t - \mu)(x_{t+h} - \mu) = c(t - [t + h]) = c(-h)$ the definition is justified.

Proposition 6.1. The auto-covariance function is at least a positive semidefnite sequence.

Proof: Let $x = \{x_t: \ t = 1, 2, \ldots n\}$ be elements of a zero mean covariance stationary sequence. Then, for an arbitrary real vector a,

$$0 \leq \mathrm{var}(a'x) = \sum_{i=0}^{n} \sum_{j=0}^{n} a_i c(i, j) a_j = \sum_{i=0}^{n} \sum_{j=0}^{n} a_i c(i-j) a_j, \quad \text{with} \quad c(i-j) = c(j-i).$$

Thus, the matrix $C = [c(i - j)]$ is symmetric and obeys, for arbitrary a

$$a'Ca \geq 0,$$

i.e. it is **positive semidefinite** and so is the sequence.

Definition 6.7. A stochastic sequence $\{\epsilon_t: \ t \in T\}$ is said to be a **white noise** sequence if it is a sequence of **uncorrelated random variables** with mean 0 and variance σ^2, and is denoted by $WN(0, \ \sigma^2)$.

Moving Average Sequences

Definition 6.8. A stochastic sequence $\{X_t: \ t \in T\}$ is said to be a **moving average sequence of order** n if and only if

$$X_t = \sum_{j=0}^{n} a_j u_{t-j}, \quad a_0 = 1, \quad u_t \sim WN(0, \ \sigma^2). \tag{6.44}$$

The condition $a_0 = 1$ is an **identification condition** and is imposed solely in order to enable the unique identification of the parameters a_j, σ^2. Alternatively, we **could have imposed the condition** $\sigma^2 = 1$.

This is an opportune time to introduce the terminology of **time invariant filter**, very common in this literature. The reader would find useful the communications engineering simile between an **input** which is transformed through the filter to produce an **output**. Thus,

Definition 6.9. A linear time invariant filter (LTIF) is a function which is linear and time invariant and when applied to an input, g_t, produces as output $y_t = f(g_t)$, where $t \in T$. Linearity means that $f(g_t + h_t) = f(g_t) + f(h_t)$ and time invariance means that $f(\cdot)$ does not depend on t directly.

Remark 6.5. Notice that all lag polynomials we utilized in our discussion are indeed **linear time invariant filters**.

An important property of LTIF which we shall not prove here is,

Assertion 2. Let $y_t = f(u_t)$ be the output (y_t) of a LTIF, f , with input u_t . If $\{u_t: t \in T\}$ is a **stationary sequence, then so is** $\{y_t: t \in T\}$, **the output sequence.** This holds for strict stationarity as well as for weak stationarity.

Definition 6.10. A stochastic sequence X_t is said to be a **linear process (or sequence)** if and only if

$$X_t = \sum_{j=-\infty}^{\infty} c_j u_{t-j}, \quad \sum_{j=-\infty}^{\infty} |c_j| < \infty, \quad u_t \sim WN(0,1), \quad t \in T. \qquad (6.45)$$

Taking the variance of the WN to be one is solely done so that the requisite identification condition is simply stated. We could also have put $c_0 = 1$.

Finally, in connection to moving average sequences, we have

Definition 6.11. Let X_t be a sequence as in Eq. (6.44); such a sequence is said to be $n-dependent$ meaning that X_t and X_s are correlated if $|t-s| \leq n$ and otherwise they are **not correlated.** Note that if, in Definition 6.8, the u - sequence is not merely $WN(0, \sigma^2)$ but rather $iid(0, \sigma^2)$, then for $|t-s| > n$, X_t and X_s are **mutually independent.**

Autoregressive Sequences

Definition 6.12. Let $\{X_t: \quad t \in T\}$, be a (scalar) stochastic sequence obeying

$$X_t = \sum_{i=1}^{m} a_i X_{t-i} + u_t, \quad t \in T, \qquad (6.46)$$

$a_i \in R$, for all i , and $u_t \sim WN(0, \sigma^2)$ [5] Then, the sequence is said to be an **autoregression of order** m **and denoted by** $AR(m)$.

Example 6.4. A common application of the first order autoregression is in the context of the general linear model (GLM), (regression model), to be considered extensively in a later chapter. There one writes, in its simplest form,

$$y_t = \alpha + \beta x_t + u_t,$$

where y_t is (in the econometric terminology) the dependent variable (the regressand in the statistical terminology), x_t is the independent or explana-

---

[5]Here R is the real line. Although the coefficients can be allowed to be complex as well, this is very uncommon in economics. The notation $\sim$ generally means "is equivalent"; in this particular use it is best for the reader to read it as "behaves like" or "has the distribution of".

tory variable (in the econometric terminology) or the regressor (in the statistical terminology)[6] and u_t is the error term of the model. The standard assumption is that the error sequence is either an i.i.d.- or at least a WN-sequence. But since a great many economic data exhibit autoregressive properties it is very common to test the alternative that the error sequence is an $AR(1)$ sequence,

$$u_t = \rho u_{t-1} + \epsilon_t,$$

where ϵ_t is at least $WN(0, \sigma^2)$, although most of the testing procedures typically used by econometricians require that the sequence be i.i.d. and/or $N(0, \sigma^2)$, i.e. normal with mean zero and variance σ^2.

Other forms of extensive use are the numerous "dynamic" models where at least one lag of the dependent variable, y_t, typically the first, y_{t-1}, is used as an explanatory variable.

If in the error term case discussed above we bring to bear the analysis provided earlier in the chapter we may write

$$(I - \rho L)u_t = \epsilon_t. \tag{6.47}$$

The condition for "stability", here better thought of as invertibility, requires that the root of $1 - \rho\psi = 0$, say ψ^* obey $|\psi^*| > 1$. It is clear that $\psi^* = 1/\rho$, satisfies this condition if $|\rho| < 1$. Consequently, we may express[7]

$$u_t = \sum_{j=0}^{\infty} \rho^j \epsilon_{t-j}. \tag{6.48}$$

A similar discussion regarding the general $AR(m)$ sequence of Definition 6.12, Eq. (6.46), using the same concept of stability, i.e. the requirement that the roots of

$$1 - \sum_{i=1}^{m} a_i \psi^i = 0, \quad \text{obey} \quad \psi_i^* > 1, \ i = 1, 2, \ldots, m, \tag{6.49}$$

suggests, in view of a previous section of this chapter, that we have the representation

$$X_t = \left(\prod_{i=1}^{m} \left[\frac{I}{I - \lambda_i L} \right] \right) u_t = \sum_{j=0}^{\infty} \phi_j u_{t-j},$$

---

[6]In the future we shall use exclusively the econometric terminology.

[7]The convergence in the right member of the equation below is, depending on the assumptions made, either absolute convergence of the sequence of the coefficients of the errors and finiteness of their first absolute moment, $E|\epsilon_t|$, for all $t \in T$, or is **convergence in mean square, or quadratic mean** a concept that will be introduced at a later chapter dealing with the underlying probability foundations of econometrics.

where convergence of the right member is in quadratic mean, as amplified in Footnote 6, $\lambda_i = (1/\psi_i^*)$, and the ϕ_j are complicated functions of the λ_i. In the literature of time series this phenomenon is not termed "stability". Instead, we have the following

Definition 6.13. Let X_t be the stochastic sequence of Definition 6.12, Eq. (6.46), and suppose (the roots of) its characteristic polynomial equation

$$1 - \sum_{i=1}^{m} a_i \psi^i = 0, \quad \text{obey} \quad \psi_i^* > 1, \ i = 1, 2, \ldots, m, \tag{6.50}$$

so that it has the representation

$$X_t = \left(\prod_{i=1}^{m} \left[\frac{I}{I - \lambda_i L} \right] \right) u_t = \sum_{j=0}^{\infty} \phi_j u_{t-j}; \tag{6.51}$$

then the $AR(m)$ is said to be **causal**.

Remark 6.6. When time series methods entered fully into the mainstream of the econometric literature (late 1970s, early 1980s) the use of the term "causal", to describe econometric models framed as autoregressions caused a great deal of confusion and a flurry of papers arguing that such models are not intrinsically, or in a philosophical sense, dealing with causality. While perhaps the term "causal" to describe the $AR(m)$ of Definition 6.12 is inopportune, we remind the reader that in mathematics a definition means neither more nor less than what it states. Any other attribution to the term defined is both harmful to the reader's comprehension as well as irrelevant. Thus, a "causal" $AR(m)$ means nothing more than it may be expressed totally and solely in terms of its constituent WN sequence as in Eq. (6.51).

Autoregressive Moving Average Sequences

Definition 6.14. Let $\{X_t: \ t \in T\}$, be a (scalar) stochastic sequence obeying

$$\sum_{j=0}^{m} b_j X_{t-j} = \sum_{i=1}^{n} a_i u_{t-i}, \quad t \in T, \quad \text{or} \quad B(L)X_t = A(L)u_t, \tag{6.52}$$

$a_i, b_j \in R$, for all i, j, $u_t \sim WN(0, \sigma^2)$, the polynomials $A(L)$, $B(L)$ have no common factors and $a_0 = b_0 = 1$ (identification conditions). Then, the sequence is said to be an **Autoregressive Moving Average sequence of order** (m, n) **and is denoted by** $ARMA(m, n)$.

If the characteristic polynomial equation of the (autoregressive) lag polynomial $B(L)$ has roots **greater than one**, i.e. the roots, ψ_j^*, of

$$B(\xi) = 1 + \sum_{j=1}^{m} b_j \xi^j = 0, \quad \text{obey} \quad |\xi_j^*| > 1,$$

the sequence $ARMA(m, n)$ is said to be **causal**.

If the roots, ϕ_i^*, of the polynomial equation

$$A(\phi) = 1 + \sum_{i=1}^{n} a_i \phi^i = 0, \quad \text{obey} \quad |\phi_i^*| > 1,$$

the sequence $ARMA(m, n)$ is said to be **invertible**.

Most of the issues related to $ARMA(m, n)$ have already been examined in the previous two sections dealing with moving average and autoregressive sequences, as well as in the discussion of the GLSEM.

Thus, observe that what we called the "final form" of the GLSEM,[8] in this case is given by

$$y'_{t\cdot} = \frac{G(L)}{b(L)} u'_{t\cdot}, \tag{6.53}$$

where the symbols are as defined in Eqs. (6.29) and (6.30). It should be noted, however, that in the GLSEM $B^*(L)$, $C(L)$ represent **matrix polynomials** in the lag operator L, while in the $ARMA(m, n)$ case we were dealing above they represent **scalar polynomials**. However, since $b(L) = |B^*(L)|$, it is a **scalar polynomial of degree** mr. Its inverse lag polynomial

$$\frac{I}{b(L)} = \prod_{i=1}^{mr} \left(\frac{I}{I - \lambda_i L} \right) \tag{6.54}$$

is well defined provided the autoregressive part of the GLSEM is causal, or in the terminology of econometrics **stable.** The representation of $y_{t\cdot}$ solely in terms of the $WN(0, \sigma^2)$ u-sequence is given thus by

$$y'_{t\cdot} = \Psi(L) u'_{t\cdot}, \quad \text{or} \quad y'_{t\cdot} = \sum_{k=0}^{\infty} \Psi_k u'_{t-k\cdot}, \tag{6.55}$$

where the Ψ_k are appropriate $m \times m$ matrices. This (matrix) representation, as well as in the case of the scalar $AR(m)$ considered earlier, is often referred to as a **moving average of infinite extent, denoted by** $MA(\infty)$.

---

[8]In the interest of simplicity of presentation we have omitted or ignored the exogenous variable component, containing $p_{t\cdot} C_0$, and lags thereof; the latter, however, can be easily added at the cost of a bit more complexity in the presentation.

Returning now to the $ARMA(m,n)$ sequence of Eq. (6.52), which we repeat below

$$\sum_{j=0}^{m} b_j X_{t-j} = \sum_{i=1}^{n} a_i u_{t-i}, \quad t \in T, \quad \text{or} \quad X_t = \frac{A(L)}{B(L)} u_t = \sum_{k=0}^{\infty} \psi_k u_{t-k}$$

we need to determine whether we could estimate the coefficients ψ_k if we estimated or knew the coefficients of the lag polynomials $A(L)$, $B(L)$. This is easily accomplished if we note that $A(L) = B(L)\psi(L)$ and match coefficients of equal powers of L on both sides of the equation, thus obtaining

$$a_s = \sum_{j=0}^{\min(s,m)} b_j \psi_{s-j}, \quad \text{for } 0 \le s \le n$$

$$0 = \sum_{j=0}^{\min(s,m)} b_{s-j} \psi_{s-j}, \quad \text{for } s > n. \tag{6.56}$$

Example 6.5. For practice, we compute recursively a few coefficients to find, $a_0 = b_0 \psi_0$, which implies $\psi_0 = 1$ in view of the normalizations; $a_1 = b_0 \psi_1 + \psi_0 b_1$, or $\psi_1 = a_1 - b_1$; $a_2 = \psi_2 b_0 + c_1 b_1 + \psi_0 b_2$, or $\psi_2 = a_2 - b_2 - \psi_1 b_1$ and so on. Finally (assuming $n \ge m$) $a_n = \sum_{j=0}^{m} b_j \psi_{n-j}$, or $\psi_n = a_n - \sum_{j=1}^{m} b_j \psi_{n-j}$. For $s > n$ the term a_n will disappear and the summation will be adjusted accordingly.

6.4.4 An $AR(m)$ with Nonlinearities; ARCH Models

The purpose of this section is to provide an accessible discussion of the ARCH model (**Autoregressive Conditional Heteroskedasticity**), which, together with its many variants has found extensive applications in modeling the rates of return of risky assets. We shall routinely refer to such models as $ARCH(m)$, rather than $AR(m)$ with nonlinearities. Our chief focus will be the scalar ARCH(m).

Definition 6.15. Let $\{X_t: \quad t \in T\}$ be a stochastic sequence obeying

$$B(L)X_t = u_t, \quad B(L) = \sum_{j=0}^{m} b_j L^j, \quad u_t = f_t^{1/2} \epsilon_t, \quad f_t = \alpha_0 + \alpha_1 u_{t-1}^2, \quad |\alpha_1| < 1,$$

$$\tag{6.57}$$

where $\{\epsilon_t: \quad t \in T\}$ is an iid$(0, 1)$ sequence, and the usual identification condition $b_0 = 1$ holds. Such a sequence is termed $ARCH(m)$.

Remark 6.7. As is evident from Eq. (6.57) the difference between $ARCH(m)$ and the standard AR(m) model lies in the specification of the stochastic sequence u_t. In the standard case, u_t is a white noise sequence $WN(0, \sigma^2)$, while in the ARCH case it is defined by

$$u_t = f_t^{1/2} \epsilon_t, \quad f_t = \alpha_0 + \alpha_1 u_{t-1}^2, \quad |\alpha_1| < 1, \tag{6.58}$$

and ϵ_t is an iid(0,1) sequence. Thus, the ARCH specification of the basic sequence (u_t) generating X_t **involves nonlinearities, which is what separates the ARCH model from the standard AR models discussed earlier.**

The properties of the basic sequence, u_t, are not specified directly and thus **must be deduced from the properties assigned to the function** f_t **and the sequence** ϵ_t. We shall now determine these properties.[9] First, we determine the mean and variance and then determine whether the sequence u_t is also **weakly stationary.**

$$E(u_t|u_{t-1}) = 0, \text{ and thus } Eu_t = 0;$$
$$\sigma^2(t) = \text{var}(u_t) = Eu_t^2 = E_u E[u_t^2|u_{t-1}] = E_u[\alpha_0 + \alpha_1 u_{t-1}^2]$$
$$= \alpha_0 + \alpha_1 \sigma^2(t-1), \tag{6.59}$$

because $E\epsilon_t^2 = 1$.

Given the second equation above we may use a heuristic approach to find the variance of the u_t sequence. Thus,

$$\sigma^2(t) = \alpha_0 + \alpha_1 \sigma^2(t-1), \text{ or } (I - \alpha_1 L)\sigma^2(t) = \alpha_0 I, \text{ or } \sigma^2(t) = \sum_{j=0}^{\infty} \alpha_1^j L^j \alpha_0$$

$$= \frac{\alpha_0}{1 - \alpha_1} < \infty. \tag{6.60}$$

Remark 6.8. While this is the correct answer to the question of what the "long term" variance of u_t is, the method used to obtain it is not rigorous, in that for fixed t we took a conditional expectation and subject to that we made a limiting argument without showing that this transposition of operations, in that context, is allowed. We shall return to this issue below.

---

[9]The notation $E_u[E[u_t|u_{t-1}]$ means that, because u_t^2 is both a function of u_{t-1}^2 **and** ϵ_t^2, in taking the expectation $E(u_t^2)$ we first take the expectation **conditional on** u_{t-1} **and then take the expectation with respect to** u_t.

Another property of the u-sequence is that it is **uncorrelated**. Specifically, for $t, h \in T$, consider

$$E(u_{t+h} u_t) = E_u E(u_{t+h} u_t | u_{t+h-1} u_{t-1}) = E_u (f_{t+h} f_t)^{1/2} E[\epsilon_{t+h} \epsilon_t] = E_u 0 = 0,$$
(6.61)

because for $h \neq 0$, ϵ_{t+h}, ϵ_t are mutually independent with mean zero and **at least ϵ_{t+h} is not related to u_{t+h-1} or u_{t-1}**. When $h = 0$ the equation above reduces to the second equation of Eq. (6.59). To recapitulate, we have established above:

i. The u_t sequence has mean zero;

ii. The u_t sequence is uncorrelated;

iii. The conditional variance of u_t **given** u_{t-1} is $E[u_t^2 | u_{t-1}] = \alpha_0 + \alpha_1 u_{t-1}^2$ [10];

iv. The unconditional variance of u_t is given by $E u_t^2 = \alpha_0 / (1 - \alpha_1)$.

We shall now provide a more rigorous solution approach to the determination of what we called the "long term" variance of the u-sequence. From the definition of the u-sequence we have

$$u_t^2 = \alpha_0 \epsilon_t^2 + \alpha_1 u_{t-1}^2 \epsilon_t^2.$$
(6.62)

Because the term $u_{t-1}^2 \epsilon_t^2$ in the equation above is **non-linear, there is no simple way of finding a solution, as we did in earlier discussions;** one possible approach is to recursively substitute for u_{t-1}^2 so as to make u_t^2 depend on more and more remote lags, whose influence on the determination of u_t^2 is **progressively weakened because** $|\alpha_1| < 1$. Doing so we obtain

$$
\begin{aligned}
u_t^2 &= \alpha_0 \epsilon_t^2 + \alpha_1 u_{t-1}^2 \epsilon_t^2 \qquad\qquad (6.63) \\
&= \alpha_0 \epsilon_t^2 + \alpha_1 [\alpha_0 \epsilon_{t-1}^2 + \alpha_1 u_{t-2}^2 \epsilon_{t-1}^2] \epsilon_t^2 \\
&= \alpha_0 [\epsilon_t^2 + \alpha_1 \epsilon_{t-1}^2 \epsilon_t^2] + \alpha_1^2 u_{t-2}^2 \epsilon_{t-1}^2 \epsilon_t^2 \\
&= \alpha_0 [\epsilon_t^2 + \alpha_1 \epsilon_{t-1}^2 \epsilon_t^2] + \alpha_1^2 [\alpha_0 \epsilon_{t-2}^2 + \alpha_1 u_{t-3}^2 \epsilon_{t-2}^2] \epsilon_{t-1}^2 \epsilon_t^2 = \ldots, \ldots \\
&= \alpha_0 \left[\sum_{j=0}^{N-1} \alpha_1^j \left(\prod_{k=0}^{j} \epsilon_{t-k}^2 \right) \right] + \alpha_1^N u_{t-N}^2 \left(\prod_{s=0}^{N-1} \epsilon_{t-s}^2 \right). \qquad (6.64)
\end{aligned}
$$

---

[10]This feature is responsible for the name of the model ARCH, which stands for autoregressive conditional heteroskedasticity.

We must now deal with the remainder (last term) which we rename

$$w_N = \alpha_1^N u_{t-N}^2 \left(\prod_{s=0}^{N-1} \epsilon_{t-s}^2 \right). \tag{6.65}$$

We will show that it converges in probability to zero with N, using Chebyshev's inequality, a fact expressed in the notation $\text{plim}_{N \to \infty} w_N = 0$. We begin by showing[11] that $\text{plim}_{N \to \infty} w_N^{1/2} = 0$, define $w_N = g(w_N^{1/2}) = (w_N^{1/2})^2$, which is evidently a continuous function, and argue that $\text{plim}_{N \to \infty} w_N = 0$. Now

$$w_N^{1/2} = \alpha_1^{N/2} u_{t-N} \prod_{s=0}^{N-1} \epsilon_{t-s}. \tag{6.66}$$

Clearly,

$$E w_N^{1/2} = \alpha_1^{N/2} E u_{t-N} \left(\prod_{s=0}^{N-1} \epsilon_{t-s} \right) = 0,$$

$$\text{var}(w_n^{1/2}) = \alpha_1^N \text{var}(u_{t-N}), \text{ because, } E \left(\prod_{s=0}^{N-1} \epsilon_{t-s} \right)^2 = 1, \tag{6.67}$$

and u_{t-N} is **independent of the** ϵ_{t-s}. Since $\text{var}(u_{t-N})$ is bounded, i.e. less than some finite number K, we have, by Chebyshev's inequality, that for any pre-assigned positive number δ, however small,

$$\Pr(|w_N^{1/2}| > \delta) \le \frac{\alpha_1^N K}{\delta^2} \tag{6.68}$$

whose right member converges to zero with N, **even if the variance is not bounded, so long as it increases at a rate less that** $(1/\alpha_1)$. Thus,

$$\text{plim}_{N \to \infty} u_t^2 = \alpha_0 \left[\sum_{j=0}^{\infty} \alpha_1^j \left(\prod_{k=0}^{j} \epsilon_{t-k}^2 \right) \right], \text{ whose expectation is}$$

---

[11]The property in question is that if ξ_n is a sequence of random variables converging in probability to ξ, written $\text{plim}_{n \to \infty} \xi_n = \xi$, and if g is a continuous (or even a measurable) function then $\text{plim}_{n \to \infty} g(\xi_n) = g(\xi)$, provided the latter is defined. For a proof of this see Dhrymes (1989), p. 144 ff. These issues will also discussed in a later chapter.

$$\text{var}(u_t^2) \quad = \quad Eu_t^2 = \alpha_0 \sum_{j=0}^{\infty} \alpha_1^j = \frac{\alpha_0}{1 - \alpha_1}. \tag{6.69}$$

This confirms the result we obtained above regarding the unconditional variance of the u-sequence, and also **it shows the X_t-sequence to be explicitly causal, i.e. it is basically expressible in terms of the fundamental iid(0, 1) ϵ-sequence, albeit nonlinearly.**

Chapter 7

Mathematical Underpinnings of Probability Theory

The purpose of this chapter is to provide a background on the results from probability and inference theory required for the study of several of the topics of contemporary econometrics.

An attempt will be made to give proofs for as many propositions as is consistent with the objectives of this chapter which are to provide the tools deemed necessary for the exposition of several topics in econometric theory; it is clearly **not our objective to provide a substitute to a mathematical textbook of modern probability theory**.

7.1 Sets and Set Operations

Let Ω be a (nonempty) collection of objects (our universe of discourse); the nature of such objects need not be specified. In most applications, the set Ω would be either the set of all possible outcomes of an **experiment**, the sample space of probability theory, the real line, or the Cartesian product of a finite number of copies of the real line; for the moment, however, we treat Ω as an abstract collection of objects and we shall refer to it as a **space** in this and subsequent sections. We denote its elements by ω and as a matter of notation we write

$$\omega \in \Omega.$$

P.J. Dhrymes, *Mathematics for Econometrics*,
DOI 10.1007/978-1-4614-8145-4_7, © The Author 2013

A subset of Ω, say A, is simply a collection of elements of Ω; we denote it, as a matter of notation, by

$$A \subset \Omega,$$

which is to be read: A is contained in Ω, or A is a subset of Ω. For completeness we also define the null set, $\emptyset$, which has no elements; **by convention, the null set is a subset of every set.** As a matter of notation we have that for every set, $A \subset \Omega$,

$$\emptyset \subset A.$$

A subset may be described either by **enumeration**, i.e. by enumerating its elements, or by some property. For example, suppose

$$\Omega = \{1, 2, 3, \ldots\}$$

One of its subsets might be $A = \{1, 2, 3, 4, 5\}$; or we may simply describe membership in a set, say B, by some property; for example

$$B = \{n : n = 2k , \ k = 1, 2, 3 \ldots\}.$$

In this case, B is the set of all even integers; it can also be easily specified by enumeration; thus, $B = \{2, 4, 6, 8, \ldots\}$. In most instances the specification of subsets is done by the specification of the properties of their element(s) rather than by enumeration, since the latter is, typically, very difficult to accomplish.

We now begin our formal discussion.

Definition 7.1. Let A, $B \subset \Omega$; then, their **union** is defined by

$$A \cup B = \{\omega : \omega \in A \text{ or } \omega \in B\}$$

which is read: the set of all points, which belong **either to A or to B, (or both)**.

Definition 7.2. Let A, B be as in Definition 7.1; then, their **intersection** is defined by

$$A \cap B = \{\omega : \omega \in A \text{ and } \omega \in B\}$$

which is read: the set of all points, which belong to **both A and B**.

Definition 7.3. Let $A \subset \Omega$, then the **complement** of A (relative to Ω) is given by

$$\bar{A} = \{\omega : \omega \in \Omega \text{ and } \omega \notin A\}.$$

In what follows we shall drop repetitive and redundant statements, such as, for example, $A \subset \Omega$; any sets we consider will be understood to be subsets of Ω, which is the universe of discourse.

A consequence of the definitions above is

Proposition 7.1. Let A, B, be any two sets; then

   i. $\overline{(A \cup B)} = \bar{A} \cap \bar{B}$

   ii. $\overline{(A \cap B)} = \bar{A} \cup \bar{B}$

Proof: If $\omega \in \overline{(A \cup B)}$, then $\omega \notin A$ and $\omega \notin B$; hence, $\omega \in \bar{A} \cap \bar{B}$; conversely, if $\omega \in \bar{A} \cap \bar{B}$, then $\omega \notin A$ and $\omega \notin B$, i.e. $\omega \in \overline{(A \cup B)}$; this proves i.

To prove ii we note that if $\omega \in \overline{(A \cap B)}$, then either $\omega \in \bar{A}$ or $\omega \in \bar{B}$; consequently, $\omega \in (\bar{A} \cup \bar{B})$. Conversely, if $\omega \in (\bar{A} \cup \bar{B})$, then either $\omega \in \bar{A}$ or else $\omega \in \bar{B}$; hence, $\omega \notin (A \cap B)$, or $\omega \in \overline{(A \cap B)}$.

                                                                  q.e.d.

Remark 7.1. The results above obviously extend by iteration to **finite** unions and intersections, i.e. the complement of a finite union is the intersection of the corresponding complements and the complement of finite intersections is the union of the corresponding complements.

It is simple to demonstrate that the results of Proposition 7.1 extend to **countable** unions and intersections, i.e. if

$$\{A_n : \ n = 1, 2, \ldots\}$$

is a sequence of sets (subsets of Ω), then

   i. $\overline{\bigcup_{n=1}^{\infty} A_n} = \bigcap_{n=1}^{\infty} \bar{A}_n$,

   ii. $\overline{\bigcap_{n=1}^{\infty} A_n} = \bigcup_{n=1}^{\infty} \bar{A}_n$.

7.2 Limits of Sequences

Let $\{a_n : \ n = 1, 2, 3, \ldots\}$ be a sequence of, say, real numbers; we recall from calculus that the limit of the sequence, if one exists, is a real number, say a, such that given any $\epsilon \geq 0$ there exists some n_0 such that for all $n \geq n_0$

$$|a_n - a| \leq \epsilon.$$

We would like to express this concept in a way that would easily enable us to generalize it to the case where the sequence in question is not a sequence of real numbers but, say, a sequence of sets (subsets of Ω).

Definition 7.4. Let $\{a_n : n \geq 1\}$ be a sequence of real numbers; then the **supremum** of the sequence, denoted by

$$\sup_{n \to \infty} a_n,$$

is the **least upper bound (l.u.b.)** of the sequence, i.e. the smallest number, say α, such that

$$a_n \leq \alpha, \; \forall \, n.$$

The **infimum** of the sequence, denoted by

$$\inf_{n \to \infty} a_n,$$

is the **greatest lower bound (g.l.b.)** of the sequence, i.e. the largest number, say, α, such that

$$a_n \geq \alpha, \; \forall \, n.$$

Remark 7.2. When dealing with a finite sequence, say $\{a_n : n = 1, 2, \ldots, N\}$ the supremum and infimum of a sequence coincide with the latter's maximum and minimum, respectively. This is so since it is possible to find the largest (maximum) and the smallest (minimum) elements of the sequence and these will obey the requirements, respectively, for the supremum and the infimum. Contrast this to the case where the sequence is infinite and the supremum and infimum need not be members of the sequence.

Example 7.1. Consider the sequence

$$\{a_n : a_n = 1 - \frac{1}{n}, \; n \geq 1\}.$$

It is easily shown that

$$\sup_{n \to \infty} a_n = 1.$$

Notice also that 1 is not a member of the sequence; on the other hand

$$\inf_{n \to \infty} a_n = 0$$

and here 0 is a member of the sequence. If we truncate the sequence at $n = N$ and consider the sequence to consist only of the first N elements, then both inf and sup are members of the sequence and correspond, respectively, to

$$\min a_n = 0, \quad \max a_n = 1 - \frac{1}{N}.$$

Consider further the sequence $\{a_n : a_n = 1 + (1/n), \; n \geq 1\}$. In this case we find

$$\inf_{n \to \infty} a_n = 1, \quad \sup_{n \to \infty} a_n = 2$$

and note that the infimum is not a member of the sequence, while the supremum is.

Definition 7.5. The sequence $\{a_n : n \geq 1\}$ is said to be a **monotone nonincreasing** sequence if

$$a_{n+1} \leq a_n, \quad \text{for all} \quad n,$$

and is said to be a **monotone nondecreasing** sequence if

$$a_{n+1} \geq a_n, \quad \text{for all} \quad n.$$

Monotone nonincreasing or nondecreasing sequences are said to be **monotone** sequences.

Remark 7.3. It is clear that if we consider limits of sequences in the extended number system, $[-\infty, \infty]$, then all monotone sequences have a limit; this is so since, in the case of monotone nonincreasing sequence either there is a (finite) greatest lower bound, or the sequence decreases to $-\infty$, while in the case of a monotone nondecreasing sequence either there is a (finite) least upper bound or the sequence increases to $+\infty$.

Monotone sequences offer an important tool in studying the limiting behavior of general sequences. This is so, since for a general sequence a limit, i.e. a point within a neighborhood of which are located all but a finite number of the elements of the sequence, may not exist. A simple example is the sequence

$$\{a_n : a_n = (-1)^n + (-1)^n \frac{1}{n}, \ n \geq 1\}.$$

Here, if we confine our attention to even numbered values of the index we have a sequence with a limit at one; on the other hand if we confine our attention to odd numbered values of the index then we have a sequence with a limit at minus one. This sequence, then, has no limit in the sense that there is no point around which are located all but a finite number of the elements of the sequence; instead, there are two such points, each corresponding, however, to distinct subsequences of the original sequence. Occasionally, such points are called **limit, or cluster** points of the sequence. Now, if we had a way in which we could determine more or less routinely the "largest" and "smallest" such point, then we would have a routine way of establishing whether the limit of a given sequence exists and, if it does, of identifying it.

Definition 7.6. Let $\{a_n : n \geq 1\}$ be a sequence of real numbers and put

$$b_n = \sup_{k \geq n} a_k, \quad c_n = \inf_{k \geq n} a_k.$$

Then, the sequences $\{b_n : n \geq 1\}$, $\{c_n : n \geq 1\}$ are, respectively, monotone nonincreasing and nondecreasing and their limits are said to be the **limit superior** and **limit inferior** of the original sequence and are denoted, respectively, by

$$\text{limsup}, \quad \text{liminf} \quad \text{or} \quad \overline{\lim} , \quad \underline{\lim}.$$

Thus, we write

$$\lim_{n \to \infty} b_n = \lim_{n \to \infty} \sup_{k \geq n} a_k,$$

$$\lim_{n \to \infty} c_n = \lim_{n \to \infty} \inf_{k \geq n} a_k.$$

We immediately have

Proposition 7.2. Let $\{a_n : n \geq 1\}$ be a sequence of real numbers; then,

$$\text{limsup } a_n \geq \text{liminf } a_n.$$

Proof: Let

$$b_n = \sup_{k \geq n} a_k, \quad c_n = \inf_{k \geq n} a_k.$$

It is evident, by construction, that

$$b_n \geq c_n, \text{ for all } \quad n. \tag{7.1}$$

Consequently,

$$\text{limsup } a_n = \lim b_n \geq \lim c_n = \text{liminf } a_n$$

The validity of the preceding rests on the validity of the middle inequality; the latter in turn is implied by Eq. (7.1). For, suppose not; then we can find $\epsilon \geq 0$, such that

$$b + \epsilon \leq c - \epsilon$$

where, of course,

$$b = \lim b_n, \quad c = \lim c_n.$$

We may now select subsequences, say,

$$\{b_{n_1} : b_{n_1} < b + \epsilon, \text{ for all } n_1 \geq N_1\},$$

$$\{c_{n_2} : c_{n_2} > c - \epsilon, \text{ for all } n_2 \geq N_2\},$$

and note that for all $n \geq N$, where $N \geq \max(N_1, N_2)$ we have, for the elements of the subsequences above,

$$b_n < b + \epsilon \leq c - \epsilon < c_n.$$

But this states that there are infinitely many elements for which

$$b_n < c_n.$$

This is a contradiction.   q.e.d.

Definition 7.7. Let $\{a_n : n \geq 1\}$ be a sequence of real numbers; then its limit exists, if and only if

$$\limsup a_n = \liminf a_n,$$

and it (the limit) is defined to be their common value.

Let us now consider sequences whose elements are sets, i.e. subsets of Ω.

Definition 7.8. Let $\{A_n : n \geq 1, \ A_n \subset \Omega\}$; define

$$B_n = \bigcup_{k=n}^{\infty} A_k, \quad C_n = \bigcap_{k=n}^{\infty} A_k$$

and note that $\{B_n : n \geq 1\}$, $\{C_n : n \geq 1\}$ are, respectively, monotone nonincreasing and monotone nondecreasing. Let

$$A^* = \lim_{n \to \infty} B_n, \quad A_* = \lim_{n \to \infty} C_n$$

where, for a monotone nonincreasing sequence

$$\lim_{n \to \infty} B_n = \bigcap_{n=1}^{\infty} B_n$$

and for a monotone nondecreasing sequence

$$\lim_{n \to \infty} C_n = \bigcup_{n=1}^{\infty} C_n.$$

Then, the **limit superior** of the sequence is defined to be A^*; the **limit inferior** of the sequence is defined to be A_* and the limit of the sequence exists, if and only if,

$$A^* = A_* = A.$$

Moreover, we have the notation

$$\lim_{n \to \infty} \sup_{k \geq n} A_k = A^*, \quad \text{or} \quad \overline{\lim_{n \to \infty}} A_n = A^*,$$

$$\lim_{n \to \infty} \inf_{k \geq n} A_k = A_*, \quad \text{or} \quad \underline{\lim}_{n \to \infty} A_n = A_*,$$

and, whenever $A^* = A_* = A$,

$$\lim A_n = A.$$

Remark 7.4. The intuitive meaning of A^* is that if $\omega \in A^*$ then ω belongs to infinitely many sets, A_n, a fact also denoted by the notation

$$A^* = \{\omega : \omega \in A_n \ , \ i.o.\},$$

the abbreviation, *i.o.*, meaning **infinitely often.** To see this, pick any element $\omega \in A^*$; evidently, ω must belong to at least one set A_n; let this occur first for $n = n_1$, and consider B_n, for $n = n_1 + 1$. Clearly, this set, B_n, does not contain A_n, for $n = n_1$; however, since it must contain ω, there must be another set, say A_n, for $n = n_2 > n_1$ which contains ω. Continuing in this fashion we can show that the elements of A^* are contained in infinitely many sets A_n.

Remark 7.5. The set A_* has the intuitive interpretation that its elements belong to all, except possibly a finite number, of the sets of the sequence. To see why this is so, note that if $\omega \in A_*$, then there exists an index, say, n_0, such that for all $n \geq n_0$, $\omega \in A_n$.

We close this section with

Proposition 7.3. $A^* \supset A_*$.

Proof: Evidently, by construction,

$$B_n \supset C_n, \quad \text{for all} \quad n.$$

Thus,

$$A^* = \lim_{n \to \infty} B_n \supset \lim_{n \to \infty} C_n = A_*.$$

q.e.d.

7.3   Measurable Spaces, Algebras and Sets

In previous sections we had introduced the abstract space Ω and have dealt with operations on sets, which are subsets of Ω. Here, we wish to impart some structure on the class of subsets under consideration. Thus, we introduce,

Definition 7.9. Let $\mathcal{A}$ be a nonempty class of subsets of Ω; then $\mathcal{A}$ is said to be an algebra if

   i. For any $A \in \mathcal{A}$, we also have $\bar{A} \in \mathcal{A}$;

   ii. For any $A_i \in \mathcal{A}$, $i = 1, 2$, $A_1 \cup A_2 \in \mathcal{A}$.

Remark 7.6. A few implications of the definition of an algebra are worth pointing out. Since an algebra is a nonempty class of subsets of Ω, it contains at least one set, say A; since it is closed under complementation it also contains the complement of A, in Ω. Since it is also closed under (finite) unions it also contains the union of A and its complement; this is of course Ω! But the complement of Ω, in Ω, is the null set, $\emptyset$. Thus, **any algebra must contain** the pair $(\Omega, \emptyset)$; moreover, one can easily verify that a class consisting solely of this pair is, indeed, an algebra.

Remark 7.7. Notice also that an algebra, $\mathcal{A}$, is closed under finite intersections as well. To see this, observe that if the sets A_i, $i = 1, 2, \ldots, n$, are in $\mathcal{A}$, then

$$\bigcup_{i=1}^{n} A_i \in \mathcal{A}$$

and, consequently, since an algebra is closed under complementation,

$$\overline{\bigcup_{i=1}^{n} A_i} = \bigcap_{i=1}^{n} \bar{A}_i \in \mathcal{A}.$$

Remark 7.8. We may render the description of an algebra, verbally, as a nonempty class of subsets of Ω which is closed under complementation, finite unions and intersections.

Definition 7.10. A nonempty class of subsets of Ω, say $\mathcal{A}$, is said to be a σ-algebra if

    i. It is an algebra and, in addition,

    ii. It is closed under countable unions, i.e. if $A_i \in \mathcal{A}$, $i \geq 1$, then $\bigcup_{i=1}^{\infty} A_i \in \mathcal{A}$.

Definition 7.11. Let Ω be a space and $\mathcal{A}$ a σ-algebra of subsets of Ω; the pair, $(\Omega, \mathcal{A})$, is said to be a **measurable space** and the sets of $\mathcal{A}$ are said to be the **measurable sets**, or $\mathcal{A}$**-measurable**.

Remark 7.9. If Ω is the real line (in this case it is typically denoted by R) and $\mathcal{A}$ the σ-algebra generated by the open intervals (a, b), where a, b are real numbers, then $\mathcal{A}$ is said to be a **Borel** σ**-algebra** and is usually denoted by $\mathcal{B}$. The sets in $\mathcal{B}$ are said to be the **Borel sets**. The measurable space $(R, \mathcal{B})$ is typically referred to as a **Borel space** or a **one dimensional Borel space**.

Definition 7.12. Let Ω_i , $i = 1, 2$, be two spaces; a function

$$X : \Omega_1 \longrightarrow \Omega_2$$

is a relation that associates to each element $\omega_1 \in \Omega_1$ an element, say $\omega_2 \in \Omega_2$, i.e. $X(\omega_1) = \omega_2$.

Definition 7.13. Let X, Ω_i, $i = 1, 2$, be as in Definition 7.12 and $A \subset \Omega_1$. The set (in Ω_2)

$$B = \{\omega_2 : \omega_2 = X(\omega_1), \ \omega_1 \in A\}$$

is said to be the **image** of A under X . Conversely, take any set $B \subset \Omega_2$. Then the set (in Ω_1)

$$A = \{\omega_1 : \omega_1 = X^{-1}(\omega_2), \ \ \omega_2 \in B\}$$

is said to be the **inverse image** of B under X , and we have the notation,

$$X(A) = B \text{ and } X^{-1}(B) = A,$$

i.e. B is the image of A under X , and A is the inverse image of B , under X .

The following question now arises: If $(\Omega_i, \ \mathcal{A}_i)$, $i = 1, 2$ are two measurable spaces and X is a function,

$$X : \Omega_1 \longrightarrow \Omega_2 ,$$

what can we say about the image of $\mathcal{A}_1$ under X and/or the inverse image of $\mathcal{A}_2$ under X ? Denoting these entities by $X(\mathcal{A}_1)$, $X^{-1}(\mathcal{A}_2)$, respectively, we have

Proposition 7.4. Let $(\Omega_i, \ \mathcal{A}_i)$, $i = 1, 2$, be measurable spaces and suppose

$$X : \Omega_1 \longrightarrow \Omega_2 .$$

Then, $X^{-1}(\mathcal{A}_2)$ is a σ -algebra, on Ω_1 , while $X(\mathcal{A}_1)$ is a σ -algebra on Ω_2 , only if X is one to one and onto.[1]

Proof: Let $\mathcal{A} = \{A : A = X^{-1}(B), \ B \in \mathcal{A}_2\}$ and suppose $A_i \in \mathcal{A}, \ i \geq 1$; we shall show that the complement and countable union of such sets are also in $\mathcal{A}$, thus showing that the latter is a σ -algebra. This, however, is quite evident,

---

[1] A function f: $\Omega_1 \longrightarrow \Omega_2$ is said to be one-to-one if and only if, for any pair $(a, b) \in \Omega_1$, $f(a) = f(b)$ implies $a = b$. A function f: $\Omega_1 \longrightarrow \Omega_2$, where Ω_i, $i = 1, 2$, are suitable spaces, is said to be **onto** if and only if **for every** $\omega_2 \in \Omega_2$ **there exists** $\omega_1 \in \Omega_1$ **such that** $f(\omega_1) = \omega_2$.

since if $A_i = X^{-1}(B_i)$, $i \geq 1$, for $B_i \in \mathcal{A}_2$, $i \geq 1$, then $\bigcup_{i=1}^{\infty} B_i \in \mathcal{A}_2$, as well. Consequently, $\bigcup_{i=1}^{\infty} A_i = \bigcup_{i=1}^{\infty} X^{-1}(B_i) = X^{-1}(\bigcup_{i=1}^{\infty} B_i)$, which shows that $\bigcup_{i=1}^{\infty} A_i \in \mathcal{A}$. Moreover, since $\overline{X^{-1}(B_i)} = X^{-1}(\bar{B}_i)$, the proof of the first part of the proposition is complete.

As for the second part, to appreciate the need for the additional conditions, consider X such that it maps Ω_1 into a set $B \subset \Omega_2$. In such a case $B \in X(\mathcal{A}_1)$, but $\bar{B} \notin X(\mathcal{A}_1)$. If X, however is **onto**, i.e. its range is Ω_2, and one to one, i.e. if for every $B \in \mathcal{A}_2$ there is one set, say, $A \in \mathcal{A}_1$ such that $X(A) = B$ and if $X(\mathcal{A}_1) = X(\mathcal{A}_2)$, then $A_1 = A_2$, we may put forth the following argument. Let

$$\mathcal{C} = \{B : B = X(A), \ A \in \mathcal{A}_1\},$$

and suppose $B_i \in \mathcal{C}$, $i \geq 1$. We show that the the countable union and complements of such sets are also in $\mathcal{C}$. For each $B_i \in \mathcal{C}$, $i \geq 1$, there exist $A_i \in \mathcal{A}_1$, $i \geq 1$, such that $X(A_i) = B_i$. Since $A = \bigcup_{i=1}^{\infty} A_i \in \mathcal{A}_1$, and since $X(A) = \bigcup_{i=1}^{\infty} X(A_i) = \bigcup_{i=1}^{\infty} B_i = B$, we conclude that $B \in \mathcal{C}$. Moreover, since $\bar{B}_i = \overline{X(A_i)} = X(\bar{A}_i)$, we conclude that $\bar{B}_i \in \mathcal{C}$.

q.e.d.

We now have the basic definition

Definition 7.14. Let $(\Omega_i, \ \mathcal{A}_i)$, $i = 1, 2$, be measurable spaces and let

$X : \Omega_1 \longrightarrow \Omega_2$.

Then, X is said to be a **measurable** function, or $\mathcal{A}_1$-measurable, if and only if

$$X^{-1}(\mathcal{A}_2) \subset \mathcal{A}_1.$$

The connection between the mathematical concepts above and econometrics is, perhaps, most obvious in the following definition

Definition 7.15. Let $(\Omega, \ \mathcal{A})$, $(R, \ \mathcal{B})$ be two measurable spaces, where R is the extended real line and $\mathcal{B}$ is the Borel σ-algebra. A random variable, X, is a function,

$X : \Omega \longrightarrow R$,

such that $X^{-1}(\mathcal{B}) \subset \mathcal{A}$, i.e. a **random variable is a real valued measurable function**.

A natural question of interest is: if X is a random variable, then what sort of "functions of X" are random variables? For example, if X is a random variable are functions like $\sin X$, X^n, $\log X$, e^X etc. also random variables? This is answered by

Proposition 7.5. Let $(\Omega, \mathcal{A})$, $(R, \mathcal{B})$ be measurable spaces and

$$X : \Omega \longrightarrow R$$

be as in Definition 7.15; let[2]

$$\phi : R \longrightarrow R$$

be a $\mathcal{B}$-measurable function. Then,

$$\psi = \phi \circ X : \Omega \longrightarrow R$$

is a random variable (i.e. a measurable function), where

$$\psi(\omega) = \phi[X(\omega)].$$

Proof: We shall show that $\psi^{-1}(\mathcal{B}) \subset \mathcal{A}$. Let C be any set in $\mathcal{B}$; since ϕ is $\mathcal{B}$-measurable there exists a set, say $B \in \mathcal{B}$ such that $\phi^{-1}(C) = B$. On the other hand since X is $\mathcal{A}$-measurable, there exists a set $A \in \mathcal{A}$ such that $X^{-1}(B) = A$. Consequently, for any set $C \in \mathcal{B}$, we have $\psi^{-1}(C) = X^{-1}[\phi^{-1}(C)] \in \mathcal{A}$. Thus, $\psi^{-1}(\mathcal{B}) \subset \mathcal{A}$.

q.e.d.

The result above is applicable, also, for functions which are defined over **sequences** of random variables. Since for the typical student in econometrics these functions are unfamiliar we present an explicit discussion of them.

Proposition 7.6. Let $(\Omega, \mathcal{A})$, $(R, \mathcal{B})$ be measurable spaces, and let

$$X_n : \Omega \longrightarrow R, \ n \geq 1$$

be random variables. Then, the following are random variables, i.e. they are $\mathcal{A}$-measurable functions from Ω to R:

i. $X_N^* = \sup_{n \leq N} X_n, \ Y^* = \sup_n X_n$;

ii. $X_{*N} = \inf_{n \leq N} X_n, \ Y_* = \inf_n X_n$;

iii. $Y_n = \sup_{k \geq n} X_k, \ Z_n = \inf_{k \geq n} X_k$;

---

[2]The function ψ, implicitly defined below, is said to be a **composition function**, of the functions ϕ and X, is denoted by $\phi \circ X$, and means that for any $\omega \in \Omega$, we first evaluate $X(\omega)$, which is an element of R and **then evaluate** $\phi[X(\omega)]$. Thus, ψ is also a measurable function transforming elements $\omega \in \Omega$ to elements in R, thus justifying the notation $\psi: \ \Omega \longrightarrow R$.

iv. $X^* = \limsup X_n, \quad X_* = \liminf X_n$;

v. $X^+ = \max(0, X), \quad X^- = \max(0, -X)$.

This proof is too complicated to produce here. However, as with other propositions presented without proof, in this and the next two chapters, proofs may be found in Dhrymes (1989).

We close this section with two clarifications, whose meaning will become clear in the ensuing sections—although this is the natural place to present them.

Remark 7.10. The results of Propositions 7.5 and 7.6, although stated in terms of (scalar) random variables, are also applicable to vector random variables. For example, if $\{X_n : n \geq 1\}$ is a sequence of random variables, and ϕ is a suitably measurable vector valued function (i.e. ϕ has, say, k components, $(\phi_i, \ i = 1, 2, \ldots, k)$, then $\phi(Z_n)$ is also a random variable (vector), where $Z_n = (X_1, X_2, \ldots, X_n)$. In the literature of probability theory, entities like random vectors are occasionally referred to as "random elements", although we shall nearly always use the term "random vectors".

Remark 7.11. The following convention will be observed throughout this volume: unless otherwise specified, a random variable will always mean an *a.c.* **finite random variable**, to be read as: "an almost certainly finite random variable". Formally, what this means is that if X is a random variable and we define

$$A = \{\omega : \mid X(\omega) \mid = \infty\},$$

then, $P(A) = 0$, where $P(\cdot)$ is the probability measure. This means, roughly speaking, that the "probability that the random variable will assume the values $\pm\infty$ is zero". **All random variables routinely dealt with in econometrics are *a.c.* finite random variables and, thus, no restriction is entailed by the adherence to the convention above.**

7.4 Measures and Probability Measures

7.4.1 Measures and Measurable Functions

Definition 7.16. Let Ω be a space and $\mathcal{A}$ be a nonempty class of subsets of Ω. A relation, μ, that associates with each set of $\mathcal{A}$ a real number is said to be a **set function**; thus,

$\mu : \mathcal{A} \to R$.

If for every $A \in \mathcal{A}$

$| \mu(A) | < \infty$

the set function, μ, is said to be finite.

Definition 7.17. Let $A \subset \Omega$ and suppose there exist pairwise disjoint sets A_i, i.e. $A_i \cap A_j = \emptyset$ for $i \neq j$, such that $A = \cup_{i=1}^{n} A_i$, then the collection,

$$C_n = \{A_i : i = 1, 2, \dots, n\},$$

is said to be a **finite partition** of A in $\mathcal{A}$. If the collection above is countably infinite, i.e.

$$C = \{A_i : i \geq 1\},$$

the constituent sets are disjoint, and $A = \cup_{i=1}^{\infty} A_i$, then C is said to be a σ-**partition** of A in $\mathcal{A}$.

Definition 7.18. Let A and its partitions be as in Definition 7.17, and let μ be a finite set function as in Definition 7.16. Then, μ is said to be **finitely additive** if for any finite partition of A, say C_n,

$$\mu(A) = \sum_{i=1}^{n} \mu(A_i).$$

If for any σ-partition of A, say C,

$$\mu(A) = \sum_{i=1}^{\infty} \mu(A_i)$$

the set function μ is said to be σ-**additive**, provided the right member above is finite.

Definition 7.19. Let Ω, $\mathcal{A}$ and μ be as in Definition 7.18; the set function μ is said to be a **measure** if and only if

   i. $\mu(\emptyset) = 0$;

   ii. $\mu(A) \geq 0$, for any $A \in \mathcal{A}$;

   iii. If C, as in Definition 7.17, is a σ-partition (in $\mathcal{A}$) of a set $A \subset \Omega$, then

$$\mu(A) = \sum_{i=1}^{\infty} \mu(A_i).$$

Moreover, if μ is a measure and in addition $\mu(\Omega) = 1$, then μ is said to be a **probability measure**, or simply a **probability**, and is denoted by $P(\cdot)$.

Remark 7.12. To summarize the essentials: if Ω is a space and $\mathcal{A}$ a nonempty collection of its subsets containing at least $\emptyset$ and Ω, then a probability measure, or simply a probability $P(\cdot)$, is a real valued nonnegative nondecreasing set function, such that $P(\emptyset) = 0$, $P(\Omega) = 1$.

Definition 7.20. Let Ω be a space, $\mathcal{A}$ a σ-algebra of subsets of Ω, and μ a measure defined on $\mathcal{A}$. Then, the triplet $(\Omega, \mathcal{A}, \mu)$ is said to be a **measure space**; if μ is a probability, then the triplet $(\Omega, \mathcal{A}, P)$, is said to be a **probability space**.

Now that we have introduced measure spaces, it is desirable to elaborate, somewhat, on the concept of measurable functions. We have already defined measurable functions in an earlier section, in the context of two measurable spaces. The requirement imposed by that definition is that the inverse image of the σ-algebra of the range space be contained in the σ-algebra of the domain space. Thus, in order to exploit or establish the measurability of a function we must rely on the fact that sets in the **range** σ-algebra have inverse images in the **domain** σ-algebra. Consequently, it would be useful to establish, under a measurable function, just what kinds of sets in the range space have inverse images which belong to the domain σ-algebra. To this end we have

Proposition 7.7. Let f be an extended real valued function, i.e. f may assume the values $\pm\infty$,

$$f : \Omega \longrightarrow R$$

where the domain space is $(\Omega, \mathcal{A})$, and the range space is $(R, \mathcal{B})$. Then, the following statements are equivalent:

i. For each $a \in R$, the set

$$A = \{\omega : f(\omega) > a\}$$

   is measurable;

ii. For each $a \in R$, the set

$$B = \{\omega : f(\omega) \geq a\}$$

   is measurable;

iii. For each $b \in R$, the set

$$C = \{\omega : f(\omega) < b\}$$

   is measurable;

iv. For each $b \in R$, the set

$$D = \{\omega : f(\omega) \leq b\}$$

is measurable.

Moreover, statements i through iv imply that for each **extended** real number, c, (one allows $c = \pm\infty$), the set

$$E = \{\omega : f(\omega) = c\}$$

is measurable.

This proof is too complicated to reproduce here. However, as with other propositions presented without proof, in this and the next two chapters, proofs may be found in Dhrymes (1989).

We give below, without proof examples of functions which are measurable.

Proposition 7.8. Let $(\Omega, \mathcal{A})$, $(R, \mathcal{B})$ be measurable spaces and let $\{f_n: \ n \geq 1\}$, be a sequence of measurable functions

$$f_n: \quad \Omega \longrightarrow R.$$

Then the following statements are true:

i. If $c \in R$, then cf_n and $c + f_n$ are measurable;

ii. For any n, m, such that f_n, f_m are measurable, then $f_n + f_m$, $f_n - f_m$, are also measurable;

iii. The functions f_n^+, f_n^-, for all n are measurable, as are $\mid f_n \mid = f_n^+ + f_n^-$, where $f_n^+ = \max(0, \ f_n)$, $f_n^- = \max(0, -f_n)$;

iv. The functions

$$\sup_{n \leq N} f_n, \ \sup_{n \geq 1} f_n, \ \limsup_{n \to \infty} f_n$$

are measurable;

v. The functions

$$\inf_{n \leq N} f_n, \ \inf_{n \geq 1} f_n, \ \liminf_{n \to \infty} f_n$$

are measurable.

Definition 7.21. Let $(\Omega, \mathcal{A})$, $(R, \mathcal{B})$ be measurable spaces and

$$f : \Omega \to R$$

be a relation. We say that f is **simple** if and only if there exists a finite partition $\{A_i : i = 1, 2, \ldots, n\}$ of Ω in $\mathcal{A}$, such that

$$f(\omega) = \sum_{i=1}^{n} x_i I_i(\omega), \; x_i \in R,$$

where $I_i(\cdot)$ is the indicator function of the set A_i, i.e.

$$I_i(\omega) \;=\; 1, \quad \text{if } \omega \in A_i, \; i = 1, 2, \ldots, n$$

$$=\; 0, \quad \text{otherwise.}$$

Proposition 7.9. Let $(\Omega, \mathcal{A}, \mu)$, $(R, \mathcal{B})$ be, respectively, a probability space and a (Borel) measurable space and suppose f is a function

$$f : \Omega \to R$$

If g is another function

$$g : \Omega \to R$$

which is $\mathcal{A}$-measurable and such that $\mu(A) = 0$, where

$$A = \{\omega : f(\omega) \neq g(\omega)\}$$

then, f is also measurable and we have the notation, $f = g$ *a.c.* [3]

Proof: Since $\mathcal{B}$ may be generated by sets of the form $(b, \infty]$, it will suffice to show that $C \in \mathcal{A}$, for all $c \in R$, where

$$C = \{\omega : f(\omega) > c, \; c \in R\}.$$

Now, since g is measurable, then for any $c \in R$, $B \in \mathcal{A}$, where

$$B = \{\omega : g(\omega) > c, \; c \in R\}.$$

Next, note that we can always write

$$C = (C \cap \bar{A}) \cup (C \cap A)$$

---

[3] The notation *a.c.* means "almost certainly"; the notations *a.s.*, read "almost surely", or *a.e.*, read "almost everywhere", are also common in this connection; in this volume, however, when dealing with probability spaces we shall use invariably the term *a.c.*

But
$$C \cap \bar{A} = (B \cap \bar{A}) \in \mathcal{A},$$
$$C \cap A \subset A$$

and we conclude that for any set in $\mathcal{B}$, say (c, ∞), its inverse image under f, consists of a set in $\mathcal{A}$, viz., $(B \cap \bar{A})$, plus a set with measure zero, viz., $(C \cap A) \in \mathcal{A}$. This is so since by the nondecreasing property of measures

$$\mu(C \cap A) \leq \mu(A) = 0.$$

q.e.d.

Another useful fact about measurable functions is worth pointing out at this stage.

Proposition 7.10. Let $(R,\ B,\ \mu)$, $(R,\ \mathcal{B})$ be, respectively, a measure space and a measurable space and let f be a measurable function

$$f : R \to R$$

such that, for $A = \{\omega : f(\omega) = \pm\infty\}$, we have $\mu(A) = 0$. Then

    i. Given any $\epsilon > 0$, however small, there exists $N > 0$ such that $|\ f\ | \leq N$, except possibly on a set of measure less than ϵ;

   ii. Given any $\epsilon_1 > 0$, however small, there exists a simple function, g, such that $|\ f(\omega) - g(\omega)\ | < \epsilon_1$, except possibly on a set, say A_1, such that $A_1 = \{\omega : |\ f(\omega)\ | > N\}$, and $\mu(A_1) < \epsilon_1$;

  iii. Given any $\epsilon_2 > 0$, however small, there exists a continuous function, h, such that $|\ f(\omega) - h(\omega)\ | < \epsilon_2$, except possibly on a set, say, $A_2 = \{\omega : |\ f(\omega) - h(\omega)\ | \geq \epsilon_2\}$ such that $\mu(A_2) < \epsilon_2$.

This proof is too complicated to produce here. However, as with other propositions presented without proof, in this and the next chapter, proofs may be found in Dhrymes (1989).

Before we proceed with the theory of Lebesgue integration it is useful to demonstrate another important property of measurable functions.

Proposition 7.11 (Egorov's Theorem). Let $(\Omega,\ \mathcal{A},\ \lambda)$, $(R,\ \mathcal{B})$, be a measure and (Borel) measurable space, respectively, and let

$$f_n : \Omega \longrightarrow R,\ n \geq 1$$

be a sequence of measurable functions such that

$$f_n \longrightarrow f, \text{ a.e.}$$

on a set A with $\lambda(A) < \infty$. Then, given any $\delta > 0$, there exists a (measurable) set $C \subset A$, with $\lambda(C) < \delta$ such that

$$f_n \longrightarrow f$$

uniformly on $A \cap \bar{C}$.

Proof: First note that the set

$$D = \{\omega : \lim_{n \longrightarrow \infty} \mid f_n(\omega) - f(\omega) \mid \neq 0\},$$

obeys $\lambda(D) = 0$, by the conditions of the Proposition; accordingly, we shall, henceforth, interpret A as $A \cap \bar{D}$. To prove the result we shall show that given any $\epsilon > 0$, then for all $\omega \in A \cap \bar{C}$ there exists N such that $\mid f_n(\omega) - f(\omega) \mid < \epsilon$, for all $n \geq N$. To see this define the sets

$$B_{k,r} = \{\omega : \mid f_k(\omega) - f(\omega) \mid \geq \frac{1}{r}, \ \omega \in A\}, \quad k \geq 1,$$

and note that

$$C_{n,r} = \{\omega : \sup_{k \geq n} \mid f_k(\omega) - f(\omega) \mid \geq \frac{1}{r}, \ \omega \in A\} = \bigcup_{k=n}^{\infty} B_{k,r}.$$

Moreover, since the sequence converges pointwise, for any r, we have

$$C_r = \bigcap_{n=1}^{\infty} C_{n,r} = \emptyset$$

Consequently,

$$\lim_{n \to \infty} \lambda(C_{n,r}) = 0$$

and given $\delta > 0$, there exists $N(r)$ such that

$$\lambda(C_{n,r}) < 2^{-r}\delta.$$

Define

$$C = \bigcup_{r=1}^{\infty} C_{N(r),r}$$

and note that

$$\lambda(C) \leq \sum_{r=1}^{\infty} \lambda(C_{N(r),r}) < \delta.$$

The construction above shows that $f_n \longrightarrow f$, uniformly on the set

$$A \cap \bar{C}.$$

For suppose not; let $\epsilon > 0$ be given and suppose there exists

$$\omega \in A \cap \bar{C}$$

for which

$$|f_n(\omega) - f(\omega)| \geq \epsilon, \text{ for all } n > N(\epsilon).$$

Now, given ϵ, there exists an $r*$ such that $(1/r*) < \epsilon$; consequently, for all $n \geq N(r*)$ we have that this ω is contained in $C_{N(r*), r*}$. This is a contradiction.

q.e.d.

Remark 7.13. The preceding discussion may be summarized loosely as follows: any function which is closely approximated by a measurable function is measurable; measurable functions which are almost bounded, i.e. the set over which they assume the values, say, $f(\omega) = \pm\infty$, has measure zero, can be closely approximated by bounded functions; measurable functions which are bounded can be approximated, arbitrarily closely, by simple functions, i.e. functions which are constant over the sets of a finite (or countable) partition of the space; finally, bounded measurable functions defined on $(R, \mathcal{B})$ are "almost" continuous, i.e. they can be arbitrarily closely approximated by continuous functions.

We are now in a position to deal with integration in measure spaces.

7.5 Integration

We begin with a brief review of the **Riemann integral**. It is assumed that the reader is thoroughly familiar with the Riemann integral, the point of the review being to set forth notation and the context of the discussion. Let

$$f : R \longrightarrow R$$

and let it be desired to find the **integral** of f over the interval $[a, b] \subset (-\infty, +\infty)$. To this effect, partition the interval

$$a = x_0 < x_1 < x_2 < \ldots < x_n = b,$$

put

$$c_i = \inf_{x \in (x_i, \ x_{i+1})} f(x), \quad C_i = \sup_{x \in (x_i, \ x_{i+1})} f(x), \ i = 0, 1, 2, \ldots, n-1,$$

and define the sums

$$s_R = \sum_{i=0}^{n-1} c_i \Delta x_{i+1}, \ S_R = \sum_{i=0}^{n-1} C_i \Delta x_{i+1}$$

with $\Delta x_{i+1} = x_{i+1} - x_i$. Take

$$\bar{s}_R = \sup s_R, \ \underline{S}_R = \inf S_R,$$

where sup and inf are taken over all possible partitions of $[a, b]$. The entities $\bar{s}_R$, $\underline{S}_R$ always exist and, evidently, $\bar{s}_R \leq \underline{S}_R$.

We say that the **Riemann integral exists**, if and only if $\bar{s}_R = \underline{S}_R$ and we denote the Riemann integral by

$$I_R = \int_a^b f(x)dx.$$

The **Riemann-Stieltjes** (RS) integral is defined similarly, except that f is weighted by another function, say G. Let it be desired to obtain the integral of f, with respect to G over the interval $[a, b]$. To this effect partition the interval as above and obtain the lower and upper sums, respectively,

$$s_{RS} = \sum_{i=0}^{n-1} c_i [G(x_{i+1}) - G(x_i)], \quad S_{RS} = \sum_{i=0}^{n-1} C_i [G(x_{i+1}) - G(x_i)].$$

Again, determine the sup and inf of these quantities over all possible partitions of the interval, thus obtaining

$$\bar{s}_{RS} = \sup s_{RS}, \quad \underline{S}_{RS} = \inf \underline{S}_{RS}.$$

If

$$\bar{s}_{RS} = \underline{S}_{RS},$$

we say that the Riemann-Stieltjes integral exists and we denote it by

$$I_{RS} = \int_a^b f(x) \, dG(x).$$

Remark 7.14. Note that if G is differentiable with derivative g, then the RS integral reduces to the ordinary Riemann integral

$$I_R = \int_a^b f(x)g(x)dx.$$

To tie the development above with the discussion to follow, let us give the RS integral a slightly different formulation. Thus, we deal with the problem of defining the integral of f over $[a, b]$ and we subdivide the interval by the points,

$$a = x_0 < x_1 < x_2 < \ldots < x_n = b.$$

On these subintervals we then define the **step functions**

$$f_n(x) = c_i, \ F_n(x) = C_i, \ x \in (x_i, \ x_{i+1}), \ i = 0, 1, 2, \ldots, n-1,$$

where c_i and C_i, are as above, i.e. they represent, respectively, the inf and sup of f over the subinterval $(x_i, \ x_{i+1})$. It is easily verified that, by construction,

$$f_n(x) \le f(x), \ F_n(x) \ge f(x), \ \text{for all} \ x \in [a, b].$$

Moreover, in terms of the definition of **any** integral it certainly makes sense to write

$$\sum_{i=0}^{n-1} c_i \Delta x_{i+1} = \int_a^b f_n(x) dx.$$

Similarly, we may put

$$\sum_{i=0}^{n-1} C_i \Delta x_{i+1} = \int_a^b F_n(x) dx.$$

Consequently, in this framework, the Riemann integral may be defined as

$$I_R = \int_a^b f(x)\,dx = \inf_{F_n \ge f} \int_a^b F_n(x)\,dx = \sup_{f_n \le f} \int_a^b f_n(x)\,dx.$$

A similar idea is employed in the construction of the **Lebesgue** integral in measure space. To simplify matters we shall take measure to be **Lebesgue outer measure**, a concept to be explained below, and we shall take Ω to be the set of real numbers. By way of clarifying the meaning of Lebesgue outer measure it is sufficient to note, for the moment, that in this context the outer measure of an interval is simply its length. Thus, if μ is a measure and $(x_i, \ x_{i+1})$ are the (sub) intervals, which we denote conveniently by D_i, then

$$\mu(D_i) = x_{i+1} - x_i.$$

Remark 7.15. The basic difference between the Riemann and Lebesgue approaches to integration is the following: in the Riemann approach we look at the domain of the function, i.e. the "x-axis", obtain a finite partition of the domain and within each (disjoint) interval we ask: what are the values assumed by the function in this interval. The integral, then, is simply the weighted sum of such values, the weights being functions of the reference (sub) intervals. In the Lebesgue approach we look at the range of the function, i.e. the "y-axis", and obtain a finite partition of it. We then ask: what is the inverse image of each (sub) interval in the range of the function, i.e. what (sub) interval in the domain of the function corresponds to the reference (sub) interval in the range. The integral is then obtained as a weighted sum of the values assumed by the function, the weights being functions of the measure of the (domain) subinterval corresponding to the reference (range) subinterval.

Remark 7.16. We note that Lebesgue measure may be defined on R as follows: let A be a subset of R; then its Lebesgue (outer) measure is given by

$$\mu(A) = \inf_{A \subset \cup_{i \in I} D_i} \sum_{i \in I} l(D_i),$$

where I is at most a countable index set, $\{D_i : i \in I\}$ is, at most, a countable collection of open sets that cover A, i.e. whose union contains A, and $l(D_i)$ indicates the length of an interval, i.e. if, for example, $D_i = (x_i, x_{i+1})$ then $l(D_i) = x_{i+1} - x_i$.

Note that Lebesgue measure, defined as (essentially) the length of an interval is not a finite measure, according to Definitions 7.16 through 7.19. In particular consider the sets $[-\infty, a)$, $a \in (-\infty, \infty)$, which generate $\mathcal{B}$. The (Lebesgue) measure of such sets is unbounded. On the other hand, if we confine our attention to, say $[-N, N]$, $N < \infty$, measure, defined as length, is clearly finite.

Example 7.2. This example will help clarify the similarities and differences between the Riemann and Lebesgue approaches to integration. Consider the function f, which is defined to be zero for $x \notin [0, 1]$ while for $x \in [0, 1]$ is defined by

$$f(x) \quad = \quad 1, \text{ if } x \text{ is irrational}$$
$$= \quad 0, \text{ otherwise}$$

One feels intuitively that the integral of this function over $[0, 1]$ must be unity since the set of rationals in $[0, 1]$ is only countably infinite, while the set of irrationals is far larger. If we apply the Riemann definition, obtaining, for example, the partition $a = 0$, $x_1 = (1/n)$, $x_2 = (2/n), \ldots, x_n = 1$ then we find, $c_i = 0$, $C_i = 1$, for all i. Thus, for all partitions $s_R \neq S_R$ and, consequently, the Riemann integral does not exist. If we follow the Lebesgue approach then we ask what is the inverse image, A, of $\{0\}$; evidently, this is the set of rationals in the interval $[0, 1]$ i.e. the set of distinct elements of the form $\{(p/q) : p \leq q, \ p, \ q \text{ positive integers}\}$. One may show that this set has measure zero, and, consequently, that the inverse image of its complement in $[0, 1]$, B, has measure one. But its complement is precisely the inverse image of $\{1\}$ under f. Thus, by the definition of a Lebesgue integral

$$I_L = 1, \quad \text{since } \mu(A) = 0, \quad \mu(B) = 1$$

where, evidently μ denotes the measure, which is here length, and A is the set of rationals in $[0, 1]$ while B is the set of all irrationals in that interval.

Let us now examine (Lebesgue) integration in measure space a bit more formally. We remind the reader, however, that when using the term measure we shall mean **Lebesgue outer measure**; to facilitate this recognition we shall, in this discussion, designate measure by λ. We shall retain the use of the symbol Ω for the space although it should be clear from the context that when dealing with the elementary concepts of integration we shall have in mind not an abstract space, but R, the set of real numbers.

Definition 7.22. Let $(\Omega,\ \mathcal{A},\ \lambda)$, $(R,\ \mathcal{B})$ be measure and measurable space, respectively. A function

$$f : \Omega \longrightarrow R$$

is said to be **simple, or elementary** if there exists a set $A \subset \Omega$, of finite measure, such that $f(\omega) = 0$, for $\omega \notin A$ and there exists a finite partition of A (in $\mathcal{A}$) such that

$$f(\omega) = y_i, \ \omega \in A_i, \ \text{and} \ y_i \neq 0, \ y_i \neq y_j \ \text{for} \ i \neq j.$$

Definition 7.23. Let A be a set in the context of Definition 7.22. The function,

$$I_A : \Omega \longrightarrow R,$$

such that

$$\begin{aligned} I_A(\omega) \ &= \ 1, \quad \text{if } \omega \in A \\ &= \ 0, \quad \text{otherwise,} \end{aligned}$$

is said to be an **indicator function**, more precisely the indicator function of the set A.

Definition 7.24. In the context of Definition 7.22, let f be a simple function such that

$$f(\omega) = 0, \ \text{if} \ \omega \notin A,$$

where A is a set of finite measure. Let $\{A_i :\ i\ =\ 1, 2, \ldots, n\}$ be a finite partition of A, in $\mathcal{A}$, let y_i be the (distinct) values assumed by f on A_i and let I_i be the indicator function for A_i. Then, the **canonical representation** of f is given by

$$f(\omega) = \sum_{i=1}^{n} y_i I_i(\omega)$$

Remark 7.17. In terms of our earlier discussion, it is evident that the Lebesgue integral of the simple function f, above, is $\sum_{i=1}^{n} y_i \lambda(A_i)$, and we have the notation

$$\int_A f \, d\lambda, \quad \text{or simply} \quad \int_A f.$$

If this volume were designed to give an extensive treatment of Lebesgue integration we would have proceeded following the usual sequence (of measure theory) outlined in Remark 7.13, which relies on Lebesgue outer measure, which is easy to grasp and we would have dealt only with sets that have finite Lebesgue outer measure. However none of the proofs for the properties of Lebesgue integration make essential use of the properties of Lebesgue outer measure **except for its finiteness**. Thus, we shall proceed directly to the properties of integration in abstract spaces that have direct relevance to econometrics. We recall that in the probability space $(\Omega, \mathcal{A}, P)$, the probability measure P is finite, in particular the probability of the sample space Ω, the universe of discourse, obeys $P(\Omega) = 1$.

7.5.1 Miscellaneous Convergence Results

In this section we present several important results involving issues of convergence of sequences of measurable functions or integrals of measurable functions. The context is still essentially that of the previous section, so that the underlying space, Ω, is simply R and measure, λ, is not necessarily such that $\lambda(\Omega) < \infty$. Otherwise, we deal with measurable functions defined on a measure space and taking values in a measurable space, specifically the one-dimensional Borel space.

Proposition 7.12 (Bounded Convergence Theorem). Let $\{f_n : n \geq 1\}$ be a sequence of measurable functions

$$f_n : \Omega \longrightarrow R, \ n \geq 1$$

defined on a set A, such that $\lambda(A) < \infty$; suppose further that they are uniformly bounded, i.e. there exists $M \in R$ such that for all n, $| f_n(\omega) | < M < \infty$. If the seqence converges to a measurable function, f, pointwise, i.e. for each $\omega \in A$

$$f_n(\omega) \longrightarrow f(\omega)$$

then

$$\int_A f d\lambda = \lim_{n \to \infty} \int_A f_n d\lambda.$$

Proof: By Proposition 7.11 (Egorov's Theorem) given $\epsilon > 0$, there exists $n(\epsilon)$ and a measurable set C with $\lambda(C) < (\epsilon/4M)$ such that for all $n \geq N$ and $\omega \in A \cap \bar{C}$, we have

$$\mid f_n(\omega) - f(\omega) \mid < \frac{\epsilon}{2}\lambda(A).$$

Define $A_1 = A \cap \bar{C}$, $A_2 = A \cap C$, and note that $A_1 \cup A_2 = A$, $\lambda(A_2) < (\epsilon/4M)$. Consequently,

$$
\begin{aligned}
\mid \int_A f_n d\lambda - \int_A f d\lambda \mid &= \mid \int_A (f_n - f)d\lambda \mid \leq \int_A \mid f_n - f \mid d\lambda \\
&= \int_{A_1} \mid f_n - f \mid d\lambda + \int_{A_2} \mid f_n - f \mid d\lambda \\
&\leq \frac{\epsilon}{2} + \frac{\epsilon}{2} = \epsilon.
\end{aligned}
$$

q.e.d.

Proposition 7.13 (Monotone Convergence Theorem). Let $\{f_n : n \geq 1\}$ be a sequence of measurable functions,

$$f_n : \Omega \longrightarrow R, \quad n \geq 1,$$

which vanish outside a set A with $\lambda(A) < \infty$. Then, the following statements are true:

i. If $f_n \geq g$, for all n, where g is an integrable function and $\{f_n: \quad n \geq 1\}$ is a sequence of monotone nondecreasing functions that converge pointwise on A, i.e. if there exists a (measurable) function, f, such that

$$\lim_{n \to \infty} f_n(\omega) = f(\omega), \text{ for } \omega \in A,$$

then,

$$\lim_{n \to \infty} \int_A f_n \, d\lambda = \int_A f \, d\lambda;$$

ii. If $\{f_n: \quad n \geq 1\}$ is a monotone nonincreasing sequence that converges pointwise to f on a set A with $\lambda(A) < \infty$, and $f_n \leq g$, where g is an integrable function, then

$$\lim_{n \to \infty} \int_A f_n \, d\lambda = \int_A f \, d\lambda.$$

In either case the convergence of integrals is monotone (in i. $\int_A f_n \, d\lambda \uparrow \int_A f \, d\lambda$, while in ii. $\int_A f_n \, d\lambda \downarrow \int_A f \, d\lambda$).

Proposition 7.14 (Fatou's Lemma). Let $\{f, g, f_n : n \geq 1\}$

$$f, \ g, \ f_n \colon \Omega \longrightarrow R$$

be a sequence of measurable functions, vanishing outside[4] a set A and such that

$$f_n \longrightarrow f, \text{a.e., on a set } A \text{ with } \lambda(A) < \infty.$$

Then, the following statements are true:

  i. If $f_n \geq g$ and $\int_A g \, d\lambda > -\infty$, then

$$\int_A \liminf_{n \to \infty} f_n \, d\lambda \leq \liminf_{n \to \infty} \int_A f_n \, d\lambda;$$

 ii. If $f_n \leq g$, $n \geq 1$, and $\int_A g \, d\lambda < \infty$, then

$$\limsup_{n \to \infty} \int_A f_n \, d\lambda \leq \int_A \limsup_{n \to \infty} f_n \, d\lambda;$$

iii. If $\mid f_n \mid \leq g$, $n \geq 1$, and $\int_A g \, d\lambda < \infty$, then

$$\int_A \liminf_{n \to \infty} f_n d\lambda \leq \liminf_{n \to \infty} \int_A f_n d\lambda \leq \limsup_{n \to \infty} \int_A f_n d\lambda \leq \int_A \limsup_{n \to \infty} f_n d\lambda.$$

Proposition 7.15 (Lebesgue Dominated Convergence Theorem). Let g, $\{f_n : n \geq 1\}$ be integrable functions over a measurable set A, such that

$$\mid f_n(\omega) \mid \leq g(\omega), \ \omega \in A,$$

and

$$\lim_{n \to \infty} f_n(\omega) = f(\omega), \text{ a.e. on } A.$$

Then,

$$\int_A f d\lambda = \lim_{n \to \infty} \int_A f_n d\lambda.$$

Proof: Consider the sequence $\{(g - f_n) : n \geq 1\}$. This is a sequence of nonnegative measurable functions that converge pointwise to $(g - f)$. Hence, by Proposition 7.14,

$$\int_A (g - f) \, d\lambda \leq \liminf_{n \to \infty} \int_A (g - f_n) \, d\lambda = \int_A g \, d\lambda - \limsup_{n \to \infty} \int_A f_n \, d\lambda.$$

---

[4]The term *a.e.* to be read **almost everywhere**, is a term in measure theory meaning in this case, for example, that the set $B \colon x \notin A$, and $f_n(x) \neq 0$ has measure zero, i.e. $\lambda(B) = 0$. *Mutatis mutandis* it has the same meaning as *a.c.*, a concept we introduced when dealing with random variables in a probability space with probability measure P.

Since f is, evidently, integrable we have

$$\int_A f d\lambda \geq \limsup_{n \to \infty} \int_A f_n d\lambda.$$

Consider now

$$\{g + f_n : n \geq 1\}.$$

This is also a sequence of nonnegative functions such that

$$g(\omega) + f_n(\omega) \longrightarrow g(\omega) + f(\omega),$$

pointwise, for $\omega \in A$. Hence, again by Proposition 7.4,

$$\int_A (g + f) d\lambda \leq \liminf \int_A (g + f_n) d\lambda = \int_A g d\lambda + \liminf \int_A f_n d\lambda.$$

Since

$$\liminf_{n \to \infty} \int_A f_n d\lambda = \int_A \liminf_{n \to \infty} f_n d\lambda \leq \int_A \limsup_{n \to \infty} f_n d\lambda = \limsup_{n \to \infty} \int_A f_n d\lambda,$$

we have the result,

$$\int_A f d\lambda \leq \liminf \int_A f_n d\lambda \leq \limsup \int_A f_n d\lambda \leq \int_A f d\lambda,$$

or,

$$\lim \int_A f_n d\lambda = \int_A f d\lambda.$$

q.e.d.

Proposition 7.16 (Continuity of Lebesgue Integral). Let f be a nonnegative measurable function, integrable over a set A. Then, given $\epsilon > 0$, there exists $\delta > 0$ such that for every $C \subset A$, with $\lambda(C) < \delta$, we have

$$\int_C f d\lambda < \epsilon.$$

Proof: Suppose not; then given $\epsilon > 0$ we can find sets C such that

$$\lambda(C) < \delta \text{ and } \int_C f d\lambda \geq \epsilon.$$

In particular, choose the sets

$$\{C_n : \lambda(C_n) < 2^{-n}\}.$$

Define

$$g_n(\omega) = f(\omega)I_n(\omega),$$

where I_n is the indicator function of the set C_n. It is clear from the definition of the functions g_n that $g_n \longrightarrow 0$, a.e., except possibly for the sets $\liminf C_n$ or $\limsup C_n$. Since

$$\limsup C_n = \bigcap_{n=1}^{\infty} \bigcup_{k=n}^{\infty} C_k$$

$$\lambda(\bigcup_{k=n}^{\infty} C_k) \le \sum_{k=n}^{\infty} 2^{-k} = 2^{-(n-1)};$$

it follows that

$$\limsup \lambda(C_n) = 0.$$

Hence, by Proposition 7.14, with $f_n = f - g_n$,

$$\int_A f d\lambda \le \liminf \int_A f_n d\lambda = \int_A f d\lambda - \limsup \int_A g_n d\lambda \le \int_A f d\lambda - \epsilon.$$

This, however, is a contradiction.

q.e.d.

In the next proposition we give a number of results that follow easily from previous discussions.

Proposition 7.17. The following statements are true:

i. Let $\{g_n: \ n \ge 1\}$ be a sequence of nonnegative measurable functions defined on a measurable set A, with $\lambda(A) < \infty$ and let $g = \sum_{n=1}^{\infty} g_n$. Then

$$\int_A g d\lambda = \sum_{n=1}^{\infty} \int_A g_n d\lambda;$$

ii. Let f be a nonnegative measurable function, and $\{A_i : i \ge 1\}$ a countable partition of the measurable set A. Then

$$\int_A f d\lambda = \sum_{i=1}^{\infty} \int_{A_i} f d\lambda;$$

iii. Let f, g, be two nonnegative measurable functions; if f is integrable and $g < f$ (both statements valid on a measurable set A), then g is also integrable and

$$\int_A (f - g) d\lambda = \int_A f d\lambda - \int_A g d\lambda.$$

Proof: Define

$$h_n = \sum_{k=1}^{n} g_k$$

and note that $\{h_n : n \geq 1\}$ is a sequence of nonnegative nondecreasing measurable functions, defined on a measurable set A such that

$$h_n \longrightarrow g, \text{ a.e., on } A.$$

Consequently, by Proposition 7.15,

$$\int_A g d\lambda = \lim_{n \to \infty} \int_A h_n d\lambda = \sum_{n=1}^{\infty} \int_A g_n d\lambda$$

which proves i. To prove ii let I_n be the indicator function of A_n and define

$$g_n(\omega) = f(\omega) I_n(\omega).$$

Note that on A

$$f = \sum_{n=1}^{\infty} g_n.$$

Consequently, by i above, we have

$$\int_A f d\lambda = \sum_{n=1}^{\infty} \int_A g_n d\lambda = \sum_{n=1}^{\infty} \int_{A_n} f d\lambda,$$

which proves ii. To deal with iii write $f = (f - g) + g$ and note that both $(f - g)$ and g are nonnegative measurable functions defined on a measurable set A; moreover $(f-g) \leq f$ on A and thus integrable over A. Consequently,

$$\int_A g d\lambda = \int_A f d\lambda - \int_A (f - g) d\lambda < \infty,$$

which shows that g is integrable over A.

q.e.d.

We close this section by introducing and (partially) characterizing a form of convergence of sequences of measurable functions that is closely related to convergence in probability, a property widely discussed in econometrics.

Definition 7.25. Let $\{f_n : n \geq 1\}$ be a sequence of measurable functions defined on a measurable set A; let f be a measurable function and, for given $\epsilon > 0$, define the set

$$C_{n,\epsilon} = \{\omega : |f_n(\omega) - f(\omega)| \geq \epsilon\}.$$

The sequence is said to **converge in measure**, to f on the set A if $C_{n,\epsilon} \subset A$ and $\lambda(C_{n,\epsilon}) < \epsilon$, for all $n \geq N(\epsilon)$. We have the following partial characterization of convergence in measure.

Proposition 7.18. Let $\{f_n : n \geq 1\}$, be a sequence of measurable functions defined on the measurable set A; let f be a measurable function and suppose that f_n converges in measure to f, on the set A. Then:

   i. Every subsequence of $\{f_n : n \geq 1\}$ converges to f in measure;

   ii. There exists a subsequence, say $\{f_{n(r)} : r \geq 1\}$ which converges to f, a.e.

The proof of this is beyond the scope of this volume.

7.6 Extensions to Abstract Spaces

In the preceding section, we have essentially used the device of dealing with **bounded** measurable functions and, in obtaining integrals, we have always operated with a set A which was specified to have finite Lebesgue measure. This has simplified the presentation considerably, but at the cost of producing results of seemingly restricted relevance; in point of fact, however, the results we have obtained in the preceding section remain valid under a broader set of conditions as well. Taking the case of bounded functions, first, we recall from Proposition 7.10 that if f is a measurable function and the set over which it assumes the values $\pm\infty$ has measure zero, then it can be closely approximated by a **bounded** measurable function, say g; moreover, g coincides with f, almost everywhere, i.e. $g = f$, a.e., and, consequently, the integrals of the two functions over a set, A, of finite measure, are equal.

We shall complete the discussion of (Lebesgue) integration by extending the definition of the Lebesgue integral to nonnegative functions, which are not necessarily bounded, and finally to unrestricted functions. In addition, we shall establish the necessary modifications, if any, to the integration results presented above.

We recall that in our initial discussion of Lebesgue integration we had dealt with measurable functions defined over a simple Borel space with measure defined as the length of the (bounded) interval in question. If we define functions over a more general measure space, it is not apparent, from our earlier discussion, under what conditions the results obtained earlier will continue to hold in the broader context.

We begin with the proper definition of integrability, where **boundedness is no longer assumed.**

Definition 7.26. Let f be a nonnegative measurable function

$$f : \Omega \longrightarrow R$$

which vanishes outside a set A, with $\lambda(A) < \infty$; let h be a bounded measurable function obeying $h(\omega) \le f(\omega)$, for $\omega \in A$ and otherwise $h(\omega) = 0$. The integral of f over A is defined by

$$I_L = \int_A f \, d\lambda = \sup_{h \le f} \int_A h \, d\lambda$$

and when $I_L < \infty$, the function f is said to be (Lebesgue) integrable over the set A.

Remark 7.18. The reader might ask: what if the function is not nonnegative? This is handled by noting that if f is an unrestricted function, then it can be written in a form involving **two nonnegative** functions as follows: define

$$f^+ = \max(f, 0), \quad f^- = \max(-f, 0),$$

note that both entities above are nonnegative and, moreover,

$$f = f^+ - f^-, \quad |f| = f^+ + f^-.$$

A direct consequence of the remark above is

Definition 7.27. Let f be a measurable function

$$f : \Omega \longrightarrow R$$

which vanishes except on a measurable set A with $\lambda(A) < \infty$. Define

$$f^+ = \max(f, 0), \quad f^- = \max(-f, 0)$$

and note that we can write

$$f = f^+ - f^-.$$

The functions f^+, f^- are nonnegative and measurable over A. If they are, also, integrable (over A), then f is integrable and its integral is defined to be

$$\int_A f \, d\lambda = \int_A f^+ \, d\lambda - \int_A f^- \, d\lambda.$$

Remark 7.19. In some contexts it is convenient to extend the notion of integrability to the case where

$$\sup_{h \leq f} \int_A h \, d\lambda = \infty.$$

In such a context, we can always approximate a nonnegative function f by a nonnegative nondecreasing sequence $\{f_n : n \geq 1\}$, such that $f_n \leq f$; for example,

$$f_n(\omega) = f(\omega), \text{ for } f(\omega) \leq n$$

$$= 0, \quad \text{ for } f(\omega) > n.$$

We may then define the integral by

$$\int_A f \, d\lambda = \lim_{n \to \infty} \int_A f_n \, d\lambda$$

since the limit, in the right member, will always exist for nonnegative functions. Note that this device "works" for (nonnegative) measurable functions, f, **even if they are unbounded and the set over which they are unbounded does not have measure zero. When it does have measure zero** then, of course, the integral will be finite and there is no need for this extension of the definition of integrability. For unrestricted functions, f, the integral will fail to exist only if we have, simultaneously,

$$\int_A f^+ \, d\lambda = \infty \quad \int_A f^- \, d\lambda = \infty.$$

If only one of the equalities above holds, then the integral of f will be either $+\infty$ or $-\infty$. As in the more restricted cases considered earlier, Lebesgue integration, in this context, is a linear operation, a fact that is made clear (without proof) in

Proposition 7.19. Let f, g be integrable functions over a set A with $\lambda(A) < \infty$. Then

   i. For any $a, b \in R$

$$\int_A (af + bg) \, d\lambda = a \int_A f \, d\lambda + b \int_A g \, d\lambda;$$

  ii. If $f \leq g$, a.e., then

$$\int_A f \, d\lambda \leq \int_A g \, d\lambda;$$

iii. If $A = A_1 \cup A_2$, and the A_i, $i = 1, 2$ are disjoint, then

$$\int_A f \, d\lambda = \int_{A_1} f \, d\lambda + \int_{A_2} f \, d\lambda.$$

Next, we ask what difference does it make, in the proofs of the results of the previous sections, whether we are dealing with Lebesgue (outer) measure, or with more general versions of measure as, simply, a nonnegative σ-additive set function defined on a σ-algebra. Even though many of the proofs had been omitted it can be shown that nowhere in these proofs does one rely on the definition of measure being length. However, there is a technical difference that is not so manifest. This is due to the fact that while Lebesgue measure is **complete**, Lebesgue measure restricted to the σ-algebra of Borel sets is not. The term is explained in

Definition 7.28. A measure space, $(\Omega, \mathcal{B}, \mu)$, is said to be complete if $\mathcal{B}$ contains all subsets of sets of measure zero, i.e. if $B \subset A$, $A \in \mathcal{B}$ and $\mu(A) = 0$, then $B \in \mathcal{B}$.

Remark 7.20. It is for the reason implied in the definition above that, in discussing integration at an earlier stage, we were somewhat vague in specifying the precise σ-algebra involved, although we had made several references to Borel space. The reason why the previous definition is necessary is that its absence may create conceptual problems. Particularly, if we fail to take into account its implications, **we may negate the measurability of a function by simply changing its values over a set of measure zero.** For example, suppose f is a measurable function on such a space and E is a measurable set of measure zero. Let $B \subset E$ and suppose the values assumed by f on B are "changed"; suppose, further, that the values assumed by f on B are not in the range σ-algebra. Then, the inverse image of the set in question would not be measurable, i.e. it will not belong to the domain σ-algebra; consequently, the measurability of the function will be negated. We can obviate this problem by insisting that if $A \in \mathcal{B}$, where $\mathcal{B}$ is the σ-algebra of the space above, and if $\mu(A) = 0$, then all subsets of A are also in $\mathcal{B}$. This would mean, of course, that when we speak of a function we will really be speaking about an equivalence class, i.e. of a collection of such functions which are identical except possibly on sets of measure zero. In this connection we have

Proposition 7.20. Let $(\Omega, \mathcal{A}, \mu)$ be a measure space; then there exists a complete measure space, say, $(\Omega_0, \mathcal{A}_0, \mu_0)$, such that

i. $\mathcal{A} \subset \mathcal{A}_0$;

ii. $A \in \mathcal{A}$, implies $\mu(A) = \mu_0(A)$;

iii. $C \in \mathcal{A}_0$ if, and only if, $C = A \cup B$, where $A \in \mathcal{A}$, $B \subset D$, $D \in \mathcal{A}$ and $\mu(D) = 0$.

The proof of this Proposition is beyond the scope of this volume. The next objective, in this discussion, is to show that although in the preceding sections we have dealt with a specific kind of measure on R, the results obtained are valid over a wide range of spaces and measures, provided certain minimal conditions are satisfied. In the course of this discussion we shall need an additional concept.

Definition 7.29. Let $(\Omega, \mathcal{A}, \mu)$ be a measure space, and let $A \in \mathcal{A}$; then A is said to be of **finite measure** if $\mu(A) < \infty$; it is said to be of σ**-finite measure**, if there exists a partition of $A \in \mathcal{A}$, say, $\{A_i: A_i \in \mathcal{A}, \mu(A_i) < \infty, i \geq 1\}$, such that $\sum_{j=1}^{\infty} \mu(A_i) < \infty$.

Definition 7.30. Let μ be a measure, as in Definition 7.2. Then μ is said to be **finite** if $\mu(\Omega) < \infty$. It is said to be σ-finite if Ω is of σ-finite measure.

Remark 7.21. Note that a probability measure is simply a normalized finite measure. Note also that every finite measure is also σ-finite, but that the converse is not true. In particular Lebesgue measure, λ, on R is σ-finite, but it, clearly, is not finite since the λ-measure of the set $(-\infty, \infty)$ is infinite.

Proposition 7.21. Let $(\Omega, \mathcal{A}, \mu)$ be a measure space and f a nonnegative measurable function

$$f : \Omega \longrightarrow R .$$

Then the following statements are true:

i. A necessary condition for $J = \int_{\Omega} f \, d\mu < \infty$, is that A is of σ-finite measure and $\mu(B) = 0$, where $A = \{\omega : f(\omega) > 0\}$, $B = \{\omega : f(\omega) = \infty\}$.

ii. If A is of σ-finite measure, but $\mu(B) = 0$ does not hold, then we can only assert that

$$\int_{\Omega} f \, d\mu = \sup_{g \leq f} \int_{\Omega} g \, d\mu,$$

where the *sup* is taken over all **bounded** measurable functions, g, such that $g \leq f$ and g vanishes outside a set, say A, of finite measure.

The proof of this Proposition is beyond the scope of this volume.

Remark 7.22. The preceding discussion concludes the demonstration that the results obtained in Propositions 7.11 through 7.19, remain valid even when dealing with abstract measure spaces.

We now turn to the discussion of another topic, which will play an important role when we consider conditional probability and conditional expectation. We have

Definition 7.31. Let μ, ν be measures defined on a measurable space $(\Omega, \mathcal{A})$; then ν is said to be **absolutely continuous** with respect to (or relative to) μ if and only if $\nu(A) = 0$, for every set $A \in \mathcal{A}$ such that $\mu(A) = 0$. This is denoted by $\nu \ll \mu$.

Definition 7.32. Let ν be a set function defined on the measurable space $(\Omega, \mathcal{A})$; then ν is said to be a **signed measure** if $\nu = \nu^+ - \nu^-$ and each of the set functions ν^+ and ν^- is a measure. Moreover, the signed measure is absolutely continuous with respect to the measure μ, if and only if the two measures, ν^+ and ν^-, are absolutely continuous with respect to μ.

The preceding discussion immediately suggests

Proposition 7.22. Let $(\Omega, \mathcal{A}, \mu)$ be a measure space and f a nonnegative $\mathcal{A}$-measurable function

$$f : \Omega \longrightarrow R$$

whose integral exists, and define the set function

$$\nu(A) = \int_A f \, d\mu.$$

Then, the following statements are true:

  i. $\nu(A)$ is a measure, and

  ii. It is absolutely continuous with respect to μ.

Proof: To prove i we note that, by the elementary properties of the Lebesgue integral, ν is a nonnegative nondecreasing function, obeying $\nu(\emptyset) = 0$. Again, using the elementary properties of the Lebesgue integral, if $A \in \mathcal{A}$ and $\{A_i, \ i \geq 1\}$ is a partition of A in $\mathcal{A}$, then $\nu(A) = \sum_{i=1}^{\infty} \nu(A_i)$, which proves i.

To prove ii, let $A \in \mathcal{A}$ and $\mu(A) = 0$; let h be a nonnegative simple function, $h(\omega) = \sum_{i=1}^{n} c_i I_i(\omega)$, where I_i are the characteristic functions of the sets of the partition of A in $\mathcal{A}$. Then,

$$\int_A h \, d\mu = \sum_{i=1}^{n} c_i \mu(A_i \cap A) = 0.$$

Accordingly, let $\{f_n : f_n \le f, \ i \ge 1\}$ be a sequence of nonnegative simple functions converging to f; by the Monotone Convergence Theorem we conclude

$$\nu(A) = \int_A f \, d\mu = \lim_{n \to \infty} \int_A f_n \, d\mu = 0.$$

q.e.d.

Corollary 7.1. Let $(\Omega, \mathcal{A}, \mu)$ be a measure space and let f be a measurable function

$$f : \Omega \longrightarrow R$$

whose integral exists, in the extended sense. Then, the set function

$$\nu(A) = \int_A f \, d\mu$$

is a signed measure which is absolutely continuous with respect to μ.

Proof: By the conditions of the Corollary, putting $f = f^+ - f^-$, we have that, at least, one of the integrals of the right member below

$$\int_A f \, d\mu = \int_A f^+ \, d\mu - \int_A f^- \, d\mu$$

is finite. Denoting the first integral, on the right, by $\nu^+(A)$ and the second by $\nu^-(A)$, we observe that each is a measure which is absolutely continuous with respect to μ. Putting

$$\nu(A) = \nu^+(A) - \nu^-(A)$$

we note that it is a well defined entity since at least one of the two right measures is finite; moreover, it is a signed measure which is absolutely continuous with respect to μ.

q.e.d.

It is remarkable that the converse of Proposition 7.22 is also valid, a fact that forms the basis of the abstract development of conditional probability and conditional expectation. We state the result without proof, since its proof will take us well beyond the objectives of this volume.

Proposition 7.23 (Radon-Nikodym Theorem). Let $(\Omega,\, \mathcal{A},\, \mu)$ be a measure space, where μ is σ-finite, and let ν be a signed measure which is absolutely continuous with respect to μ; then, there exists an extended measurable function, i.e. a measurable function taking values in $[-\infty,\, \infty]$ and such that for any $A \in \mathcal{A}$

$$\nu(A) = \int_A f \, d\mu.$$

The function f is unique up to sets of μ-measure zero, i.e. if g is another such function, then the set $C = \{\omega : f \neq g\}$ obeys $\mu(C) = 0$. If ν is a measure then the function f is nonnegative, i.e. it takes values only in $[0,\, \infty]$.

Remark 7.18. The function f, of Proposition 7.23 is said to be the **Radon-Nikodym derivative of the measure** ν **with respect to the measure** μ, and is denoted by

$$\frac{d\nu}{d\mu}.$$

Chapter 8

Foundations of Probability

8.1 Discrete Models

Consider the problem of constructing a model of the process (experiment) of throwing a die and observing the outcome; in doing so, we need to impose on the experiment a certain probabilistic framework since the same die thrown under ostensibly identical circumstances, generally, yields different outcomes. The framework represents, primarily, the investigator's view of the nature of the process, but it must also conform to certain logical rules.

Example 8.1. Suppose the "experiment" consists of throwing a single die and recording the face showing; thus, the "outcomes" of the "experiment" are contained in the set $\Omega = \{1, 2, 3, 4, 5, 6\}$, which is the sample space or universe. We may consider the collection of all possible subsets

$$(\emptyset); (1), (2), \ldots, (6);$$
$$(1, 2), (1, 3), \ldots, (1, 6); (2, 3), \ldots, (2, 6); (3, 4), \ldots, (3, 6); \ldots; (5, 6);$$
$$(1, 2, 3), (1, 2, 4), \ldots (1, 2, 6); \ldots \ldots; (4, 5, 6);$$
$$\vdots$$
$$(1, 2, 3, 4, 5, 6), \text{etc.}$$

say $\mathcal{C}$, and note that it is finite; in general the collection of all subsets of a space is too large to be considered a σ-algebra. In this particular case it merely indicates the subsets relevant for various experiments involving the tossing of the die in n trials.

If the die is only thrown once ($n = 1$), the only relevant part of this collection is the subsets of singletons $\{1\}, \{2\}, \{3\}, \{4\}, \{5\}, \{6\}$, which together with the null set constitute the relevant σ-algebra. On this space we may define the probability "measure", P, as follows: if $A = \{i\}$, $i \leq 6$, then $P(A) = 1/6$. This reflects one's (the investigator's) view that all outcomes of this experiment are equally likely. The reader can verify that P is a

P.J. Dhrymes, *Mathematics for Econometrics*,
DOI 10.1007/978-1-4614-8145-4_8, © The Author 2013

nonnegative finitely additive set function with $P(\emptyset) = 0$ and $P(\Omega) = 1$. If the die is thrown twice ($n = 2$) the relevant part of the collection is the set of (distinct) pairs $(1, 2), (1, 3), \ldots, (1, 6); (2, 3), \ldots, (2, 6); (3, 4), \ldots, (3, 6); \ldots; (5, 6)$. The probability assignment now requires us to also specify the properties of the trials (tossings); in such models it is assumed that the trials are **independent**, i.e. knowing what is the outcome of any trial does not convey any information regarding a subsequent trial. Such trials are usually termed **Bernoulli trials**. This designation implies that the die is perfectly balanced, so there is no "tilt" that tends to favor say a 4 over a 5.

The meaning of the pairs is that the first entry denotes the outcome of the first trial and the second entry the outcome of the second trial. A similar interpretation applies to all n-tuplets, i.e. the outcomes of n trials.

Example 8.2. Let the "experiment" now consist of throwing the die twice and recording the "outcomes" in the order in which they have occurred. In this case the sample space, Ω, in addition to the null set ($\emptyset$) which is part of every space, consists of the collection of pairs

$$\begin{aligned}
\Omega \quad = \quad & (1,1), (1,2), \ldots, (1,6); \\
& (2,1), (2,2), \ldots, (2,6); \\
& (3,1), (3,2), \ldots, (3,6); \\
& \vdots \\
& (6,1), (6,2), \ldots, (6,6).
\end{aligned}$$

The σ-algebra, $\mathcal{A}$, may be defined as the collection of all subsets above including the null set, their unions, intersections etc., and it may be verified that it contains a finite number of sets. On this σ-algebra we may define the probability "measure" P, as follows: if A is one of the elements of Ω, i.e., if $A = (i, j)$ with $i, j = 1, 2, 3, \ldots, 6$, then $P(A) = 1/36$. If A is a set that is made up of the union of, say, k disjoint sets of the type above, then $P(A) = k/36$. Note that the elements of Ω, i.e., the sets (i, j), $i, j = 1, 2, \ldots, 6$ are disjoint. In the language of probability these are called also, **simple events**, while any member of the σ-algebra (the collection of subsets of Ω) is said to be an **event**, or sometimes a **compound event**. For example, if we wish to calculate the probability of the event

$$A = \{\text{the sum of the faces showing is less than 10}\}$$

we may proceed as follows: first, we write A, if possible, as the union of disjoint events, and then we write the latter as unions of simple events. Executing the first step, we have: $A = \bigcup_{i=2}^{9} A_i$, where A_i is the compound event: the sum of the faces showing is i. Executing the second step we reason as follows: since A_2 consists only of the simple event $(1, 1)$,

$P(A_2) = 1/36$; $A_3 = (1,2) \cup (2,1)$, hence, $P(A_3) = 2/36$; $A_4 = (1,3) \cup (2,2) \cup (3,1)$; hence, $P(A_4) = 3/36$; $A_5 = (1,4) \cup (2,3) \cup (3,2) \cup (4,1)$; hence, $P(A_5) = 4/36$; $A_6 = (1,5) \cup (2,4) \cup (3,3) \cup (4,2) \cup (5,1)$; hence, $P(A_6) = 5/36$; $A_7 = (1,6) \cup (2,5) \cup (3,4) \cup (4,3) \cup (5,2) \cup (6,1)$; hence, $P(A_7) = 6/36$; $A_8 = (2,6) \cup (3,5) \cup (4,4) \cup (5,3) \cup (6,2)$; hence, $P(A_8) = 5/36$; finally $A_9 = (3,6) \cup (4,5) \cup (5,4) \cup (6,3)$; hence, $P(A_9) = 4/36$. Consequently, $P(A) = \sum_{i=1}^{6}(i/36) + (5/36) + (4/36)$.

Remark 8.1. The two simple examples, above, contain a great deal of the fundamental concepts of abstract probability theory. Let Ω_i, $i = 1, 2$, be two exact copies of the space Ω of Example 8.1 and notice that the space of Example 8.2, is simply the Cartesian (or direct) product of these spaces, i.e.,

$$\Omega = \{(\omega_1, \omega_2): \quad \omega_1 \in \Omega_1, \omega_2 \in \Omega_2\}.$$

The notation for a Cartesian product is $\Omega = \Omega_1 \times \Omega_2$. Similarly, if we put $\mathcal{A}_i$, $i = 1, 2$, for the σ-algebras of the two copies above, then

$$\mathcal{A} = \mathcal{A}_1 \otimes \mathcal{A}_2$$

is the σ-algebra of the measurable space of Example 8.2. The notation above is nonstandard and $\otimes$ usually denotes the direct product of two entities, such as matrices, for example. In the usage above **it simply denotes the smallest σ-algebra containing the collection**[1] where

$$\mathcal{J} = \{A: \quad A = A_1 \times A_2, \ A_i \in \mathcal{A}_i, \ i = 1, 2\}.$$

The reader ought to verify the claims just made; in doing so he ought to consider the term σ-algebra to mean, just for this space, "the class of all subsets" of the space. Nearly all of the concepts introduced through the two examples, above, generalize easily to abstract spaces, except, obviously for the manner in which the σ-algebra is generated. For a general space the "class of all subsets" of Ω is too large a collection on which to define a measure. We shall begin the discussion of such issues in the next section, beginning with the case where $\Omega = R = (-\infty, \infty)$.

---

[1] Certain other usages are also common; thus the collection $\mathcal{J}$ is also denoted by $\mathcal{J} = \mathcal{A}_1 \times \mathcal{A}_2$, which is to be distinguished from $\mathcal{A}_1 \otimes \mathcal{A}_2$, the latter being equal to $\sigma(\mathcal{J})$,

8.2   General Probability Models

8.2.1   The Measurable Space (R^n, $\mathcal{B}(R^n)$)

We consider the space $\Omega = R$ and a certain collection of subsets of R; the collection in question is one that consists of what we shall call the **basic or elementary sets** of the space; they are of the form $(a,\ b]$, where $a, b \in R$ and others, which can be expressed as a finite union of the basic intervals, together with the null set. As a matter of convention, we consider $(b,\ \infty]$ to be the same as $(b,\ \infty)$; this is necessary in order to enforce the property that the complement of a set of the form $(-\infty,\ b]$ is a set of the same form, i.e., open on the left and closed on the right. Let this collection be denoted by $\mathcal{A}$; it is easy to verify that $\mathcal{A}$ is an algebra. This is so since, if $A_i \in \mathcal{A}$, $i = 1, 2, \ldots, n$, then $A = \bigcup_{i=1}^n A_i \in \mathcal{A}$, where $n < \infty$ and $A_i = (a_i,\ b_i]$, so that the collection is closed under finite unions. Moreover, the complement of a set of the form $(a_i,\ b_i]$ is simply $(-\infty,\ a_i] \cup (b_i,\ \infty]$; consequently, the complement of any set in $\mathcal{A}$ is also in $\mathcal{A}$, so that the latter is closed under complementation and thus it is an algebra.

Remark 8.2. Given a collection of sets, say $\mathcal{J}$, there is always a smallest σ-algebra that contains $\mathcal{J}$. This is proved as follows: clearly the set of all subsets of the space is a σ-algebra that contains $\mathcal{J}$; consider now the collection of all σ-algebras containing $\mathcal{J}$. As we have just shown this collection is nonempty. Define the desired smallest σ-algebra to be the intersection of all σ-algebras containing $\mathcal{J}$. This σ-algebra is denoted by $\sigma(\mathcal{J})$ and is said to be the σ-algebra **generated by** $\mathcal{J}$. The elements of the set $\mathcal{J}$, i.e., the set "generating" the σ-algebra are said to be **the elementary sets**.

Although in general it is not possible to describe the process of constructing the σ-algebra generated by any **arbitrary** collection of sets, we may do so in particular cases. In point of fact, if $\mathcal{J}$ **is an algebra**, it means that it is already closed under complementation and finite unions. Thus, if we add to it all sets which are limits of sets in $\mathcal{J}$, we shall have the desired σ-algebra. A similar argument will describe the **algebra generated** by a (nonempty) collection of subsets of Ω as, simply the smallest algebra containing the class of subsets in question. The following proposition establishes a relation between the "size" of the collection and the "size" of the algebra or σ-algebra it generates.

Proposition 8.1. Let $\mathcal{C}$, $\mathcal{D}$, be two nonempty collections of the subsets of Ω. Denote by $\mathcal{A}(\mathcal{C})$, $\mathcal{A}(\mathcal{D})$ the algebras generated by the two collections respectively, and let $\sigma(\mathcal{C})$, $\sigma(\mathcal{D})$ be the σ-algebras generated by the two collections, respectively. If $\mathcal{C} \subset \mathcal{D}$, then $\mathcal{A}(\mathcal{C}) \subset \mathcal{A}(\mathcal{D})$ and $\sigma(\mathcal{C}) \subset \sigma(\mathcal{D})$.

Proof: We shall give a proof for the case of σ-algebras; the proof for algebras is entirely similar and is left to the reader. Let

$$Z_\mathcal{C} \;=\; \{\mathcal{B} : \mathcal{B} \supset \mathcal{C},\ \mathcal{B}\ \text{a}\ \sigma-\text{algebra}\}$$

$$Z_\mathcal{D} \;=\; \{\mathcal{G} : \mathcal{G} \supset \mathcal{D},\ \mathcal{G}\ \text{a}\ \sigma-\text{algebra}\}.$$

It is easy[2] to see that if $\mathcal{G} \in Z_\mathcal{D}$, then $\mathcal{G} \in Z_\mathcal{C}$, since $\mathcal{G}$ **is a** σ-**algebra** and $\mathcal{G} \supset \mathcal{D} \supset \mathcal{C}$. By definition,

$$\sigma(\mathcal{C}) \;=\; \bigcap_{\mathcal{B} \in Z_\mathcal{C}} \mathcal{B}$$
$$\subset\; \bigcap_{\mathcal{G} \in Z_\mathcal{D}} \mathcal{G}$$
$$=\; \sigma(\mathcal{D}).$$

This is so since $Z_\mathcal{D} \subset Z_\mathcal{C}$ and, thus, the intersection of all the elements in $Z_\mathcal{D}$ **contains** the intersection of all the elements in $Z_\mathcal{C}$.

q.e.d.

Definition 8.1. Let $\mathcal{J}$ be the collection of intervals $(a, b]$, with $a, b \in R$, as above. Then, the σ-algebra, $\sigma(\mathcal{J})$, generated by $\mathcal{J}$, is said to be the **Borel** σ-algebra and is usually denoted by $\mathcal{B}$, or $\mathcal{B}(R)$. The sets of this σ-algebra are said to be the **Borel** sets, and the pair $(R,\ \mathcal{B}(R))$ is said to be the **Borel** measurable space, or simply the one dimensional Borel space.

Now, suppose we are dealing with the Cartesian product of two real lines, which we denote, for clarity, by R_i, $i = 1, 2$. As a matter of notation put $R^2 = R_1 \times R_2$, and on this space we define, by analogy with the one dimensional case, rectangles, say, $T^2 = T_1 \times T_2$, where $T_i \in R_i$, is a set of the form $(a_i,\ b_i]$, $i = 1, 2$. If $\mathcal{J}$ is the collection of all such rectangles in R^2, then the σ-algebra generated by $\mathcal{J}$, i.e. $\sigma(\mathcal{J})$, is also denoted by $\mathcal{B}(R^2)$; this is read: the σ-algebra generated by the (half open) rectangles of R^2. As an alternative consider the collection $\mathcal{J}^*$ of rectangles with "Borel sides", i.e., sets of the form $B = B_1 \times B_2$, where $B_i \in \mathcal{B}(R_i)$, $i = 1, 2$.

Definition 8.2. Let $\mathcal{J}^* = \{B : B = B_1 \times B_2,\ B_i \in \mathcal{B}(R_i),\ i = 1, 2\}$ i.e., the set of all (two dimensional) rectangles with Borel sides; the σ-algebra generated by this collection, $\sigma(\mathcal{J}^*)$, is said to be the **direct product** of the σ-algebras $\mathcal{B}(R_i)$, $i = 1, 2$, and is often denoted by $\mathcal{B}(R_1) \otimes \mathcal{B}(R_2)$.

---

[2]In this argument it is assumed that the collections $Z_\mathcal{C}$, $Z_\mathcal{D}$, are nonempty; otherwise there is nothing to prove. Evidently, if the collections $\mathcal{C}$, $\mathcal{D}$ are **algebras** then it is easy to see that $Z_\mathcal{C}$, $Z_\mathcal{D}$ are nonempty collections.

Remark 8.3. Note that if B, C are any two sets in $\mathcal{J}^*$, their union is not necessarily in $\mathcal{J}^*$; this is so since $B \cup C \neq (B_1 \cup C_1) \times (B_2 \cup C_2)$, and consequently, it is not necessarily a set of the form $(D_1 \times D_2)$ with $D_i \in \mathcal{B}(R_i)$, $i = 1, 2$. On the other hand, $B \cap C = (B_1 \cap C_1) \times (B_2 \cap C_2) \in \mathcal{J}^*$. Considering the complement of B, a little reflection will show that

$$\bar{B} = (\bar{B}_1 \times B_2) \cup (B_1 \times \bar{B}_2) \cup (\bar{B}_1 \times \bar{B}_2),$$

which is, evidently, the union of disjoint sets (in $\mathcal{J}^*$) and as such it is in $\mathcal{J}^*$.

The observation above leads to

Definition 8.3. Let $\mathcal{J}$ be a collection of subsets of a space Ω; if $\Omega \in \mathcal{J}$, the complement of a set in $\mathcal{J}$ is the union of **disjoint sets** in $\mathcal{J}$, and $\mathcal{J}$ is closed under (finite) intersections, then $\mathcal{J}$ is said to be a **semi-algebra**.

Remark 8.4. Note first that the collection $\mathcal{J}^*$, of Remark 8.3 is a semi-algebra. Note, also, that if, in general, $\mathcal{H}$ is a semi-algebra and is augmented by adding to it the null set and all sets which are finite disjoint unions of sets in $\mathcal{H}$, the resulting collection, $\mathcal{H}^*$, may be shown to be an **algebra**. This is so since, if $A \in \mathcal{H}$, then its complement, $\bar{A}$, is the union of disjoint sets in $\mathcal{H}$, and hence $\bar{A} \in \mathcal{H}^*$; if $A \in \mathcal{H}^*$, but $A \notin \mathcal{H}$, then it is the union of disjoint sets in $\mathcal{H}$, and by a similar argument we may establish that its complement is also in $\mathcal{H}^*$. Evidently, the augmented set is closed under finite unions. Moreover, $\sigma(\mathcal{H}^*) = \sigma(\mathcal{H})$. The argument for this is quite simple. Since $\mathcal{H}^* \supset \mathcal{H}$ it follows, by Proposition 8.1, that $\sigma(\mathcal{H}^*) \supset \sigma(\mathcal{H})$. Let $\mathcal{A}(\mathcal{H})$ be the algebra generated by $\mathcal{H}$. If $A \in \mathcal{H}^*$, then it is the union of disjoint sets in $\mathcal{H}$ and, hence, $A \in \mathcal{A}(\mathcal{H})$; this shows that $\mathcal{H}^* \subset \mathcal{A}(\mathcal{H})$. Thus, again by Proposition 8.1, $\sigma(\mathcal{H}^*) \subset \sigma(\mathcal{A}(\mathcal{H}))$; but $\sigma(\mathcal{A}(\mathcal{H})) = \sigma(\mathcal{H})$. We, thus, conclude that $\sigma(\mathcal{H})^* = \sigma(\mathcal{H})$.

Referring the contents of Remark 8.4 to the earlier discussion, we note that the two collections of elementary sets, say,

$$\mathcal{J}_1 = \{T^2 : \quad T^2 = T_1 \times T_2, \ T_i = (a_i, \ b_i], \ i = 1, 2\}$$

and

$$\mathcal{J}_2 = \{B^2 : \quad B^2 = B_1 \times B_2, \ B_i \in \mathcal{B}(R_i), \ i = 1, 2\}$$

are both semi-algebras. Thus, to show that their respective σ-algebras are the same it will be sufficient to show that that the elementary sets of one are contained in the σ-algebra of the other. Now, it is evident that $\mathcal{J}_1 \subset \mathcal{J}_2$, since, evidently, $T_i \in \mathcal{B}(R_i)$, $i = 1, 2$. By Proposition 8.1, then, $\sigma(\mathcal{J}_1) \subset \sigma(\mathcal{J}_2)$. Conversely, consider $\sigma(\mathcal{J}_1)$ and note that it contains the sets $B_1 \times R_2$, $R_1 \times B_2$, for arbitrary $B_i \in \mathcal{B}(R_i)$, $i = 1, 2$. Hence, it contains, also, their intersection, which is nothing more than $B_1 \times B_2$, for arbitrary $B_i \in \mathcal{B}(R_i)$, $i = 1, 2$. This implies that $\sigma(\mathcal{J}_2) \subset \sigma(\mathcal{J}_1)$; we, thus, conclude that $\sigma(\mathcal{J}_1) = \sigma(\mathcal{J}_2)$. In fact it may be shown that

$$\mathcal{B}(R^2) = \sigma(\mathcal{J}_1) = \sigma(\mathcal{J}_2) = \mathcal{B}(R_1) \otimes \mathcal{B}(R_2).$$

Remark 8.5. The import of the preceding discussion is that given n identical unidimensional Borel (measurable) spaces, we can routinely construct the n-dimensional Borel (measurable) space, $(R^n, \mathcal{B}(R^n))$, where $R^n = R_1 \times R_2 \times \ldots \times R_n$, and $\mathcal{B}(R^n) = \mathcal{B}(R_1) \otimes \mathcal{B}(R_2) \otimes \ldots \otimes \mathcal{B}(R_n)$.

We close this section by considering the infinite dimensional Borel space. This is an extremely important space, in that it is the space of (infinite) ordered sequences; as such, or in suitably generalized fashion, it plays an important role in the asymptotic theory of econometrics. Note that the space in question is

$$R^\infty = \{x : x = (x_1, x_2, x_3, \ldots)\}, \text{ where } x_i \in R_i, \ i = 1, 2, 3, \ldots,$$

i.e. the ith real line R_i, is the space of the ith coordinate of the infinite sequence. To complete the construction of the infinite dimensional Borel space $(R^\infty, \mathcal{B}(R^\infty))$, we need to specify its measurable sets, i.e., its σ-algebra. From our previous discussion it is clear that this "should" be $\bigotimes_{i=1}^\infty \mathcal{B}(R_i)$; as pedagogical reinforcement, let us proceed to this task from first principles, i.e., by first specifying a certain collection of "elementary sets", usually a semi-algebra, and then obtaining the σ-algebra it generates. On the real line this is the collection of intervals $\{T : (a, b], a, b \in R\}$. On R^2, it is the collection of rectangles $T_1 \times T_2$, and on R^n it is the collection

$$\mathcal{T}_n = \{T^n : T^n = T_1 \times T_2 \ldots \times T_n, \ T_i = (a_i, b_i], \ a_i, b_i \in R_i, i = 1, 2, \ldots, n\}.$$

The obvious extension of this procedure is to specify the collection $T_1 \times T_2 \times \ldots$, but this does not offer an operational framework, i.e. it does not afford us the means of carrying out the required operations. Instead, we define the collection of basic or elementary sets by

$$\mathcal{G}(T^n) = \{x : x = (x_1, x_2, \ldots, x_n, \ldots), \ x_i \in T_i, \ T_i = (a_i, \ b_i]\},$$

i.e., the elementary sets consist of all infinite sequences, the first n elements of which lie in the intervals T_i, $i = 1, 2, \ldots, n$, for $a_i, b_i \in R_i$. Such sets, i.e. sets that require a finite number of elements (of an infinite sequence) to lie in certain subsets of the appropriate coordinate space and leave all others free are said to be **cylinder sets**. Thus, the typical cylinder set above could, more carefully, be specified as $T_1 \times T_2 \ldots \times T_n \times R \times R \ldots$. The σ-algebra generated by the cylinder sets above is denoted by $\mathcal{B}(R^\infty)$. As before we have the alternative of considering cylinder sets, where the first n elements of the infinite sequence are required to lie in the Borel sets of the appropriate coordinate space. Thus, we consider the elementary sets to be the collection

$$\mathcal{G}(B^n) = \{x : x = (x_1, x_2, \ldots, x_n, \ldots), \quad x_i \in B_i \in \mathcal{B}(R_i), \quad i = 1, 2, \ldots n\}.$$

The smallest σ-algebra that contains this collection, i.e. the σ-algebra generated by $\mathcal{G}(B^n)$, for arbitrary n, is the (infinite) direct product of the constituent σ-algebras, viz. $\mathcal{B}(R_1) \otimes \mathcal{B}(R_2) \otimes \mathcal{B}(R_3) \otimes \ldots$. Finally, we may consider the elementary sets to be the collection of cylinders $\mathcal{G}(B^n) = \{x : x = (x_1, x_2, x_3, \ldots, x_n, \ldots), (x_1, x_2 \ldots, x_n) \in B^n \in \mathcal{B}(R^n)\}$.

It may be shown that the σ-algebras generated by all three such collections of elementary sets are the same. The formal proof of this is somewhat tedious but an intuitive understanding can be easily obtained by noting that if T^n is an n dimensional rectangle with (basic) interval sides, then clearly it is a special form of an n dimensional rectangle with Borel sides, and the latter is clearly a special case of a set in $\mathcal{B}(R^n)$. On the other hand any rectangle with Borel sides can be approximated by unions and/or intersections of rectangles with interval sides. As to the unspecified components of the sequences, note that the σ-algebra generated by the sets $T^n = T_1 \times T_2 \ldots \times T_n$, i.e., the collection $\mathcal{G}(T^n)$, is the same as that generated by the collection, $T^n \times R$, is the same as that generated by the collection $T^n \times R \times R$, etc. This is so since the "character" of the set is determined by the intervals T_i, $i = 1, 2, \ldots n$, while the additional dimensions occupied by the real lines only determine the "position" of the set in the higher dimension. This is so whether we are dealing with the first or the second or the third type of elementary cylinder sets.

8.2.2   Specification of Probability Measures

The purpose of this section is to elucidate some of the basic properties of the probability measure (whose definition was given in the preceding chapter) and to show its connection with distribution functions. We recall that a distribution function

$$F : R \longrightarrow [0, \ 1]$$

has the following properties:

   i. $F(-\infty) = 0$;

  ii. $F(\infty) = 1$;

 iii. It is nondecreasing;

iv. It is right continuous, i.e., if $x_n \downarrow x$, then $\lim_{x_n \downarrow x} F(x_n) = F(x)$, and moreover, for each $x \in R$ the $\lim_{x_n \uparrow x} F(x_n) = F(x-)$ exists.

We repeat, for convenience, the definition of a probability measure given in the preceding chapter.

Definition 8.4. Let $(\Omega, \mathcal{A})$ be a measurable space; the set function

$$P : \mathcal{A} \longrightarrow R$$

is said to be a probability measure if and only if

  i. $P(\emptyset) = 0$;

 ii. $P(\Omega) = 1$;

iii. If $\{A_i : i \geq 1, \ A_i \in \mathcal{A}\}$ is a collection of pairwise disjoint sets, then

$$P(\bigcup_{i=1}^{\infty} A_i) = \sum_{i=1}^{\infty} P(A_i);$$

i.e. if and only if it is nonnegative, σ-additive and satisfies property ii.

A few basic properties of the probability measure follow immediately.

Proposition 8.2. Let $(\Omega, \mathcal{A}, P)$ be a probability space;

  i. If $A, \ B \in \mathcal{A}$, then $P(A \cup B) = P(A) + P(B) - P(A \cap B)$;

 ii. If $A, \ B \in \mathcal{A}$ and $A \subset B$, then $P(A) \leq P(B)$;

iii. If $A_i \in \mathcal{A}$, $i \geq 1$, and $A = \bigcup_{i=1}^{\infty} A_i$; then $P(A) \leq \sum_{i=1}^{\infty} P(A_i)$.

Proof: For the proof of i we note that since $(A \cup B) = A \cup (\bar{A} \cap B)$, and the two sets in the right member above are disjoint,

$$P(A \cup B) = P(A) + P(\bar{A} \cap B).$$

On the other hand, $B = (\bar{A} \cap B) \cup (A \cap B)$ and, again because the two sets on the right are disjoint, $P(B) = P(\bar{A} \cap B) + P(A \cap B)$. Thus, $P(A \cup B) = P(A) + P(B) - P(A \cap B)$. For ii, suppose $A \subset B$; then, we can write $B = A \cup (\bar{A} \cap B)$, so that the two components of B (in the right member of the equation above) are disjoint; consequently,

$$P(B) = P(A) + P(\bar{A} \cap B) \geq P(A).$$

For iii, we employ essentially the same construction as above, viz. we define: $B_1 = A_1$, $B_2 = A_2 \cap \bar{A}_1$, $B_3 = A_3 \cap \bar{A}_2 \cap \bar{A}_1 \ldots$, so that the sequence $\{B_i : i \geq 1\}$ consists of disjoint sets. Then,

$$P(\bigcup_{i=1}^{\infty} A_i) = P(\bigcup_{i=1}^{\infty} B_i) = \sum_{i=1}^{\infty} P(B_i) \leq \sum_{i=1}^{\infty} P(A_i).$$

q.e.d.

A number of other fundamental properties, which will be needed in subsequent discussion, are most conveniently exposited at this juncture.

Proposition 8.3. Let $(\Omega, \mathcal{A})$ be a measurable space and P a nonnegative, finitely additive set function defined on $\mathcal{A}$, with $P(\Omega) = 1$; then, the following four conditions are equivalent:

i. P is σ-additive, i.e., P is a probability;

ii. P is continuous at $\emptyset$, i.e., if $A_i \supset A_{i+1}$ and $\bigcap_{i=1}^{\infty} A_i = \emptyset$, then $\lim_{i \to \infty} P(A_i) = 0$;

iii. P is continuous from above, i.e., for any sets $A_i \in \mathcal{A}$, $i \geq 1$ such that $A_i \supset A_{i+1}$, $\lim_{i \to \infty} P(A_i) = P(\bigcap_{i=1}^{\infty} A_i)$;

iv. P is continuous from below, i.e., for any sets $A_i \in \mathcal{A}$, $i \geq 1$ such that $A_i \subset A_{i+1}$, $\lim_{i \to \infty} P(A_i) = P(\bigcup_{i=1}^{\infty} A_i)$;

Proof: We shall show that i implies iv; iv implies iii; iii implies ii; and finally, that ii implies i, thus completing the proof. To show that i implies iv, let $\{A_i : i \geq 1\}$ be a nondecreasing sequence and define as before

$$B_1 = A_1, \quad B_i = A_i \bigcap_{j=1}^{i-1} \bar{A}_j, \quad i > 1.$$

Since the sets are nondecreasing, we may simplify the expression above to

$$B_1 = A_1, \quad B_i = A_i \cap \bar{A}_{i-1}, \quad i > 1$$

and still preserve the disjointness of the B_i. Moreover, we find

$$\sum_{i=1}^{n} P(B_i) = P(A_1) + P(A_2) - P(A_1) + P(A_3) - P(A_2)$$

$$+ \ldots + P(A_n) - P(A_{n-1}) = P(A_n).$$

Given the σ-additivity of P we obtain

$$P(\bigcup_{i=1}^{\infty} A_i) = \sum_{i=1}^{\infty} P(B_i) = \lim_{n \to \infty} \sum_{i=1}^{n} P(B_i)$$

$$= \lim_{n \to \infty} P(A_n),$$

which proves that i implies iv; to show that iv implies iii, define

$$B_n = A_1 \cap \bar{A}_n, \ n \geq 1,$$

noting that $B_1 = \emptyset$, $B_n \subset B_{n+1}$ and, as required by iii, the sequence $\{A_n : n \geq 1\}$ is nonincreasing. Since $\{B_n : n \geq 1\}$ is, evidently, a nondecreasing sequence we have, by iv,

$$P(\bigcup_{i=1}^{\infty} B_i) = \lim_{n \to \infty} P(B_n).$$

But from the definition of B_n, above, we easily ascertain that $P(B_n) = P(A_1) - P(A_n)$, or more usefully, $P(A_n) = P(A_1) - P(B_n)$. Thus,

$$\lim_{n \to \infty} P(A_n) = P(A_1) - \lim_{n \to \infty} P(B_n)$$

and, moreover,

$$\lim_{n \to \infty} P(B_n) = P(\bigcup_{n=1}^{\infty} (A_1 \cap \bar{A}_n))$$

The set whose probability measure is taken in the right member of the equation above, may also be rendered as

$$\bigcup_{n=1}^{\infty} (A_1 \cap \bar{A}_n) = A_1 \cap (\bigcup_{n=1}^{\infty} \bar{A}_n).$$

Since we can always write

$$A_1 = [A_1 \cap (\bigcup_{n=1}^{\infty} \bar{A}_n)] \cup [A_1 \cap (\bigcap_{n=1}^{\infty} A_n)]$$

we have the relation

$$P(A_1) = P(A_1 \cap [\bigcup_{n=1}^{\infty} \bar{A}_n]) + P(\bigcap_{n=1}^{\infty} A_n).$$

Thus,

$$\lim_{n \to \infty} P(A_n) = P(A_1) - \lim_{n \to \infty} P(B_n) = P(A_1) - P(A_1) + P(\bigcap_{n=1}^{\infty} A_n).$$

To show that iii implies ii is quite simple since

$$\lim_{n \to \infty} P(A_n) = P(\bigcap_{n=1}^{\infty} A_n) = P(\emptyset) = 0.$$

Finally, to show that ii implies i define $B_i \in \mathcal{A}$, $i \geq 1$ to be pairwise disjoint and further define $A_n = \bigcup_{k=n}^{\infty} B_k$. Note that

$$P(\bigcup_{k=1}^{\infty} B_k) = \sum_{k=1}^{n-1} P(B_k) + P(A_n)$$

and moreover that $\{A_n : n \geq 1\}$ is a monotone nonincreasing sequence obeying, $\bigcap_{n=1}^{\infty} A_n = \emptyset$. Thus we have by ii, and rearranging an argument just used above,

$$\sum_{k=1}^{\infty} P(B_k) = \lim_{n \to \infty} \sum_{k=1}^{n} P(B_k) = P(\bigcup_{k=1}^{\infty} B_k) - \lim_{n \to \infty} P(A_n) = P(\bigcup_{k=1}^{\infty} B_k).$$

q.e.d.

To see the connection between probability measures and distribution functions in the context of the measurable space $(R, \mathcal{B})$, let P be a probability measure defined on $\mathcal{B}$; let $A = (-\infty, x]$ and define

$$F(x) = P(A).$$

Clearly the function F is nonnegative; it is also nondecreasing by iv of Proposition 8.3; $F(-\infty) = P(\emptyset) = 0$ by i and $F(\infty) = P(R) = 1$, by property iv of that proposition. Moreover, it is right continuous by iii of Proposition 8.3. Thus, it is a distribution function as claimed. Conversely, if F is a distribution function defined on the measurable space in question, there exists a unique measure, say $P : \mathcal{B}(R) \longrightarrow R$ such that for any set of $\mathcal{B}$, say $A = (x, y]$, x, $y \in R$, it obeys

$$P(A) = F(y) - F(x).$$

By way of explanation consider the collection of intervals on the real line, i.e., sets of the form $T = (a, b]$, and suppose

$$A = \bigcup_{i=1}^{n} (a_i, \ b_i],$$

where the intervals involved are disjoint. Define the probability measure

$$P_0(A) = \sum_{i=1}^{n} [F(b_i) - F(a_i)],$$

and note that this defines uniquely a set function on the collection (semi-algebra) of the intervals of the real line, which is finitely additive. It turns out that P_0 is σ-additive on this semi-algebra and, moreover, that it can be extended to $\mathcal{B}(R)$. This demonstration involves arguments that are too technical for our purposes and we give the result, below, without a complete proof.

Proposition 8.4. Let F be a distribution function defined on the real line. Then, there exists a unique probability measure on $(R, \ \mathcal{B}(R))$ such that for any $x, \ y \in R$

$$P((x, \ y]) = F(y) - F(x)$$

Proof: For the basic interval collection, i.e., for sets of the form $(a, b]$ define

$$P_0((a, \ b]) = F(b) - F(a),$$

and for unions of disjoint such sets define

$$P_0(\bigcup_{i=1}^{n}(a_i, \ b_i]) = \sum_{i=1}^{n}[F(b_i) - F(a_i)].$$

One easily verifies that the preceding defines, uniquely, a nonnegative, nondecreasing finitely additive set function, on the semi-algebra of the elementary sets of R. Moreover, $P_0(R) = 1$. The remainder of the proof makes use of Caratheodory's extension theorem which is given below (in a somewhat generalized form).

Proposition 8.5 (Caratheodory Extension Theorem). Let Ω be a space, let $\mathcal{C}$ be a semi-algebra of its subsets and let $\sigma(\mathcal{C})$ be the smallest σ-algebra containing $\mathcal{C}$. Let P_0 be a measure defined on $(\Omega, \ \mathcal{C})$; then, there exists a unique measure, P, on $(\Omega, \ \sigma(\mathcal{C}))$ which is an extension of P_0 to $\sigma(\mathcal{C})$, i.e., if $A \in \mathcal{C}$, then

$$P(A) = P_0(A);$$

the measure P_0 is said to be be the restriction of P to $\mathcal{C}$, denoted by $P \mid \mathcal{C}$. Moreover, if P_0 is a probability or a σ-finite measure then so is P.

Proof: See Chow and Teicher [Theorem 1, Chap. 6.1].

Example 8.3. Consider the distribution function

$$\begin{aligned} F(x) \ &= \ 0 \text{ if } x < 0 \\ &= \ \tfrac{x}{N} \text{ if } 0 \leq x \leq N < \infty \\ &= \ 1 \text{ if } x > N \end{aligned}$$

Applying Proposition 8.3, we can assert that if a, $b \in [0, N]$, then there exists a measure say P such that

$$P((a,\ b]) = b - a$$

Here, the space is $\Omega = [0, N]$ and the σ-algebra of (Borel) subsets of the space is given by

$$\mathcal{B}([0,\ N]) = \{B \cap [0,\ N],\ B \in \mathcal{B}(R)\}.$$

Clearly, for sets in $\mathcal{B}([0, N])$, P essentially defines the simple Lebesgue (outer) measure on $[0, N]$.

Remark 8.6. If the distribution function of Proposition 8.3 is **absolutely continuous**, i.e. if there exists an integrable function, say f, such that F is the (indefinite) integral of f, then the measure of that proposition is definable by

$$P((a,\ b]) = \int_a^b f dx = F(a) - F(b),$$

and it should be apparent to the reader that sets of the form $(a,\ b)$, $(a,\ b]$, $[a,\ b)$, $[a,\ b]$, all have the same measure.

Generalization of such results to the measurable space $(R^n,\ \mathcal{B}^n)$, for finite n, is straightforward. Incidentally, $\mathcal{B}^n$ is shorthand for $\mathcal{B}(R^n)$. Thus, if P is a probability measure on this space, let

$$T = T_1 \times T_2 \times \ldots \times T_n$$

where $T_i = (-\infty,\ x_i]$, $i = 1, 2, \ldots, n$, and define

$$F(x) = P(T),\ x = (x_1,\ x_2, \ldots, x_n).$$

It can be routinely verified that F is indeed a distribution function. The converse result is also valid and can be proved essentially in the same manner as Proposition 8.4. In particular, if F is absolutely continuous and if T is as above, then the probability measure obeys

$$P(T) = \int_a^b f dx = \int_{a_1}^{b_1} \int_{a_2}^{b_2} \cdots \int_{a_n}^{b_n} f dx_n \ldots dx_1,$$

except that, now, we take $T_i = (a_i,\ b_i]$. The extension of these results to the space $(R^\infty,\ \mathcal{B}(R^\infty))$ is due to Kolmogorov and essentially involves the idea that if we can specify measures on $\mathcal{B}(R^n)$ in a consistent fashion, then, in fact, we have established a measure on $\mathcal{B}(R^\infty)$. Proceeding in the same fashion as before, let us ask ourselves, what do we want to take as the elementary sets in this space. A natural choice would be the **cylinder sets** with base on the Borel sets of $\mathcal{B}(R^n)$. Let A be a Borel set in R^n; we recall that the cylinder set with base A is defined by

$$C_n(A) = \{x : x \in R^\infty, \ (x_1, x_2, \ldots, x_n) \in A \in \mathcal{B}(R^n)\}.$$

A less precise, but perhaps more revealing, way of representing this (cylinder) set is to write it as $A \times R \times R \ldots \ldots$ Consider now another set, say $A^* = A \times R$, and the cylinder set associated with it, viz., $C_{n+1}(A^*)$; if we use the more revealing notation to represent it we see that the two sets are indeed identical. Hence, we would expect that

$$P(C_n(A)) = P(C_{n+1}(A^*)).$$

Indeed, the basic character of such (cylinder) sets is determined by the set A, and the fact that they are "infinitely dimensional", creates only "bookkeeping" problems of properly dealing with the dimensionality of the sets to which various probability measures apply. Specifically, if P were a probability measure on $(R^\infty, \mathcal{B}(R^\infty))$ we would want it to satisfy the property above. However, in the context of $(R^\infty, \mathcal{B}(R^\infty))$, the operation $P(A)$ does not make any sense, since A is strictly speaking not in that σ-algebra; if we want to "place" A therein, we have to represent it as $A \times R \times R \ldots$, i.e., as the cylinder set $C_n(A)$. If we denote by P_n a probability measure on $(R^n, \mathcal{B}(R^n))$, then we would want to have

$$P(C_n(A)) = P_n(A). \tag{8.1}$$

Thus, if we construct a sequence of probability measures, P_i, on the space $(R^i, \mathcal{B}(R^i))$ we would want them to satisfy the following **consistency property** for any set $A \in \mathcal{B}(R^i)$, $i \geq 1$:

$$P_{i+1}(A \times R) = P_i(A). \tag{8.2}$$

It is remarkable that the converse of this result also holds and, moreover, that it is valid for abstract measurable spaces as well.

Proposition 8.6 (Kolmogorov Extension Theorem). Let P_i, $i \geq 1$, be a sequence of probability measures defined on the measurable spaces, $(R^i, \mathcal{B}^i)$, $i = 1, 2, \ldots$, respectively, and satisfying the consistency property of Eq. (8.2). Then, there exists a unique probability measure, say P, on $(R^\infty, \mathcal{B}(R^\infty))$, such that for any cylinder set $C_n(A)$ with $A \in \mathcal{B}(R^n)$

$$P(C_n(A)) = P_n(A).$$

Proof: See Chow and Teicher [Theorem 2, Chap. 6.4].

Remark 8.7. The reader may wonder: are the preceding results, which were developed for the case where $\Omega = R$, restricted to that space alone? In fact they are not. The real number system possesses two types of properties: algebraic, i.e., those that have something to do with notions of addition, multiplication etc., and **metric**, or **topological** properties, i.e., those that have something to do with the distance between two numbers or sets of numbers as well as those dealing with the concept of the limit. These metric or topological properties are not confined to R alone. In fact, we have the following generalization of the version of Kolmogorov's consistency theorem given above.

Proposition 8.7. Let $(\Omega_i,\ \mathcal{G}_i,\ P_i)$, $i \geq 1$, be a sequence of probability spaces; then there exists a unique probability measure, say P, on the (infinite dimensional product) measurable space $(\Omega,\ \mathcal{G})$, where

$$\Omega = \Omega_1 \times \Omega_2 \times \ldots$$

$$\mathcal{G} = \mathcal{G}_1 \otimes \mathcal{G}_2 \otimes \ldots,$$

such that if $A \in \mathcal{G}_1 \otimes \ldots \otimes \mathcal{G}_n$, then

$$P(C_n(A)) = (P_1 \times P_2 \times \ldots \times P_n)(A).$$

Proof: See Chow and Teicher [Theorem 1, Chap. 6.4].

Remark 8.8. While the results of interest to us are valid for copies of the measurable space $(R,\ \mathcal{B})$ as well as copies of $(\Omega,\ \mathcal{G})$, it must not be supposed that there are no differences between the two sets of structures. For example, if we consider the sequence of probability spaces $(R^i,\ \mathcal{B}(R^i),\ P_i)$ and $(\Omega_i,\ \mathcal{G}_i,\ P_i)$ such that $\mathcal{G}_i \subset \mathcal{G}_{i+1}$, and $P_i = P_{i+1} \mid \mathcal{G}_i$ (the equivalent of the consistency property for the infinite dimensional Borel space) then, defining

$$P(A) = \lim_{n \to \infty} P_n(A)$$

for $A \in \bigcup_{n=1}^{\infty} \mathcal{G}_n$, does not yield a σ-additive measure on that algebra; on the other hand if we consider the Borel probability spaces, the measure so defined will be σ-additive, as Proposition 8.6 asserts. In any event, the import of the discussions above, is that if $(\Omega_i,\ \mathcal{G}_i,\ P_i)$, $i \geq 1$, is a sequence of probability spaces, as described in the proposition, and if A is a set of the form

$$A = A_1 \times A_2 \times \ldots \times A_n \times \Omega \times \Omega \ldots \ldots,$$

then there exists a unique probability measure, say P, such that

$$P(A) = \prod_{i=1}^{n} P_i(A_i).$$

8.2.3 Fubini's Theorem and Miscellaneous Results

We begin by repeating the definition of a product measure (space). If $(\Omega_i, \mathcal{G}_i, \mu_i)$, $i = 1, 2$ are two measure spaces, consider the space

$$\Omega = \Omega_1 \times \Omega_2$$

and the semi-algebra

$$\mathcal{G} = \mathcal{G}_1 \times \mathcal{G}_2.$$

We recall that the notation above indicates that the sets in $\mathcal{G}$, say A, are of the form $A = A_1 \times A_2$, such that $A_i \in \mathcal{G}_i$, $i = 1, 2$. On this semi-algebra we may define a measure by the operation

$$\mu_0(A) = \prod_{i=1}^{2} \mu_i(A_i),$$

and for unions of disjoint such sets we may further require that

$$\mu_0\left(\bigcup_{i=1}^{n} B_i\right) = \sum_{i=1}^{n} \mu_0(B_i).$$

If by $\sigma(\mathcal{G})$ we denote the σ-algebra generated by the semi-algebra $\mathcal{G}$, we may extend μ_0 to $\sigma(\mathcal{G})$, using Caratheodory's extension theorem. We finally recall that if μ_0 is a probability then so would be its extension. Thus, let μ be the desired extension and consider the space $(\Omega, \sigma(\mathcal{G}), \mu)$; this is a product space and it is in the context of this space that we shall discuss Fubini's theorem. First, however, a few preliminaries.

Consider the set $A = A_1 \times A_2$, in the product space above, and let $\omega_i \in A_i$, $i = 1, 2$. Define now the sets

$$A_{\omega_2}^{(1)} = \{\omega_1 : (\omega_1, \omega_2) \in A, \text{ for fixed } \omega_2\}$$

$$A_{\omega_1}^{(2)} = \{\omega_2 : (\omega_1, \omega_2) \in A, \text{ for fixed } \omega_1\}.$$

Definition 8.5. Consider the product measure space $(\Omega, \sigma(\mathcal{G}), \mu)$ and the set $A = A_1 \times A_2$, with $A_i \in \mathcal{G}_i$, $i = 1, 2$. Then, the sets $A_{\omega_2}^{(1)}$, $A_{\omega_1}^{(2)}$ as defined above are said to be the **sections** of A at ω_2 and ω_1, respectively. More generally, if $A \in \mathcal{G}$ (and not necessarily as above), sets of the form $A_{\omega_2}^{(1)}$, $A_{\omega_1}^{(2)}$, are said to be **sections** of A at ω_2 and ω_1, respectively, with $\omega_i \in \Omega_i$, $i = 1, 2$.

Remark 8.9. Note that, in the general definition of sections, if $A = A_1 \times A_2$ and $\omega_2 \in A_2$, then $A_{\omega_2}^{(1)} = A_1$; otherwise it is the null set, and if $\omega_1 \in A_1$, $A_{\omega_1}^{(2)} = A_2$, otherwise it is the null set. The development above leads to

Proposition 8.8. Let $(\Omega, \mathcal{G}, \mu)$ be the product measure space, where $\Omega = \Omega_1 \times \Omega_2$, $\mathcal{G} = \sigma(\mathcal{G}_1 \times \mathcal{G}_2)$, $\mu = \mu_1 \times \mu_2$ and all measures are σ-finite. Then:

i. For any measurable set A, the sections $A_{\omega_2}^{(1)}$, $A_{\omega_1}^{(2)}$ are $\mathcal{G}_1$-, $\mathcal{G}_2$-measurable, respectively;

ii. If $\mu(A) = 0$, then $\mu_1(A_{\omega_2}^{(1)}) = 0$ and $\mu_2(A_{\omega_1}^{(2)}) = 0$;

iii. If f is a measurable function from $(\Omega, \mathcal{G}, \mu)$ to $(R, \mathcal{B})$, then for every $\omega_1 \in \Omega_1$, $f(\omega_1, \omega_2)$ defines a measurable function from $(\Omega_2, \mathcal{G}_2, \mu_2)$ to $(R, \mathcal{B})$, and for every $\omega_2 \in \Omega_2$, $f(\omega_1, \omega_2)$ defines a measurable function from $(\Omega_1, \mathcal{G}_1, \mu_1)$ to $(R, \mathcal{B})$.

Proof: To prove i, let

$$\mathcal{A} = \{A : A \in \mathcal{G}, A_{\omega_2}^{(1)} \in \mathcal{G}_1, \text{ for every } \omega_2 \in \Omega_2\}$$

$$\mathcal{C} = \{C : C = C_1 \times C_2, C_i \in \mathcal{G}_i, i = 1, 2\}$$

and note that by Remark 8.9, $\mathcal{C} \subset \mathcal{A}$. Moreover, if $A_i \in \mathcal{A}$, $i \geq 1$, consider $A = \bigcup_{i=1}^{\infty} A_i$ and its section at ω_2, which we denote by $A^{(1)}$, for simplicity. An easy calculation will show that $A^{(1)} = \bigcup_{i=1}^{\infty} A_i^{(1)} \in \mathcal{G}_1$. Similarly, if $A \in \mathcal{A}$ its complement also belongs to $\mathcal{A}$, since it can be written as the countable union of sets of the form $B_1 \times B_2$, with $B_i \in \mathcal{G}_i$, $i = 1, 2$. Thus, by the previous argument, the section of this union is the union of sections and thus the complement of $A \in \mathcal{A}$. But this shows that $\mathcal{A}$ is a σ-algebra that contains $\sigma(\mathcal{C}) = \mathcal{G}$.

To prove ii, let A be of the form $A = A_1 \times A_2$; clearly, for such a set $\mu(A) = \mu_1(A_1)\mu_2(A_2)$ and its indicator (or characteristic) function obeys,

$$I_{12}(\omega_1, \omega_2) = I_1(\omega_1)I_2(\omega_2)$$

where $I_i(\omega_i), i = 1, 2$, are the indicator functions of the sets A_i, $i = 1, 2$, respectively, i.e.,

$$I_{12}(\omega_1, \omega_2) = 1, \text{ if } (\omega_1, \omega_2) \in A_1 \times A_2$$

$$= 0, \text{ otherwise.}$$

Moreover,

$$\mu(A) = \int_{\Omega} I_{12}(\omega_1, \omega_2) \, d\mu = \int_{\Omega_2} \mu_1(A^{(1)}) d\mu_2 = \int_{\Omega_1} \mu_2(A^{(2)}) d\mu_1.$$

Hence, on the semi-algebra, say $\mathcal{A}$, of sets of the form $A = A_1 \times A_2$, the result in ii holds, since evidently, $\mu(A) = 0$ implies both $\mu_1(A^{(1)}) = 0$, a.e., and $\mu_2(A^{(2)}) = 0$, *a.e.*. Next, consider the restriction $\mu \mid \mathcal{A}$. This is a σ-finite measure on the semi-algebra $\mathcal{A}$ and thus, by the Caratheodory extension theorem there exists a unique extension to the σ-algebra $\sigma(\mathcal{A})$. By uniqueness, this extension is μ since $\sigma(\mathcal{A}) = \mathcal{G}$, which completes the proof of ii.

To prove iii, note that setting

$$f(\omega_1, \omega_2) = I_{12}(\omega_1, \omega_2),$$

where $I_{12}(\cdot, \cdot)$ is, as in the proof of ii, the indicator function of the set $A = A_1 \times A_2$, with $A_i \in \mathcal{G}_i$, $i = 1, 2$, we may conclude, by the discussion immediately preceding, that

$$f(\omega_1, \omega_2) = 1, \text{ if } (\omega_1, \omega_2) \in A$$
$$= 0, \text{ otherwise.}$$

In particular, treating ω_2 as fixed, we have the indicator function for A_1, while treating ω_1 as fixed we have the indicator function for A_2. Thus, clearly the result in iii holds for indicator functions of sets of the form $A = A_1 \times A_2$. But any measurable function on the space $(\Omega_1 \times \Omega_2, \sigma(\mathcal{G}_1 \times \mathcal{G}_2))$ can be approximated by simple functions which are constant on (disjoint) sets of the form $A_1 \times A_2$, i.e., by

$$f_n(\omega_1, \omega_2) = \sum_{i=1}^{n} c_i I_{12i}(\omega_1, \omega_2)$$

where I_{12i} is the indicator function of a set of the form $A_{1i} \times A_{2i}$. Thus, for fixed ω_1, f_n is $\mathcal{G}_2$-measurable and for fixed ω_2 it is $\mathcal{G}_1$-measurable. The conclusion then follows by the convergence of such functions to f.

q.e.d.

Proposition 8.9 (Fubini's Theorem). Let $(\Omega, \mathcal{G}, \mu)$ be the product measure space above (with μ σ-finite) and let $(\Psi, \mathcal{C})$ be a measurable space. Let

$$f : \Omega \longrightarrow \Psi$$

be a measurable function which is $\mu(= \mu_1 \times \mu_2)$-integrable. Then, the following statements are true:

i. The integrals

$$\int_{\Omega_1} f(\omega_1, \omega_2) \, d\mu_1, \int_{\Omega_2} f(\omega_1, \omega_2) \, d\mu_2$$

are well defined for all ω_2, ω_1, respectively;

ii. The integrals in i are $\mathcal{G}_2$-, $\mathcal{G}_1$-measurable, respectively, and moreover $\mu_2(D_2) = \mu_1(D_1) = 0$, where

$$D_2 = \{\omega_2 : \int_{\Omega_1} f(\omega_1, \ \omega_2) \, d\mu_1 = \infty\}$$

$$D_1 = \{\omega_1 : \int_{\Omega_2} f(\omega_1, \ \omega_2) \, d\mu_2 = \infty\}$$

iii.

$$
\begin{aligned}
\int_\Omega f \, d\mu &= \int_\Omega f(\omega_1, \ \omega_2) \, d(\mu_2 \times \mu_1) \\
&= \int_{\Omega_1} [\int_{\Omega_2} f(\omega_1, \ \omega_2) \, d\mu_2] \, d\mu_1 \\
&= \int_{\Omega_2} [\int_{\Omega_1} f(\omega_1, \ \omega_2) \, d\mu_1] \, d\mu_2.
\end{aligned}
$$

Proof: The proof of i is an immediate consequence of Proposition 8.8, since we have shown there that for fixed ω_2, f is $\mathcal{G}_1$-measurable and for fixed ω_1, it is $\mathcal{G}_2$-measurable. To prove ii and iii we begin with the case of nonnegative (measurable) functions. Thus, consider the set $A = A_1 \times A_2$ with $A_i \in \mathcal{G}_i$, $i = 1, 2$, take

$$f(\omega_1, \ \omega_2) = I_{12}(\omega_1, \ \omega_2),$$

where I_{12} is the indicator function of the set A above, and observe that, in the obvious notation,

$$I_{12}(\omega_1, \ \omega_2) = I_1(\omega_1) I_2(\omega_2).$$

Consequently,

$$\int_{\Omega_2} f \, d\mu_2 = I_1(\omega_1)\mu_2(A_2)$$

and

$$\int_{\Omega_1} f \, d\mu_1 = I_2(\omega_2)\mu_1(A_1)$$

which are, evidently, $\mathcal{G}_1$- and $\mathcal{G}_2$-measurable functions, respectively. Now, every nonnegative measurable function f, can be approximated by a sequence of simple (nondecreasing) functions, converging pointwise to f. As we recall, the simple functions are linear combinations of indicator functions of the type examined above, i.e.,

$$f_n = \sum_{i=1}^n c_i I_{12i}(\omega_1, \ \omega_2),$$

such that $f_n \leq f$, and I_{12i} is the indicator function of a set of the form $A_{1i} \times A_{2i}$. Notice that in view of the inequality above, we must also have, for fixed ω_1, or ω_2,

$$f_n(\omega_1, \ \omega_2) \leq f(\omega_1, \ \omega_2).$$

Hence, by the Monotone Convergence Theorem

$$H_{ni} = \int_{\Omega_i} f_n \, d\mu_i \longrightarrow \int_{\Omega_i} f \, d\mu_i = H_i, \ i = 1, 2$$

and similarly, since $H_{ni} \leq H_i$, again by the Monotone Convergence Theorem we obtain

$$\int_{\Omega_2} H_{n1}(\omega_2) \, d\mu_2 \longrightarrow \int_{\Omega_2} H_1(\omega_2) \, d\mu_2.$$

Moreover, the integral of H_{n2} converges to that of H_2. But this demonstrates the validity of iii. It is obvious, then that ii must be valid as well. This is so, since the functions displayed there have finite integrals; a necessary condition for this to be so is that the set over which the function(s) become unbounded must have measure zero. Having shown the validity of the proposition for non-negative (measurable) functions and noting that any (measurable) function, f, can be written as $f^+ - f^-$, the proof of the proposition is completed.

q.e.d.

8.3   Random Variables

8.3.1   Generalities

In this section we shall gather a number of results, regarding random variables, some of which have been dealt with in the previous sections and some of which are entirely new. The purpose is to assemble in one location a number of useful characterizations and other pertinent information about random variables. First, we recall the definition that a random variable is **a real valued measurable function, defined on a probability space.** Thus, given the extensive discussion of measurable functions in the previous chapter, we already know a great deal about random variables, since nearly everything discussed in that chapter dealt with measurable functions defined on general, or at worst σ-finite, measure spaces. Since a random variable is a real valued function defined on a probability space, which is certainly σ-finite, all results obtained therein are immediately applicable to random variables.

If a function is given, how can we determine whether it is measurable? This is answered unambiguously by

Proposition 8.10. Let $(\Omega, \ \mathcal{A})$, $(R, \ \mathcal{B})$ be measurable spaces and

$$f : \Omega \longrightarrow R$$

be a relation. Let $\mathcal{C}$ be a collection of subsets of R such that

$$\sigma(\mathcal{C}) = \mathcal{B}(R).$$

Then, f is measurable if and only if

$$A \in \mathcal{A}, \; A = \{\omega : f(\omega) \in C\}$$

for all $C \in \mathcal{C}$.

Proof: If the condition holds, then evidently $A \in \mathcal{A}$; thus, consider the sufficiency part. Let

$$\mathcal{H} = \{B : B \in \mathcal{B}(R) \text{ and } f^{-1}(B) \in \mathcal{A}\}$$

and consider the sequence B_i, $i \geq 1$, such that $B_i \in \mathcal{H}$. Since

$$f^{-1}(\bar{B}_i) = \overline{f^{-1}(B_i)}$$

$$f^{-1}\left(\bigcup_{i=1}^{\infty} B_i\right) = \bigcup_{i=1}^{\infty} f^{-1}(B_i)$$

we conclude that $\mathcal{H}$ is a σ-algebra. Clearly $\mathcal{C} \subset \mathcal{H} \subset \mathcal{B}(R)$. Therefore,[3] $\sigma(\mathcal{C}) \subset \mathcal{H} \subset \mathcal{B}(R)$. But, by the condition of the proposition

$$\sigma(\mathcal{C}) = \mathcal{B}(R).$$

q.e.d.

Corollary 8.1. A necessary and sufficient condition for X to be a random variable, is that the set

$$A_x \in \mathcal{A}, \text{ where } A_x = \{\omega : X(\omega) \leq x\} \text{ or, } \{\omega : X(\omega) < x\}$$

for every $x \in R$, i.e. that such sets be $\mathcal{A}$-measurable.

Proof: Let $\mathcal{C}$ be the collection of intervals of the form $(-\infty, x)$, $x \in R$ and $\mathcal{C}^*$ the collection of intervals of the form $(-\infty, x]$. From previous discussion and Proposition 8.10 we know that $\sigma(\mathcal{C}) = \sigma(\mathcal{C}^*) = \mathcal{B}(R)$.

q.e.d.

Remark 8.11. Notice that, in the course of the proof above, if $\sigma(\mathcal{C}) = \mathcal{B}(R)$, then putting

$$\mathcal{H} = \{H : H = X^{-1}(C), \; C \in \mathcal{C}\}$$

we easily conclude that $\mathcal{H}$ is a σ-algebra. This σ-algebra is often denoted by $\sigma(X)$ and is said to be the σ-**algebra induced by** X.

---

[3]This is a consequence of the fact that if we have two (collections of) sets obeying $\mathcal{C}_1 \subset \mathcal{C}_2$ then the σ-algebras they generate obey $\sigma(\mathcal{C}_1) \subset \sigma(\mathcal{C}_2)$.

Remark 8.12. When dealing with random variables it is occasionally convenient to allow the range of measurable functions to be

$$\bar{R} = [-\infty, \ \infty], \text{ instead of } R = (-\infty, \ \infty).$$

Since the set $[-\infty, \ \infty]$ is said to be the **extended** real line, such random variables are said to be **extended** random variables. **When dealing with such random variables it is crucial to bear in mind a number of important conventions as follows: if** $a \in R$, $a \pm \infty = \pm\infty$; **if** $a > 0$, $a \cdot \infty = \infty$; **if** $a < 0$, $a \cdot \infty = -\infty$; **if** $a = 0, a \cdot \pm\infty = 0$. **Moreover,** $\infty + \infty = \infty$; $-\infty + (-\infty) = -\infty$. **We must also recognize that, despite the conventions above, we are still left with the following indeterminate forms:** ∞/∞, $\infty - \infty$, $0/0$.

In most of our discussion we shall be dealing with a.c. finite random variables, i.e., if X is a random variable, then

$$P(A) = 0, \ A = \{\omega : X(\omega) = \pm\infty\}.$$

Note further that if X is integrable, i.e., if

$$\int X \, dP < \infty,$$

then for A, as defined above, we must also have $P(A) = 0$.

Definition 8.6. The expectation of a random variable, defined over the probability space $(\Omega, \ \mathcal{A}, \ P)$, is given by the integral above, whenever the latter exists; thus, the expectation of a random variable is

$$\int_\Omega X \, dP$$

and is denoted by $E(X)$, or $E X$, E being the expectation operator.[4]

Remark 8.13. Note that the use of the expectation operator, combined with the notation for indicator functions, virtually eliminates the need to write an integral sign. For purposes of notational ease we shall frequently employ this procedure. For example, suppose we wish to take the integral of the square of a zero mean random variable over a set A. Instead of the notation

$$\int_A X^2 dP,$$

we can write simply $E[X^2 I_A]$.

---

[4]When the context is clear and no confusion is likely to arise we shall generally use the notation $E X$.

The reader is no doubt very familiar with statements of the form: Let X be a random variable with distribution function, F; moreover he is no doubt well aware of the fact that F is defined on the real line. This terminology might create the erroneous impression that random variables are defined on the real line which is incompatible with our previous discussion. The following few words are meant to clarify these issues. Thus,

Definition 8.7. Let X be a random variable as in Definition 8.6. Then, its **probability distribution** is a **set function**

$$P_x : \mathcal{B}(R) \longrightarrow R$$

such that for all sets $B \in \mathcal{B}$

$$P_x(B) = P(X^{-1}(B)).$$

Remark 8.14. The probability function, P_x, of a random variable, X, is to be distinguished from its **distribution function**, say F, sometimes also termed the p.d.f. (probability distribution function), or c.d.f (cumulative distribution function), which is a point function

$$F : R \longrightarrow [0, 1]$$

and is defined, for all $x \in R$, by

$$F(x) = P(X^{-1}((-\infty, \ x])) = P_x((-\infty, \ x]).$$

8.3.2 Moments of Random Variables and Miscellaneous Inequalities

Let $(\Omega, \ \mathcal{A}, \ P)$, $(R, \ \mathcal{B})$ be a probability and measurable space, respectively, and let

$$X : \Omega \longrightarrow R$$

be a random variable.[5] We recall that the **expectation** or **mean** of a random variable, say X, is denoted by $E(X)$ and is given by

$$E(X) = \int_\Omega X(\omega) \, dP.$$

In all subsequent discussion it will be understood that all random variables are integrable in the sense that the relevant integrals exist and are finite, i.e. we

---

[5]In order to avoid this cumbersome phraseology in the future, when we say that X is a random variable, it is to be understood that we have predefined the appropriate probability and measurable spaces. Thus, mention of them will be suppressed.

shall always be dealing with a.c. finite random variables. If X is a random variable then so is X^k, and the k **th moment** of the random variable X is defined by

$$\int_\Omega X^k(\omega)\, dP = E(X^k) = \mu_k, \quad k = 1, 2, \ldots$$

provided the integrals exist and are finite. The second moment about the mean, (μ_1), is of special significance; it is termed the **variance** of the random variable and is given by

$$\mathrm{var}(X) = E(X - \mu_1)^2.$$

If X is a **random vector** then μ_1 is a vector of means[6] and the concept of variance is generalized to that of the **covariance matrix**

$$\mathrm{Cov}(X) = E(X - \mu_1)(X - \mu_1)' = \Sigma,$$

which is usually denoted by the capital Greek letter Σ.

Proposition 8.12 (Generalized Chebyshev Inequality). Let X be a nonnegative integrable random variable; then, given $\epsilon > 0$,

$$P(A) \le \frac{E(X)}{\epsilon}$$

where $A = \{\omega : X(\omega) \ge \epsilon\}$.

Proof: Let I_A be the indicator function of the set A, above, and note that

$$X \ge XI_A \ge \epsilon I_A.$$

Taking expectations we find

$$E(X) \ge E(XI_A) \ge \epsilon E(I_A) = \epsilon P(A).$$

Noting that $P(A)$ is the proper notation for $P(X \ge \epsilon)$ we have the standard result for the general case.

q.e.d.

Corollary 8.2. If ξ is an unrestricted random variable with mean μ and variance σ^2, then

$$P(|\xi - \mu| \ge \epsilon) \le \frac{\sigma^2}{\epsilon^2}.$$

---

[6]It is a common practice that **the subscript for the first moment is omitted for both scalar and vector random variables.** We shall follow this practice unless reasons of clarity require otherwise.

Proof: Let $X = | \, \xi - \mu \, |^2$ and note that $E(X) = \sigma^2$; by Proposition 8.12, $P(X \geq \epsilon^2) \leq \sigma^2/\epsilon^2$. Next, consider the sets $A = \{\omega : X \geq \epsilon^2\}$ and $A^* = \{\omega : X^{1/2} \geq \epsilon\}$. We shall show that $A = A^*$. This is so since, if $\omega \in A$, then we must, also, have that $[X(\omega)]^{1/2} \geq \epsilon$, so that $\omega \in A^*$, which shows that $A \subset A^*$; similarly, if $\omega \in A^*$, then we must have that $X(\omega) \geq \epsilon^2$, so that $\omega \in A$, which shows that $A^* \subset A$. The latter, in conjunction with the earlier result, implies $A = A^*$ and thus, $P(A) = P(A^*)$.

q.e.d.

Corollary 8.3. Let ξ be a vector random variable with mean μ and covariance matrix Σ. Then

$$P(\| \, \xi - \mu \, \|^2 \geq \epsilon^2) \leq \frac{\mathrm{tr}\Sigma}{\epsilon^2}.$$

Proof: Note that $\| \, \xi - \mu \, \|^2 = X$ is a nonnegative integrable random variable; hence by Proposition 8.12,

$$P(X \geq \epsilon^2) \leq \frac{\mathrm{tr}\Sigma}{\epsilon^2},$$

where, evidently, $\mathrm{tr}\Sigma = E(\| \, \xi - \mu \, \|^2)$.

q.e.d.

Proposition 8.13 (Cauchy Inequality). Let X_i, $i = 1, 2$, be zero mean random variables and suppose that $\mathrm{var}(X_i) = \sigma_{ii} \in (0, \infty)$; then

$$(E \, | \, X_1 X_2 \, |)^2 \leq \sigma_{11}\sigma_{22}.$$

Proof: Since $\sigma_{ii} > 0$, $i = 1, 2$, define the variables

$$\xi_i = \frac{X_i}{\sigma_{ii}^{\frac{1}{2}}},$$

and note that $\mathrm{var}(\xi_i) = 1$. Moreover, since

$$(| \, \xi_1 \, | - | \, \xi_2 \, |)^2 \geq 0,$$

we have that

$$2E(| \, \xi_1 \xi_2 \, |) \leq E(\xi_1^2 + \xi_2^2).$$

But this implies $E(| \, X_1 X_2 \, |) \leq (\sigma_{11}\sigma_{22})^{1/2}$.

q.e.d.

Corollary 8.4. The correlation between any two square integrable random variables, lies in $[-1, 1]$.

Proof: Let X_i, $i = 1, 2$, be any two square integrable random variables as in the proposition and put $\sigma_{12} = E(X_1 X_2) = \text{Cov}(X_1, X_2)$. We recall that the correlation (or correlation coefficient) between two random variables is given by

$$\rho_{12} = \frac{\sigma_{12}}{(\sigma_{11}\sigma_{22})^{1/2}}.$$

Since from Proposition 8.13 we have $\sigma_{12}^2 \leq \sigma_{11}\sigma_{22}$, the result follows immediately.

q.e.d.

Proposition 8.14 (Jensen's Inequality). Let h be a measurable function

$$h : R^n \longrightarrow R$$

and X be an integrable random vector, i.e., $\| E(X) \| < \infty$; then

i. If h is a convex function, $h[E(X)] \leq E(h[X])$;

ii. If h is a concave function, $h[E(X)] \geq E(h[X])$.

Proof: If h is a convex function,[7] then we can write for any point x_0 [8]

$$h(x) \geq h(x_0) + s(x_0)(x - x_0),$$

where x is an appropriate (column) vector valued function. Consequently, for $x = X$ and $x_0 = E(X)$, we have the proof of i, upon taking expectations. As for ii, we note that if h is a convex function then $-h$ is concave. The validity of ii is, then, obvious.

q.e.d.

Proposition 8.15 (Liapounov's Inequality). Let X be a suitably integrable random variable and $0 < s < r$ be real numbers; then

$$(E \mid X \mid^s)^{\frac{1}{s}} \leq (E \mid X \mid^r)^{\frac{1}{r}}.$$

Proof: Define $\mid X \mid^s = \xi$, and consider ξ^k, where $k = (r/s)$; since ξ^k is a convex function, by Jensen's inequality, we have

$$[E(\xi)]^k \leq E(\xi^k).$$

---

[7]For twice differentiable convex functions the matrix of the second order partial derivatives is positive semidefinite; for concave functions, it is negative semidefinite; in both cases this is to be understood in an *a.e.* sense.

[8]This result is obtained by using a Taylor series expansion around the point x_0, retaining terms up to and including the second derivative (third term of the expansion).

Now, reverting to the original notation this gives the result

$$(E \mid X \mid^s)^{\frac{1}{s}} \leq (E \mid X \mid^r)^{\frac{1}{r}}.$$

q.e.d.

Corollary 8.5. Let X be a suitably integrable random variable; then for any integer n, for which $\mid X \mid^n$ is integrable,

$$E \mid X \mid \leq (E \mid X \mid^2)^{\frac{1}{2}} \leq \ldots \ldots \leq (E \mid X \mid^n)^{\frac{1}{n}}.$$

Proof: Obvious by repeated application of Liapounov's inequality.

q.e.d.

Proposition 8.16 (Holder's Inequality). Let X_i, $i = 1, 2$, be suitably integrable random variables and let $p_i \in (1, \infty)$, such that $(1/p_1) + (1/p_2) = 1$. Then, provided the $\mid X_i \mid^{p_i}$ are integrable,

$$E \mid X_1 X_2 \mid \leq (E \mid X_1 \mid^{p_1})^{\frac{1}{p_1}} (E \mid X_2 \mid^{p_2})^{\frac{1}{p_2}}.$$

Proof: Evidently, if $E \mid X_i \mid^{p_i} = 0$, then $X_i = 0$, a.c., and consequently the result of the proposition is valid. Thus, we suppose that $E \mid X_i \mid^{p_i} > 0$. Define, now $\xi_i = \mid X_i \mid /c_i$, where $c_i = (E \mid X_i \mid^{p_i})^{1/p_i}$, $i = 1, 2$. Since the logarithm is a concave function, it is easy to show that

$$\ln[ax + by] \geq a \ln x + b \ln y$$

for $x, y, a, b > 0$ such that $a + b = 1$. But this means that the following inequality is also valid

$$x^a y^b \leq ax + by.$$

Applying this inequality with $x = \xi_1^{p_1}$, $y = \xi_2^{p_2}$, $a = 1/p_1$, $b = 1/p_2$, we find

$$E(\xi_1 \xi_2) \leq \frac{1}{p_1} E(\xi_1^{p_1}) + \frac{1}{p_2} E(\xi_2^{p_2})$$

and, reverting to the original notation, we have

$$E \mid X_1 X_2 \mid \leq (E \mid X_1 \mid^{p_1})^{\frac{1}{p_1}} (E \mid X_2 \mid^{p_2})^{\frac{1}{p_2}}.$$

q.e.d.

Proposition 8.17 (Minkowski's Inequality). Let X_i, $i = 1, 2$, be random variables, $p \in (1, \infty)$ such that $\mid X_i \mid^p$ is integrable. Then

$$\mid X_1 + X_2 \mid^p \text{ is integrable}$$

and, moreover,

$$(E \mid X_1 + X_2 \mid^p)^{\frac{1}{p}} \leq (E \mid X_1 \mid^p)^{\frac{1}{p}} + (E \mid X_2 \mid^p)^{\frac{1}{p}}.$$

Proof: Consider the function $F(x) = (x+a)^p - 2^{p-1}(x^p + a^p)$ and note that $F'(x) = p(x+a)^{p-1} - 2^{p-1}px^{p-1}$. From this we easily deduce that the function has a maximum at $x = a$, provided a and x are restricted to be positive. But we note that $F'(a) = 0$; consequently, we have

$$(x+a)^p \leq 2^{p-1}(x^p + a^p).$$

Since

$$(\mid X_1 + X_2 \mid)^p \leq (\mid X_1 \mid + \mid X_2 \mid)^p \leq 2^{p-1}(\mid X_1 \mid^p + \mid X_2 \mid^p),$$

the validity of the first part of the proposition is evident. For the second part, note that

$$(\mid X_1 + X_2 \mid)^p \leq \mid X_1 \mid \mid X_1 + X_2 \mid^{p-1} + \mid X_2 \mid \mid X_1 + X_2 \mid^{p-1}.$$

Applying Holder's inequality to the two terms of the right member above, we find

$$E(\mid X_1 \mid \mid X_1 + X_2 \mid^{p-1}) \leq (E \mid X_1 \mid^p)^{\frac{1}{p}}(E \mid X_1 + X_2 \mid^{(p-1)q})^{\frac{1}{q}}$$

$$E(\mid X_2 \mid \mid X_1 + X_2 \mid^{p-1}) \leq (E \mid X_2 \mid^p)^{\frac{1}{p}}(E \mid X_1 + X_2 \mid^{(p-1)q})^{\frac{1}{q}},$$

where q is such that $(1/p) + (1/q) = 1$; this being so, note that $(p-1)q = p$. Adding the two inequalities above we find

$$(E \mid X_1 + X_2 \mid^p)^{\frac{1}{p}} \leq (E \mid X_1 \mid^p)^{\frac{1}{p}} + (E \mid X_2 \mid^p)^{\frac{1}{p}}.$$

q.e.d.

8.4 Conditional Probability

8.4.1 Conditional Probability in Discrete Models

The reader is no doubt familiar with the general notion of conditional probability. Thus, for example, if $(\Omega, \mathcal{A}, P)$ is a probability space and $A, B \in \mathcal{A}$, then the conditional probability of A, given B is given by the relation

$$P(A \mid B) = \frac{P(A \cap B)}{P(B)},$$

provided $P(B) \neq 0$. The underlying principle is that by conditioning **we shift the frame of reference, from the general space Ω to the conditioning entity**, in this case the event B. Thus, the probability of the event A given the event B is the probability assigned by the (probability) measure P to that part of A which is also in the new frame of reference, viz., B ; this, of course, is simply the intersection $A \cap B$; division by $P(B)$ is simply a bookkeeping device to ensure that the probability (measure) assigned to the "new" space is unity. This basic idea is easily transferable to discrete random variables but its extension to general random variables, i.e., the case where the conditioning entity is a σ-algebra, is somewhat less transparent; conditional probability is also one of the most frequently misunderstood concepts in econometrics.

We begin by considering conditional probability in the case of discrete probability models. First, an informal discussion by example. Suppose we have two random variables, X_i, $i = 1, 2$, which are independent, identically distributed and assume the values, $1, 2, \ldots, 6$, with equal probability, viz., $1/6$. Define, a new random variable, say $X = X_1 + X_2$. This is, evidently, the two independent dice model we had discussed earlier, and the random variable X is simply the sum of the faces showing at each throw; it is clear that X assumes the values $2, 3, \ldots, 12$ with the probabilities determined at an earlier stage. We note that if we condition on X_2, say by requiring that $X_2 = i$, $1 \le i \le \min(k - 1, \ 6))$, then

$$P(X = k \mid X_2 = i) = P(X_1 = k - i).$$

The preceding has resolved the problem of calculating the probability that X will assume a specific value given that X_2 assumes a specific value, which is really a special case of conditioning one "event" in terms of another. But what would we want to mean by $X \mid X_2$, i.e. by conditioning the random variable X in terms of the random variable X_2. Clearly, if we can determine the probability with which this new variable "assumes" values in its range, subject to the condition, we would have accomplished our task. Now, define the sets

$$D_{ij} = \{\omega_j : X_j(\omega_j) = i\}, \ i = 1, 2, \ldots, 6, \text{ and } j = 1, 2,$$

and denote the indicator functions of the sets D_{ij} by $I_{ij}(\omega)$. Both collections of sets just defined are **finite partitions of the spaces** $\Omega_j, j = 1, 2$, **respectively**. The sample space of X is $\Omega = \Omega_1 \times \Omega_2$ [9]; the probability defined thereon is the product of the two probabilities, but since the two spaces are independent we shall, for notational convenience, use the symbol P to denote the probability measure on all three spaces. Next, construct the finite partition of the space Ω by defining

$$D = (D_2, D_3, \ldots, D_{10}, D_{11}, D_{12}),$$

---

[9]Note that Ω_j, $j = 1, 2$, are exact copies of the same space.

where $D_k = \{\omega: \quad X(\omega) = k.\}$ If we were dealing solely with discreet models, then perhaps there would be no need to abstract the salient aspects of the problem and seek to generalize the solution obtained. Unfortunately, however, this is not the case. By way of motivation consider the following "experiment": choose a point, x, at "random" in the unit interval $(0, 1)$; then toss a coin whose "probability" of showing heads is x and of showing tails is $1 - x$. If we engage in such an experiment n times what is the conditional probability that exactly k of the tosses result in heads, conditional on the fact that the probability (of heads) is x. Since, in the context of the uniform distribution over the unit interval, the probability of choosing x is zero, the usual approach fails. Nonetheless, it makes perfectly good sense to expect that the required probability is given by the binomial distribution with $p = x$. More formally, the elementary definition of conditional probability of an event A, given another, B, requires that $P(B) > 0$. We have just given a very real problem in which this condition is violated. Thus, there is need for abstraction and we now turn to this task.

A careful examination of the solution we have given to the two dice example indicates that the random variables as such do not play a direct intrinsic role; rather, the result depends on certain collections of sets which are, of course, determined by the random variables in question. From the discussion above, the (unconditional) probability that $X = k$ is evidently $P(D_k)$ and since X is a discrete random variable it has the representation

$$X(\omega) = \sum_{k=2}^{12} k I_k(\omega),$$

where $I_k(\omega)$ is the indicator function of the set D_k. Thus, the unconditional expectation of X is evidently given by

$$EX(\omega) = \sum_{k=2}^{12} k E I_k(\omega) = \sum_{k=2}^{12} k P(D_k).$$

Now what would we want to mean by the **conditional expectation** $E(X \mid X_2)$? By the previous discussion, and what we did above, we should define the random variable

$$X \mid X_2 = \sum_{k=2}^{12} \{ \sum_{s+r=k} (s+r) I_{s1}(\omega) I_{r2}(\omega) \},$$

where $I_{ij}(\omega)$ are the indicator functions of the sets D_{ij}, $i = 1, 2, \ldots, 6$, $j = 1, 2$, defined above. Thus the probability of the event $X = k \mid X_2 = j$ is given by

$$P(X = k \mid X_2 = j) = E I_{k-j,1}(\omega) I_{j2}(\omega) = P(D_{k-j,1}) P(D_{j,2}),$$

and consequently

$$E(X \mid X_2) = \sum_{k=2}^{12} k\{ \sum_{s+j=k} P(D_{s1})EI_{j2}(\omega)\}.$$

Let us now examine the abstract elements of this procedure. First, note that the discrete random variable X_2, i.e. the conditioning variable, gives rise to the collection of sets: $\mathcal{D}_2 = \{D_{i2} : i = 1, 2, \ldots 6\}$. Moreover, note that this is a **partition** of the space Ω_2, in the sense that the D_{i2} are **disjoint** and $\bigcup_{i=1}^{6} D_{i2} = \Omega_2$. Notice also that there is another partition, of Ω_1, in terms of the collection $\mathcal{D}_1 = \{D_{i1} : i = 1, 2, \ldots, 6\}$, and in fact that X is defined over the space $\Omega = \Omega_1 \times \Omega_2$; the σ-algebra of this space is generated by sets of the form $D_{i1} \times D_{j2}$ with $i, j = 1, 2, \ldots, 6$; the probability measure on this space is simply the product (probability) measure. Thus, we can express the random variable X as

$$X(\omega) = \sum_{k=2}^{12} kI_k(\omega)$$

where I_k is the indicator set of D_k. Since by definition

$$EX = \sum_{k=2}^{12} kP(D_k)$$

it is natural, and it conforms to the earlier discussion, to define

$$E(X \mid X_2 = j) = \sum_{k=2}^{12} kP(D_k \mid X_2 = j).$$

But of course $P(D_k \mid X_2 = j) = P(D_k \mid D_{j2})$. In this context, what would we want to mean by the notation $P(D_k \mid X_2)$? The notation alludes to the probability to be assigned to the set D_k **given** or **conditionally on** the variable X_2. Since the latter is a random variable, so should be the former. Moreover, for every value assumed by the random variable X_2, say j, a corresponding value ought to be assumed by $P(D_k \mid X_2)$, viz., $P(D_k \mid D_{j2})$. This leads us to define[10]

$$P(D_k \mid X_2) = \sum_{j=1}^{6} P(D_k \mid D_{j2})I_{j2}(\omega),$$

---

[10]To connect this with the earlier discussion of the topic note that

$$P(D_k \mid D_{j2}) = \frac{P(D_k \cap D_{j2})}{P(D_{j2})} = P(D_{k-j,1} \cap D_{j2})/P(D_{j2}).$$

again it being understood that I_{j2} is the indicator function of the set D_{j2} and the conditional probability is defined to be zero whenever $k < j + 1$. With the help of these redefinitions of the steps we had taken earlier based on elementary probability considerations, we can now write

$$E(X \mid X_2) = \sum_{k=2}^{12} kP(D_k \mid X_2) = \sum_{k=2}^{12} k \left\{ \sum_{s+j=k} P(D_{s-j,1})I_{j2}(\omega) \right\}.$$

In addition, note that in some sense the notion of conditional expectation is somewhat more "fundamental" than the notion of conditional probability, in the sense that conditional probability can always be expressed as conditional expectation. For example, given any set A, we may define its indicator set, to be the random variable associated with it; by analogy with the standard definition we may then set

$$E(A \mid X_2) = \sum_{j=1}^{m} P(A \mid D_{j2})I_{j2}(\omega),$$

as the conditional expectation of A with respect to the random variable X_2. In this framework, what would one want to mean by the conditional probability of $A \mid X_2$? Presumably one would want the random variable that rearranges the mass assigned to A, over the constituent (elementary) sets of the partition induced by X_2. This is simply

$$P(A \mid X_2) = \sum_{j=1}^{6} P(A \mid D_{j2})I_{j2}(\omega).$$

If we take the expectation of this random variable (i.e. if we take the expectation of the conditional probability) we have the standard formula for what is known as **total probability**. Specifically, we have

$$E[P(A \mid X_2)] = \sum_{j=1}^{6} P(A \mid D_{j2})P(D_{j2}) = P(A),$$

and, moreover,

$$E[E(A \mid X_2)] = P(A),$$

which exhibits the probability of an event A, as the expectation of a conditional probability, as well as the expectation of a conditional expectation. We now undertake the formal development of the subject.

Definition 8.10. Let $(\Omega, \mathcal{A}, P)$ be a probability space, $A \in \mathcal{A}$ and $\mathcal{D} = \{D_i : i = 1, 2, \ldots, n\}$ be a finite partition of Ω (i.e., the D_i are disjoint sets whose union is Ω). The conditional probability of the "event" A given the partition $\mathcal{D}$ is defined by

$$P(A \mid \mathcal{D}) = \sum_{i=1}^{n} P(A \mid D_i) I_i(\omega)$$

where, evidently, I_i is the indicator function of D_i.

Remark 8.16. It is important to realize just what the operation of conditioning with respect to a partition involves. It does not involve holding anything "constant" although this notion is useful in operations involving integration. What it does involve conceptually, however, is the rearrangement of the probability assigned to an event (or more generally a random variable) in terms of the conditioning entity. In terms of the definition above, the event A has probability in terms of the measure assigned to A by P, in the context of the space Ω. Conditioning in terms of the partition $\mathcal{D}$ shifts attention from Ω to $\mathcal{D}$. In this new framework the probability of A is distributed over the constituent parts (i.e., the sets D_i of $\mathcal{D}$) and the **random variable** $P(A \mid \mathcal{D})$ takes on the value $P(A \mid D_i)$, whenever $\omega \in (D_i \cap A)$. Notice, further that the expectation of this random variable yields the probability of the event A! Specifically, since $E(I_i) = P(D_i)$, we easily establish

$$E[P(A \mid \mathcal{D})] = \sum_{i=1}^{n} P(A \cap D_i) = P(A).$$

Thus, loosely speaking, **conditioning an event A in terms of a partition $\mathcal{D}$ means distributing the probability assigned to A over the constituent elements (sets) of the partition.** It is evident from the definition above, that the conditional probability of an event A, with respect to a decomposition is a simple random variable which assumes a constant value over the elements of the decomposition, this value being simply the conditional probability of A given the element in question, say D_i. It follows also immediately that

$$P(A \mid \Omega) = P(A),$$

and that if A, B are two disjoint sets, then

$$P(A \cup B \mid \mathcal{D}) = P(A \mid \mathcal{D}) + P(B \mid \mathcal{D}).$$

Remark 8.17. It is now straightforward to apply to random variables the notion of conditional expectation with respect to a partition. Thus, let $\mathcal{D}$ be a partition as above, and let X be a discrete (simple) random variable, say

$$X(\omega) = \sum_{j=1}^{m} x_j I_j(\omega),$$

where I_j is the indicator function of the set $B_j = \{\omega : X(\omega) = x_j\}$, $j = 1, 2, \ldots, m$. Since

$$E(X) = \sum_{j=1}^{m} x_j P(B_j),$$

it is natural to define the conditional expectation of X with respect to the partition $\mathcal{D}$ as

$$E(X \mid \mathcal{D}) = \sum_{j=1}^{m} x_j P(B_j \mid \mathcal{D}).$$

Notice that, as we have sought to explain above, the operation of conditioning the expectation with respect to a partition simply involves the rearrangement of the probability mass of the random variable X, in terms of the conditioning entity. Particularly, the random variable X, originally, assumes constant values over the sets B_j and these values are x_j, respectively. Its conditional expectation with respect to $\mathcal{D}$, on the other hand, redistributes (or perhaps one should say rearranges) its probability mass over the constituent elements of the partition, so that $E(X \mid \mathcal{D})$ assumes constant values over the (elementary) constituent sets of $\mathcal{D}$, say D_i, $i = 1, 2, \ldots, n$; these values are, respectively, $\sum_{j=1}^{m} x_j P(B_j \mid D_i)$. It should also be apparent that taking a second expectation, over the partition, restores us to the basic notion of expectation, since the expectation of the indicator functions of the elements of the partition yields $P(D_i)$. This discussion introduces

Definition 8.11. Let $(\Omega, \mathcal{A}, P)$ be a probability space, X a random variable defined thereon and $\mathcal{D}$ a finite partition of Ω as above. Then, the conditional expectation of X with respect to the partition $\mathcal{D}$ is given by

$$E(X \mid \mathcal{D}) = \sum_{j=1}^{m} x_j P(B_j \mid \mathcal{D}) = \sum_{i=1}^{n} E(X \mid D_i).$$

Finally, we formally introduce two common terms.

Definition 8.12. Let X be a (simple) random variable on the probability space above and define the sets

$$D_i = \{\omega : X(\omega) = x_i\}.$$

The collection

$$\mathcal{D}_x = \{D_i : i = 1, 2, \ldots, n\}$$

is a finite partition (or a decomposition) of the space Ω and is said to be the decomposition (or partition) **induced by the random variable** X.

Definition 8.13. Let X be a random variable as in Definition 8.11 and suppose $\mathcal{D}$ is a finite partition of Ω; we say that X is $\mathcal{D}$-measurable, if and only if $\mathcal{D}$ is finer than $\mathcal{D}_x$, i.e. if X can be represented as

$$X(\omega) = \sum_{j=1}^{r} z_j I_j(\omega)$$

where some of the z_j may be repeated, and I_j, $j = 1, 2, \ldots, r$ are the indicator functions of the sets of $\mathcal{D}$.

The following elementary properties for conditional expectations, follow almost immediately.

Proposition 8.18. Let $(\Omega, \mathcal{A}, P)$ be a probability space, X_i, $i = 1, 2$, be discrete (simple) random variables, define the sets

$$D_{i1} = \{\omega : X_1(\omega) = x_{i1}, \ i = 1, 2, \ldots, n\}$$

$$D_{j2} = \{\omega : X_2(\omega) = x_{j2}, \ j = 1, 2, \ldots, m\},$$

and note that the collections, $\mathcal{D}_1 = \{D_{i1} : i = 1, 2, \ldots, n\}$, $\mathcal{D}_2 = \{D_{j2} : j = 1, 2, \ldots, m\}$ are the finite partitions of Ω induced by the variables X_i, $i = 1, 2$, respectively. Suppose, further, that $\mathcal{D}_2$ is a finer partition than $\mathcal{D}_1$, in the sense that every set in $\mathcal{D}_1$ can be expressed as a union of sets in $\mathcal{D}_2$. Then the following statements are true:

  i. $E(X_i \mid X_i) = X_i$, $i = 1, 2$;

  ii. $E[E(X_1 \mid X_2)] = E(X_1)$;

  iii. If X_3 is another simple random variable with an induced partition $\mathcal{D}_3$, then
$$E(aX_1 + bX_2 \mid X_3) = aE(X_1 \mid X_3) + bE(X_2 \mid X_3);$$

  iv. If X_3 is any other random variable, as in iii, i.e. with an induced partition $\mathcal{D}_3$, then, given $\mathcal{D}_2$ is finer than $\mathcal{D}_1$,
$$E[E(X_3 \mid X_2) \mid X_1] = E(X_3 \mid X_1);$$

  v. Let $\mathcal{D}$ be a decomposition of Ω, and suppose it, $\mathcal{D}$, is finer than $\mathcal{D}_1$; let X_3 be another random variable as in iv., i.e. with an induced partition $\mathcal{D}_3$, then
$$E(X_1 X_3 \mid \mathcal{D}) = X_1 E(X_3 \mid \mathcal{D}).$$

Proof: By definition $E(X_1 \mid X_1) = \sum_{j=1}^{n} \sum_{i=1}^{n} x_{i1} P(D_{i1} \mid D_{j1}) I_{j1}(\omega)$. Since the constituent sets of a partition are disjoint, $P(D_{i1} \mid D_{j1}) = 0$, if $i \neq j$ and is equal to unity if $i = j$. Thus, we may write

$$E(X_1 \mid X_1) = \sum_{i=1}^{n} x_{i1} I_{i1}(\omega) = X_1.$$

The proof for X_2 is entirely similar.

The proof of ii is as follows: by definition

$$E(X_1 \mid X_2) = \sum_{j=1}^{m} E(X_1 \mid D_{j2}) I_{j2}(\omega).$$

Consequently, $E[E(X_1 \mid X_2)] = \sum_{j=1}^{m} E(X_1 \mid D_{j2}) P(D_{j2})$. But

$$E(X_1 \mid D_{j2}) P(D_{j2}) = \sum_{i=1}^{n} x_{i1} P(D_{i1} \mid D_{j2}) P(D_{j2}).$$

Thus,

$$
\begin{aligned}
E[E(X_1 \mid X_2)] &= \sum_{i=1}^{n} x_{i1} \sum_{j=1}^{m} P(D_{i1} \cap D_{j2}) \\
&= \sum_{i=1}^{n} x_{i1} P(D_{i1}) = E(X_1)
\end{aligned}
$$

which concludes the proof of ii; the result above may be looked upon as a generalization of the formula for total probability.

The proof of iii is immediate from the definition of conditional expectation. To prove iv we note that by definition

$$E[E(X_3 \mid X_2) \mid X_1] = \sum_{s=1}^{n} E[E(X_3 \mid X_2) \mid D_{i1}] I_{i1}(\omega).$$

Since

$$E(X_3 \mid X_2) = \sum_{j=1}^{m} E(X_3 \mid D_{j2}) I_{j2}(\omega)$$

we conclude that

$$E[E(X_3 \mid X_2) \mid D_{i1}] = \sum_{j=1}^{m} \sum_{s=1}^{k} x_{s3} P(D_{s3} \mid D_{j2}) P(D_{j2} \mid D_{i1}).$$

Since $\mathcal{D}_2$ is finer than $\mathcal{D}_1$, it follows that $P(D_{j2} \cap D_{i1})/P(D_{j2})$ is either 1, when D_{j2} is one of the sets that make up D_{i1}, or else it is zero. Hence, summing over those indices, j, for which $D_{j2} \subset D_{i1}$, we may rewrite the right member of the equation above as

$$\sum_{s=1}^{k} x_{s3} P(D_{s3} \mid D_{i1}) = E(X_3 \mid D_{i1}).$$

Thus,

$$E[E(X_3 \mid X_2) \mid X_1] = \sum_{i=1}^{n} E(X_3 \mid D_{i1})I_{i1}(\omega) = E(X_3 \mid X_1).$$

which proves iv, thus showing that the conditioning over the coarser partition prevails.

The proof for v is entirely similar; thus, let D_i, $i = 1, 2, \ldots, r$, be the elementary sets of $\mathcal{D}$; since X_1 is $\mathcal{D}$-measurable it has the representation

$$X_1(\omega) = \sum_{i=1}^{r} y_{i1} I_i(\omega),$$

where some of the y_{i1} may be repeated and I_i, $i = 1, 2, \ldots, r$, are the indicator functions of the elementary sets of $\mathcal{D}$. By definition, then we have

$$E(X_1 X_3 \mid \mathcal{D}) = \sum_{i=1}^{r} E(X_1 X_3 \mid D_i)I_i(\omega).$$

But,

$$E(X_1 X_3 \mid D_i) = \sum_{m=1}^{r} \sum_{j=1}^{k} y_{m1} x_{j3} P(D_m \cap D_{j3} \mid D_i).$$

In view of the fact that the elementary sets of a partition are disjoint, we must have

$$E(X_1 X_3 \mid D_i) = y_{i1} \sum_{j=1}^{k} P(D_{j3} \mid D_i) = y_{i1} E(X_3 \mid D_i).$$

Thus, we conclude that

$$E(X_1 X_3 \mid \mathcal{D}) = \sum_{i=1}^{r} y_{i1} I_i(\omega) E(X_3 \mid D_i).$$

On the other hand, using again the disjointness of the elementary sets of $\mathcal{D}$, we have that $I_i I_m = 0$, if $m \neq i$, and is equal to I_i, for $i = m$. Consequently,

$$E(X_1 X_3 \mid \mathcal{D}) = \sum_{i=1}^{r} y_{i1} I_i(\omega) \sum_{m=1}^{r} E(X_3 \mid D_m) I_m(\omega),$$

or that $E(X_1 X_3 \mid \mathcal{D}) = X_2 E(X_3 \mid \mathcal{D})$.

q.e.d.

8.4.2 Conditional Probability in Continuous Models

In this section we shall extend the notions of conditional probability and conditional expectation to continuous models; or, more precisely, we shall examine the concepts of conditioning with respect to a σ-algebra. We begin with

Definition 8.14. Let $(\Omega, \mathcal{A}, P)$ be a probability space, let X be a (nonnegative extended) random variable defined thereon and let $\mathcal{G}$ be a σ-algebra contained in $\mathcal{A}$. The conditional expectation of X with respect to the σ-algebra $\mathcal{G}$, denoted by $E(X \mid \mathcal{G})$ is a (nonnegative extended) random variable such that

i. $E(X \mid \mathcal{G})$ is $\mathcal{G}$-measurable;

ii. For every set $B \in \mathcal{G}$

$$\int_B X \, dP = \int_B E(X \mid \mathcal{G}) \, dP.$$

Remark 8.18. When dealing with extended random variables, the question always arises as to when expectations exist. This is true as much in the standard case as it is in the case of conditional expectations. Thus, let X be a random variable in the context of the definition above, except that we do not insist that it is nonnegative. How do we know that conditional expectation, as exhibited above, is well defined? Note that, since we are dealing with nonnegative random variables, the problem is not that some expectation (integral) is unbounded, but rather whether the definition leads to one of the indeterminate forms, such as, e.g., $\infty - \infty$. This is resolved by the convention: The conditional expectation of any random variable X, with respect to the σ-algebra $\mathcal{D}$, which is contained in $\mathcal{A}$, exists if and only if

$$\min(E(X^+ \mid \mathcal{G}), \ E(X^- \mid \mathcal{G})) < \infty.$$

It is evident that the definition of conditional expectation above is not a vacuous one. In particular, note that the Radon-Nikodym (RN) theorem of the previous chapter guarantees that the conditional expectation exists. Thus, recall that setting

$$Q(B) = \int_B X \, dP,$$

where X is a nonnegative random variable and $B \in \mathcal{G}$, we have that Q is a measure which is absolutely continuous with respect to P. The RN theorem then asserts the existence of a $\mathcal{G}$-measurable function, $E(X \mid \mathcal{G})$, unique up to sets of P-measure zero such that

$$Q(B) = \int_B E(X \mid \mathcal{G}) \, dP,$$

which, therefore, establishes the existence of conditional expectation; its salient properties are given below in

Proposition 8.19. Let X, X_i, $i = 1, 2$, be random variables defined on the probability space $(\Omega, \mathcal{A}, P)$; let $\mathcal{G}$, $\mathcal{G}_i$, $i = 1, 2$ be σ-(sub)algebras contained in $\mathcal{A}$, and suppose all random variables are extended and that their expectations exist. Then the following statements[11] are true:

   i. If K is a constant, and $X_1 = K$, a.c., then $E(X \mid \mathcal{G}) = K$;

   ii. If $X_1 \leq X_2$, then $E(X_1 \mid \mathcal{G}) \leq E(X_2 \mid \mathcal{G})$;

   iii. For any random variable X, $\mid E(X \mid \mathcal{G}) \mid \leq E(\mid X \mid \mid \mathcal{G})$;

   iv. For any scalars, a_i, $i = 1, 2$, such that $\sum_{i=1}^{2} a_i E(X_i)$ is defined,

$$E(a_1 X_1 + a_2 X_2 \mid \mathcal{G}) = a_1 E(X_1 \mid \mathcal{G}) + a_2 E(X_2 \mid \mathcal{G});$$

   v. $E(X \mid \mathcal{A}) = X$;

   vi. If $\mathcal{G}_0 = (\emptyset, \Omega)$, $E(X \mid \mathcal{G}_0) = E(X)$;

   vii. If X is a random variable which is independent of the σ-algebra $\mathcal{G}$. Then,

$$E(X \mid \mathcal{G}) = E(X);$$

   viii. If Y is $\mathcal{G}$-measurable, with $E(\mid Y \mid) < \infty$, $E(\mid X \mid) < \infty$, then

$$E(YX \mid \mathcal{G}) = Y E(X \mid \mathcal{G});$$

   ix. If $\mathcal{G}_1 \subseteq \mathcal{G}_2$, (i.e., if $\mathcal{G}_2$ is finer), then

$$E[E(X \mid \mathcal{G}_2) \mid \mathcal{G}_1] = E(X \mid \mathcal{G}_1);$$

   x. $\mathcal{G}_2 \subseteq \mathcal{G}_1$, (i.e., if $\mathcal{G}_1$ is finer), then

$$E[E(X \mid \mathcal{G}_2) \mid \mathcal{G}_1] = E(X \mid \mathcal{G}_2);$$

   xi. $E[E(X \mid \mathcal{G})] = E(X)$.

Proof: To prove i we note that $X = K$ is both $\mathcal{A}$- and $\mathcal{G}$-measurable since it can be given the trivial representation $X(\omega) = K I(\omega)$, where I is the indicator function of Ω. Hence for any set $B \in \mathcal{G}$

$$\int_B X \, dP = \int_B E(X \mid \mathcal{G}) \, dP \quad \text{implies} \quad K P(B) = K P(B),$$

---

[11]These statements are to be understood in the a.c. sense, where appropriate.

which completes the proof of i.

To prove ii we note that, for any set $A \in \mathcal{G}$,

$$\int_A E(X_1 \mid \mathcal{G}) \, dP = \int_A X_1 \, dP \leq \int_A X_2 \, dP = \int_A E(X_2 \mid \mathcal{G}) \, dP,$$

which implies that

$$E(X_1 \mid \mathcal{G}) \leq E(X_2 \mid \mathcal{G}).$$

To see that, define

$$C = \{\omega : E(X_1 \mid \mathcal{G}) > E(X_2 \mid \mathcal{G})\}.$$

If $P(C) = 0$, then the proof of ii is complete; if not, consider

$$\int_C [E(X_1 \mid \mathcal{G}) - E(X_2 \mid \mathcal{G})] \, dP,$$

which is unambiguously positive; this is a contradiction and, hence, $P(C) = 0$, which proves ii.

To prove iii we note that for any random variable, X, $-\mid X \mid \leq X \leq \mid X \mid$; consequently, by ii we have

$$-E(\mid X \mid \mid \mathcal{G}) \leq E(X \mid \mathcal{G}) \leq E(\mid X \mid \mid \mathcal{G}),$$

which completes the proof of iii.

To prove iv we note that, by definition, for any set $A \in \mathcal{G}$

$$a_1 \int_A X_1 \, dP = a_1 \int_A E(X_1 \mid \mathcal{G}) \, dP$$

$$a_2 \int_A X_2 \, dP = a_2 \int_A E(X_2 \mid \mathcal{G}) \, dP.$$

Summing, and using the fundamental properties of the integral, we establish the validity of iv.

The proof of v is trivial since, evidently, X is $\mathcal{A}$-measurable; consequently, for **every** $A \in \mathcal{A}$

$$\int_A X \, dP = \int_A E(X \mid \mathcal{A}) \, dP.$$

But this means that $E(X \mid \mathcal{A}) = X$.

For vi we note that the integral of any measurable function over the null set is zero and, moreover,

$$\int_\Omega X \, dP = \int_\Omega E(X) \, dP.$$

An argument similar to that used in connection with the proof of ii, will then show the validity of vi.

To prove vii we note that $E(X)$ is $\mathcal{G}$-measurable, and using the fundamental definition of independence (which will be also be discussed in the next section) we establish for any $A \in \mathcal{G}$

$$\int_A X \, dP = \int_\Omega I_A X \, dP = E(I_A) E(X) = P(A) E(X) = \int_A E(X) \, dP.$$

The proof of viii is as follows: clearly, $Y E(X \mid \mathcal{G})$ is $\mathcal{G}$-measurable; let $B \in \mathcal{G}$, $Y = I_B$, the latter being the indicator function of B, and let A be any set in $\mathcal{G}$. Then,

$$\int_A Y X \, dP = \int_{A \cap B} X \, dP = \int_{A \cap B} E(X \mid \mathcal{G}) \, dP = \int_A I_B E(X \mid \mathcal{G}) \, dP.$$

Hence,

$$E(Y X \mid \mathcal{G}) = Y E(X \mid \mathcal{G})$$

for $Y = I_B$, $B \in \mathcal{G}$; consequently, the result holds for nonnegative simple $\mathcal{G}$-measurable random variables. Thus, by the Lebesgue dominated convergence theorem (Proposition 7.15 of Chap. 7), if Y is a nonnegative random variable and $\{Y_n : n \geq 1, Y_n \leq Y\}$ is a sequence of simple random variables converging to Y, we have that

$$\lim_{n \to \infty} E(Y_n X \mid \mathcal{G}) = E(Y X \mid \mathcal{G}), \text{ a.c.}$$

Moreover, since $E(\mid X \mid) \leq \infty$, it follows that $E(X \mid \mathcal{G})$ is a.c. finite. Consequently,

$$\lim_{n \to \infty} E(Y_n X \mid \mathcal{G}) = \lim_{n \to \infty} Y_n E(X \mid \mathcal{G}) = Y E(X \mid \mathcal{G}),$$

which shows the result to hold for nonnegative Y. The proof for general Y is established by considering $Y = Y^+ - Y^-$.

For the proof of ix let $Z = E(X \mid \mathcal{G}_2)$; thus Z is a $\mathcal{G}_2$-measurable random variable, and we wish to show that

$$E(Z \mid \mathcal{G}_1) = E(X \mid \mathcal{G}_1).$$

For any $B \in \mathcal{G}_1$, we have, by definition,

$$\int_B E(Z \mid \mathcal{G}_1) \, dP = \int_B Z \, dP.$$

Since $\mathcal{G}_1 \subseteq \mathcal{G}_2$, $B \in \mathcal{G}_2$ as well and, therefore, bearing in mind the definition of Z, we find that for all sets $B \in \mathcal{G}_1$,

$$\int_B E(X \mid \mathcal{G}_2) \, dP = \int_B X \, dP = \int_B E(X \mid \mathcal{G}_1) \, dP.$$

This, in conjunction with the preceding result shows that

$$E[E(X \mid \mathcal{G}_2) \mid \mathcal{G}_1] = E(Z \mid \mathcal{G}_1) = E(X \mid \mathcal{G}_1).$$

To prove x we must show that $E(Y \mid \mathcal{G}_1) = E(X \mid \mathcal{G}_1)$, where $Y = E(X \mid \mathcal{G}_2)$. Since Y is $\mathcal{G}_2$-measurable and $\mathcal{G}_2 \subseteq \mathcal{G}_1$, Y is also $\mathcal{G}_1$-measurable; moreover, if $B \in \mathcal{G}_2$, then $B \in \mathcal{G}_1$ as well. Consequently, for every $B \in \mathcal{G}_2$

$$\int_B E(Y \mid \mathcal{G}_1)\, dP = \int_B Y\, dP = \int_B E(X \mid \mathcal{G}_2)\, dP.$$

But this shows that

$$E[E(X \mid \mathcal{G}_2) \mid \mathcal{G}_1] = E(Y \mid \mathcal{G}_1) = E(X \mid \mathcal{G}_2),$$

which completes the proof of x.

The proof of xi is a simple consequence of vi and ix. Thus, let $\mathcal{G}_1 = (\emptyset, \Omega)$ and $\mathcal{G}_2 = \mathcal{G}$. Then, clearly, $\mathcal{G}_1 \subseteq \mathcal{G}$ and, by ix,

$$E[E(X \mid \mathcal{G}_2)] = E[E(X \mid \mathcal{G}_2) \mid \mathcal{G}_1) = E(X).$$

q.e.d.

8.4.3 Independence

It is well established in elementary probability theory that two events are "independent" if the probability attached to their intersection, i.e., their joint occurrence, is the product of their individual probabilities. In the preceding sections we have also seen another possible interpretation of independence; this is the intuitively very appealing description that holds that if two events, say A and B are independent, the probability attached to A is the same whether or not we condition on the event B. Another way of expressing this concept is that being told that event B has occurred does not convey any implication regarding the probability of A's occurrence.

In this section we shall formalize these notions and apply them to the case of random variables and families of random variables defined on suitable probability spaces. We begin by noting that if A, B are two independent events, i.e.

$$P(A \mid B) = P(A),$$

then this, in conjunction with the definition (and provided that $P(B) > 0$),

$$P(A \mid B) = \frac{P(AB)}{P(B)}$$

implies
$$P(AB) = P(A)P(B),$$
which is the operational characterization of independence. In the preceding, the intersection operator ($\cap$) was omitted; **we shall follow this practice in this section for notational simplicity, so that the notation AB will always mean $A \cap B$; another notational simplification that will be observed in this section is the following: if $A \supset B$, then we shall write**
$$A - B = A \cap \bar{B}.$$

These two operations will occur sufficiently frequently in the ensuing discussion so as to make the notational conventions above quite useful.

Definition 8.15. Let $(\Omega, \mathcal{A}, P)$ be a probability space and let $\mathcal{C}_i$, $i = 1, 2$, be two classes of events, contained in $\mathcal{A}$. The two classes are said to be independent classes if and only if any events $C_i \in \mathcal{C}_i$, $i = 1, 2$ are independent i.e., $P(C_1 C_2) = P(C_1)P(C_2)$.

Definition 8.16. Let $\mathcal{C}_\pi$, $\mathcal{C}_\lambda$ be two classes of subsets of Ω.

  i. $\mathcal{C}_\pi$ is said to be a π-class if and only if $A, B \in \mathcal{C}_\pi$ implies $AB \in \mathcal{C}_\pi$;

  ii. $\mathcal{C}_\lambda$ is said to be a λ-class if and only if

   a. $\Omega \in \mathcal{C}_\lambda$;

   b. If $A_i \in \mathcal{C}_\lambda$, $i = 1, 2$ and $A_1 A_2 = \emptyset$, then $A_1 \cup A_2 \in \mathcal{C}_\lambda$;

   c. If $A_i \in \mathcal{C}_\lambda$, $i = 1, 2$, and $A_1 \subset A_2$, then $A_2 - A_1 \in \mathcal{C}_\lambda$;

   d. If $A_n \in \mathcal{C}_\lambda$, $n \geq 1$ and $A_n \subset A_{n+1}$, then $\lim_{n \to \infty} A_n \in \mathcal{C}_\lambda$.

A simple consequence of the definition is

Proposition 8.20. If a λ-class, $\mathcal{C}$, is also a π-class, then it is a σ-algebra.

Proof: Let $A_i \in \mathcal{C}$, $i \geq 1$, and recall that $\Omega \in \mathcal{C}$; since the complement of A_i, is given by $\Omega - A_i$ and this is in $\mathcal{C}$, due to the fact that the latter is a λ-class, it follows that $\mathcal{C}$ is closed under complementation. Next we show that $\mathcal{C}$ is closed under countable unions; thus consider $A_1 \cup A_2$; if the two sets are distinct, this union lies in $\mathcal{C}$ since it is a λ-class; if not distinct, write

$$A_1 \cup A_2 = A = A_1 \cup (A_2 \bar{A}_1).$$

Since $\mathcal{C}$ is also a π-class, the second component of the union above is in $\mathcal{C}$; since the two components of the union are disjoint and belong to $\mathcal{C}$, their union also belongs to $\mathcal{C}$ because $\mathcal{C}$ is a λ-class. Finally, define

$$C_n = \bigcup_{i=1}^{n} A_i.$$

By the preceding discussion, $C_n \subset C_{n+1}$, and $C_n \in \mathcal{C}$; since $\mathcal{C}$ is a λ-class $\lim_{n \to \infty} C_n \in \mathcal{C}$.

q.e.d.

An interesting consequence of the preceding discussion is

Proposition 8.21. If a λ-class $\mathcal{C}$, contains a π-class $\mathcal{D}$, then it also contains $\sigma(\mathcal{D})$, the σ-algebra generated by $\mathcal{D}$, i.e. the minimal σ-algebra containing $\mathcal{D}$.

Proof: It suffices to show that the minimal λ-class, $\mathcal{A}$, containing $\mathcal{D}$ also contains $\sigma(\mathcal{D})$. Define $\mathcal{A}_1 = \{A : AD \in \mathcal{A}, \forall D \in \mathcal{D}\}$; evidently, $\mathcal{A}$ contains $\mathcal{D}$ and thus $\mathcal{A}_1 \supset \mathcal{A}$, since it is a λ-class. From this argument we conclude that for all $A \in \mathcal{A}$ and $D \in \mathcal{D}$, $AD \in \mathcal{A}$. Next, define $\mathcal{A}_2 = \{B : BA \in \mathcal{A}, \forall A \in \mathcal{A}\}$; clearly, $\mathcal{A}_2$ is a λ-class, it contains $\mathcal{D}$ and thus $\mathcal{A}$. But this implies that if A, $B \in \mathcal{A}$, then $AB \in \mathcal{A}$, so that $\mathcal{A}$ is a π-class, as well. By Proposition 8.20, we conclude that $\mathcal{A}$ is a σ-algebra which contains $\mathcal{D}$; consequently, it contains $\sigma(\mathcal{D})$.

q.e.d.

We may now use these basic concepts to characterize independence among random variables.

Proposition 8.22. Let $\mathcal{G}_i$, $i = 1, 2$, be independent classes (of subsets of Ω) and suppose further that $\mathcal{G}_2$ is also a π-class. Then $\mathcal{G}_1$ and $\sigma(\mathcal{G}_2)$ are independent.

Proof: For any $A \in \mathcal{G}_1$, define $\mathcal{A} = \{B : B \in \sigma(\mathcal{G}_2), P(AB) = P(A)P(B)\}$ Clearly, $\mathcal{A} \supset \mathcal{G}_2$; moreover, $\mathcal{A}$ is a λ-class, since

i. If $B_1 B_2 = \emptyset$, then

$$
\begin{aligned}
P[(B_1 \cup B_2)A] &= P(B_1 A) + P(B_2 A) \\
&= P(B_1)P(A) + P(B_2)P(A) \\
&= [P(B_1) + P(B_2)]P(A) = P(B_1 \cup B_2)P(A);
\end{aligned}
$$

ii. If $B_2 \supset B_1$,then $B_2 = B_1 \cup (B_2 - B_1)$ and the two components of the union are disjoint; thus $P(B_2 A) = P(B_1 A) + P[(B_2 - B_1)A]$; rearranging we have $P[(B_2 - B_1)A] = P(B_2 - B_1)P(A)$;

iii. $\Omega \in \mathcal{A}$ since $\Omega \in \sigma(\mathcal{G}_2)$ and $P(\Omega A) = P(\Omega)P(A)$;

iv. If $B_i \subset B_{i+1}$, $B_i \in \mathcal{A}$, $i \geq 1$, then $P(B_i A) = P(B_i)P(A)$, $\lim_{n \to \infty} B_i = B \in \sigma(\mathcal{G})$, and thus $P(BA) = \lim_{n \to \infty} P(B_i A) = \lim_{n \to \infty} P(B_i)P(A) = P(B)P(A)$.

This concludes the demonstration that $\mathcal{A}$ is a λ-class, containing the π-class $\mathcal{G}_2$; hence by Proposition 8.21, it contains $\sigma(\mathcal{G})$; but this means that if B is any set in $\sigma(\mathcal{G}_2)$ and $\mathcal{A}$ is any set in $\mathcal{G}_1$, then $P(AB) = P(A)P(B)$.

q.e.d.

With these preliminaries aside we may now turn our attention to the question of independence (of sets) of random variables. We begin with

Definition 8.17. Let T be a nonempty index set (generally the real line) and let $\{X_t : t \in T\}$ be a family of random variables indexed by the set T. This family of random variables is said to be a **stochastic process, if T is continuous and a stochastic sequence if T is discrete.**

The reader no doubt has an intuitive view as to what it means for a **set of random variables to be independent, or to be independent of another set.** No matter what intuitive meaning one ascribes to this concept, the latter would not be very useful unless we can attach to it a specific operational meaning. The question raised here is this: what does it mean, operationally in this context, for random variables to be independent. This is answered in

Definition 8.18. Let $\{X_i : i = 1, 2, \ldots, n\}$ be a set of random variables defined on the probability space $(\Omega, \mathcal{A}, P)$; they are said to be independent (of one another), or mutually independent if and only if $\sigma(X_i)$, $i = 1, 2, \ldots, n$ are independent classes.

For stochastic processes, we have the obvious extension.

Definition 8.19. Let $\{X_t : t \in T\}$ be a stochastic process defined on the probability space $(\Omega,\ \mathcal{A},\ P)$; let T_i, $i \geq 1$ be distinct subsets of the index set T. The (stochastic) subprocesses, $\{X_t : t \in T_i\}$, $i \geq 1$, are said to be independent, if and only if $\mathcal{C}_i = \sigma(X_t,\ t \in T_i)$, $i \geq 1$, are independent classes.

Proposition 8.23. Let $\{X_t : t \in T\}$ be a family of random variables indexed by the nonempty index set T and defined on the probability space $(\Omega,\ \mathcal{A},\ P)$; let T_i, $i = 1, 2$, be disjoint subsets of T and suppose $t_j^{(i)}$, $j = 1, 2, \ldots, m$ are m distinct elements of the the subsets T_i, $i = 1, 2$, respectively. Define the sets $D_{im} = \{\omega : X_{ji} \leq x_j,\ j = 1, 2, \ldots, m,\ x_j \in R\}$, for all $x \in R$, and all integers m. Define

$$\mathcal{D}_i = \{D_{im} : m \geq 1\},\ i = 1, 2.$$

If the $\mathcal{D}_i$, are independent classes, so are $\sigma(\mathcal{D}_i)$, $i = 1, 2$.

Proof: It is evident that $\mathcal{D}_i$, $i = 1, 2$ are π-classes, since if D_{im} and D_{in} are two sets in $\mathcal{D}_i$, $i = 1, 2$, their intersection is a similar set, i.e., a set that describes the region of the domain over which a group of variables indexed by the set T_i assume values in certain intervals of their range, of the form $(-\infty,\ x]$. By Proposition 8.22, $\mathcal{D}_1$ is independent of $\sigma(\mathcal{D}_2)$; applying Proposition 8.22, again, and noting that $\mathcal{D}_1$ is also a π-class, we conclude that the two σ-algebras, $\sigma(\mathcal{D}_i)$, $i = 1, 2$, are independent.

q.e.d.

Corollary 8.6. Let $\{X_t : t \in T\}$ be a family of random variables as in the Proposition above; suppose, further, that the random variables are independent, in the sense that for any indices $j(i) \in T_i$, $i = 1, 2, \ldots, n$, and any integer n, $\sigma(X_{j(i)})$ are independent classes. Let T_1, T_2, be disjoint nonempty subsets of T, then $\sigma(\mathcal{D}_1) = \sigma(X_t, t \in T_1)$, and $\sigma(\mathcal{D}_2) = \sigma(X_t, t \in T_2)$, are independent classes.

Proof: Obvious, since constructing the classes of sets $\mathcal{D}_i$, $i = 1, 2$, of the proposition above, we conclude that they are independent classes; by Proposition 8.23, so are $\sigma(\mathcal{D}_i)$, $i = 1, 2$; but it is apparent from the construction of these σ-algebras that $\sigma(\mathcal{D}_i) = \sigma(X_t, t \in T_i)$, $i = 1, 2$, i.e., they amount to the σ-algebras generated by the random variables indexed by the elements of the set T_i.

q.e.d.

We have, finally, the fundamental characterization of independence of sequences of random variables as follows:

Proposition 8.24. Let $\{X_i : i = 1, 2, \ldots, n\}$ be a sequence of random variables defined on the probability space $(\Omega, \mathcal{A}, P)$; define their (joint) distribution function by

$$F_{(n)}(x_1, \ldots, x_n) = P(A_1 A_2 \ldots A_n),$$

and their individual (marginal) distribution functions by

$$F_i(x_i) = P(A_i), \; i = 1, 2, \ldots, n,$$

where $A_i = \{\omega : X_i(\omega) \in (-\infty, x_i]\}$. These random variables are independent if and only if

$$F_{(n)}(x_1, \ldots, x_n) = \prod_{i=1}^{n} F_i(x_i).$$

Proof: Necessity is obvious, since if $F_n = \prod_{i=1}^{n} F_i$, then we must have

$$P(A_1 A_2 \ldots A_n) = \prod_{i=1}^{n} P(A_i),$$

which shows the classes $\sigma(X_i)$, $i = 1, 2, \ldots, n$ to be independent. To prove sufficiency note that $A_i \in \sigma(X_i)$ and note, also, that if the random variables are independent, then the $\sigma(X_i)$ are independent classes. Hence, that

$$P(A_1 A_2 \ldots A_n) = \prod_{i=1}^{n} P(A_i),$$

and the conclusion follows from the definition of the distribution functions.

q.e.d.

The following corollaries simply rephrase or articulate more explicitly some of the preceding results.

Corollary 8.7. Let $\{X_t : t \in T\}$ be a stochastic process; the random variables of the stochastic process are mutually independent if and only if for any finite number of indices $t_i \in T, i = 1, 2, \ldots, n$ the joint distribution of the variables X_{t_i}, $i = 1, 2, \ldots, n$, $F_{(n)}$, is equal to the product of their marginal distributions, $\prod_{i=1}^{n} F_{t_i}$.

Corollary 8.8. If $\{X_i : i = 1, 2, \ldots, n\}$ is a sequence of independent random variables, and if, similarly, $\{Z_i : i = 1, 2, \ldots, n\}$ is a sequence of independent random variables; moreover, if X_i and Z_i are **identically distributed**, then the joint distribution of the first sequence is identical to the joint distribution of the second sequence, i.e., if $F_{(n)}$ and $G_{(n)}$ are the joint distributions of the two sequences, respectively, then

$$F_{(n)}(a_1, a_2, \ldots, a_n) = G_{(n)}(a_1, a_2, \ldots, a_n).$$

We close this section with a definition that will play an important role in examining questions of convergence of sequences of random variables, in subsequent discussions.

Definition 8.20. Let $\{X_i : i \geq 1\}$ be a sequence of random variables defined on the probability space $(\Omega, \mathcal{A}, P)$; the **tail** σ-algebra of this sequence is given by

$$\bigcap_{n=1}^{\infty} \sigma(X_i, \; i \geq n),$$

where $\sigma(X_i, \; i \geq n)$ is the σ-algebra generated by the semi-algebra

$$\mathcal{J}_n = \sigma(X_n) \times \sigma(X_{n+1}) \times \sigma(X_{n+2}) \times \ldots\ldots.$$

Proposition 8.25 (Kolmogorov Zero-One Law). Let $\{X_n : n \geq 0\}$ be a sequence of independent random variables defined on the probability space $(\Omega, \mathcal{A}, P)$; in this sequence, tail events have probability either zero or one.

Chapter 9

LLN, CLT and Ergodicity

9.1 Review and Miscellaneous Results

We recall from Chap. 8 that discussion of random variables (r.v.) takes place in a probability space $(\Omega,\ \mathcal{A},\ \mathcal{P})$, where Ω is the **sample space**, $\mathcal{A}$ is the σ-**algebra** and $\mathcal{P}$ is the **probability measure**.[1]

If X is an a.c. finite random variable,[2] i.e. if $P(A) = 0$ for $A = \{\omega : X(\omega) = \infty\}$, then

$$\frac{X}{b_n} \overset{a.c}{\to} 0,$$

where b_n is a sequence such that $\lim_{n\to\infty} |b_n| = \infty$.

Let $\{X_n:\ n \geq 1\}$ be a sequence of random variables defined on the probability space $(\Omega,\ \mathcal{A},\ \mathcal{P})$ and suppose that

$$S_n = \sum_{i=1}^{n} X_i, \quad \frac{S_n}{n} \overset{\text{P or a.c.}}{\longrightarrow} 0,$$

then

$$\frac{X_n}{n} \overset{\text{P or a.c.}}{\longrightarrow} 0.$$

See pp. 148–150.

---

[1] Part of this chapter is an adaptation of a set of Lectures given in the Spring of 2005, at the University of Cyprus, whose purpose was stated as "The purpose of these Lectures is to set forth, in a convenient fashion, the essential results from probability theory necessary to understand classical econometrics." All page references, unless otherwise indicated, are to: Dhrymes (1989).

[2] The notation a.c. means **almost certainly**; an alternative notation is a.s. which means **almost surely**. More generally, the notation, say, $X_n \overset{a.c}{\to} X_0$, or $X_n \overset{P}{\to} X_0$ means that the sequence of random variables $\{X_n:\ n \geq 1\}$ converges to the random variable X_0, respectively, almost certainly or in probability. These concepts will be defined more precisely below.

P.J. Dhrymes, *Mathematics for Econometrics*,
DOI 10.1007/978-1-4614-8145-4_9, © The Author 2013

Definition 9.1. Let X be a random variable, defined on a probability space $(\Omega, \mathcal{A}, \mathcal{P})$; the σ-algebra induced by X, say $\sigma(X) \subset \mathcal{A}$, is the smallest σ-algebra for studying X, and is defined by

$$\sigma(X) = \{A: A = X^{-1}(B), B \in \mathcal{B}(R)\},$$

where R is the real line and $\mathcal{B}(R)$ is the Borel σ-algebra.

Remark 9.1. The connection between random variables as used here and random variables as usually taught in elementary statistics courses is the following: for each random variable, X, defined on the probability space $(\Omega, \mathcal{A}, \mathcal{P})$, consider its range, R, and the σ-algebra associated with it, the Borel σ-algebra, $\mathcal{B}(R)$. Perform the identity transformation (so that X is identified solely by the values it assumes)

$$X : R \to R$$

and deal exclusively with the values assumed by the random variable; we assign probabilities to such sets (of values assumed by the random variable) by the rule: if $B \in \mathcal{B}(R)$, assign a probability to the event that X assumes a value in B, by the rule
$$P(B) = \mathcal{P}(A), \quad A = X^{-1}(B).$$
It may be shown that P (as in $P(B)$) is a proper probability measure $P : \mathcal{B}(R) \to [0, 1]$, and that $(R, \mathcal{B}(R), P)$ **is the probability space induced by** X, in the context employed in standard statistics discussions.

In this context what distinguishes one variable from another is the probability measure they induce, P. Hence, the usual statement: let X be a random variable with distribution, G. The connection between P and G is given by $G(x) = P(B)$ in the case of a set $B = (-\infty, x]$.

9.1.1 Limits of sets

Let $\{A_n : n \geq 1\}$ be a sequence of sets defined on some σ-algebra $\mathcal{A}$. Define[3]

$$B_n = \bigcup_{i=n}^{\infty} A_i, \quad C_n = \bigcap_{i=n}^{\infty} A_i, \quad B^* = \lim_{n \to \infty} B_n, \quad C_* = \lim_{n \to \infty} C_n,$$

and note that
$$B_n \supseteq A_n, \quad C_n \subseteq A_n,$$

and both are monotonic (sequences) obeying $B_n \supseteq B_{n+1}$ and $C_n \subseteq C_{n+1}$, respectively. This means that the B-sequence is **non-increasing** while the

---

[3]This is a summary of relevant results presented earlier in Chap. 8.

C-sequence is **non-decreasing**. This justifies the definition of B^*, C^* as above since

$$B^* = \bigcap_{n=1}^{\infty} B_n, \quad C^* = \bigcup_{n=1}^{\infty} C_n.$$

The set B^* is said to be the **limit superior**, and the set C_* is said to be the **limit inferior** of the sequence, and one writes

$$B^* = A^* = \limsup_{n \to \infty} A_n, \quad C_* = A_* = \liminf_{n \to \infty} A_n.$$

Evidently, $A^* \supseteq A_*$. **If $A^* = A_*$**, the common value of the limit inferior and limit superior is said to be the limit of the sequence and is denoted by $A = \lim_{n \to \infty} A_n$. (pp. 8–10)

Note the notational equivalence:

$$\overline{A^*} = \bigcup_{n=1}^{\infty} \bigcap_{k=n}^{\infty} \bar{A}_k = \bar{A}_*,$$

and conversely with the limit inferior, i.e. the complement of the limit inferior of the sequence $\{A_n : n \geq 1\}$ is the limit superior of the sequence $\{\bar{A}_n : n \geq 1\}$.

Remark 9.2. The intuitive meaning of the limit superior, A^*, is that it contains elements that belong to members of the sequence above *infinitely often*, a fact that is denoted by (the notation)

$$A^* = \{\omega : \omega \in A_n, \ i.o.\}$$

In connection with the preceding we have the following important result.

Proposition 9.1 (Borel-Cantelli Lemma and extension). : Let $\{A_n : n \geq 1\}$ be a sequence of events defined on the probability space $(\Omega, \mathcal{A}, \mathcal{P})$. If

$$\sum_{i=1}^{\infty} \mathcal{P}(A_i) < \infty, \quad \text{then} \quad \mathcal{P}(A_n, i.o.) = 0.$$

If the events are **mutually independent** and

$$\sum_{i=1}^{\infty} \mathcal{P}(A_i) = \infty, \quad \text{then} \quad \mathcal{P}(A_n, i.o.) = 1.$$

As will become clear below, this result has potential applications in showing the convergence of estimators with probability one. (See pp. 136–137)

9.1.2 Modes of Convergence

Let $\{X_n : n \geq 0\}$ be a sequence of random variables defined on the probability space $(\Omega, \mathcal{A}, \mathcal{P})$, we want to define statements like: The sequence converges to X_0. To this end define, for arbitrary integer r, the sets

$$A_{n,r} = \{\omega : |X_n(\omega) - X_0(\omega)| \geq \frac{1}{r}\}. \tag{9.1}$$

Definition 9.2 (Convergence with probability one). Let $\{X_n : n \geq 0\}$ be a sequence of random variables defined on the probability space $(\Omega, \mathcal{A}, \mathcal{P})$; we say that the sequence $\{X_n : n \geq 1\}$ converges to X_0 **with probability one, or almost surely (a.s), or almost certainly (a.c.)** if and only if for given r, denoting by A_r^* the limit superior of the sequence of sets in Eq. (9.1), the latter obeys

$$\mathcal{P}(A^*) = \lim_{r \to \infty} \mathcal{P}(A_r^*) = 0, \quad \text{because} \quad A^* = \bigcup_{r=1}^{\infty} A_r^*, \quad \text{and} \quad A_r^* \subseteq A_{r+1}^*. \tag{9.2}$$

For convenience in use we shall frequently denote this fact by

$$X_n \overset{\text{a.c. (or a.s.)}}{\to} X_0,$$

which means that for arbitrary r the probability attached to the limit superior A_r^* can be made arbitrarily close to zero with proper choice of r. (pp. 134–136)

9.1.3 Convergence in Probability

Definition 9.3 (Convergence in probability). Consider the sequence in Definition 9.2. We say that the sequence converges in probability to X_0, denoted by

$$X_n \overset{\text{P}}{\to} X_0,$$

if and only if for any r there exists a number $q(r)$, such that

$$\lim_{n \to \infty} \mathcal{P}(A_{nr}) \leq q(r) \tag{9.3}$$

and $q(r)$ can be made arbitrarily close to zero by making the proper choice of r. Evidently, convergence a.c. implies convergence in probability. (pp. 134–136)

9.1.4 Convergence in Mean of Order p, or L^p

Definition 9.4 (Convergence in Mean of order p). Let $\{X_n : n \geq 0\}$ be a sequence of random variables as in Definition 9.2 and suppose

$$E|X_n|^p < \infty, \quad n \geq 0, \quad p > 0.$$

The sequence converges to X_0 , **in mean of order p**, denoted by

$$X_n \overset{\mathrm{L^p}}{\to} X_0 \qquad\qquad (9.4)$$

if and only if

$$\lim_{n\to\infty} E|X_n - X_0|^p = 0.$$

See p. 156ff.

9.1.5 Convergence in Distribution

Definition 9.5 (Convergence in Distribution). Let $\{X_n : n \geq 0\}$ be a sequence of random variables as in Definition 9.2 with respective cumulative distribution functions (cdf) F_n ; it converges **in distribution** to a random variable X_0 if and only if

$$F_n \overset{\mathrm{c}}{\to} F_0, \qquad\qquad (9.5)$$

where c indicates **complete convergence** as follows. Let G_n, $n \geq 1$ be a sequence of non-decreasing functions; it is said to **converge weakly** to a non-decreasing function G , denoted by $G_n \overset{\mathrm{w}}{\to} G$ if and only if

$$\lim_{n\to\infty} G_n(x) = G(x), \quad x \in C(G),$$

where $C(G)$ is the **set of continuity points** of G . If, in addition

$$\lim_{n\to\infty} G_n(\pm\infty) = G(\pm\infty) = \lim_{x\to\pm\infty} G(x),$$

the sequence is said to **converge completely**, a fact that is denoted by $G_n \overset{\mathrm{c}}{\to} G$.

Since the cdf are non-decreasing functions, the explanation is complete, and convergence in distribution is denoted by

$$X_n \overset{\mathrm{d}}{\to} X_0.$$

Remark 9.3. Note that, essentially, convergence in distribution **does not involve convergence to a random variable**, even though common usage of the term indicates that this is so. In fact convergence in distribution (or convergence in Law as is termed in a more antique usage) indicates that the cdf of the sequence converge completely and, thus, to a distribution function (cdf). Since to each cdf there corresponds (non-uniquely) a random variable, (cf.: let X be a sequence of i.i.d. random variables) what we mean by the terminology is that the sequence converges **in an equivalence class** of random variables having the distribution F . (pp. 152–153)

9.2   Relationship Among Modes
of Convergence

Let $\{X_n : n \geq 0\}$ be a sequence of random variables defined on the probability space $(\Omega, \mathcal{A}, \mathcal{P})$, then[4]

1. $X_n \overset{\text{a.c.}}{\to} X_0$ implies $X_n \overset{\text{P}}{\to} X_0$,

2. $X_n \overset{\text{P}}{\to} X_0$ implies $X_n \overset{\text{d}}{\to} X_0$

3. $X_n \overset{\text{L}^p}{\to} X_0$ implies $X_n \overset{\text{P}}{\to} X_0$,

4. $X_n \overset{\text{P}}{\to} X_0$ **does not imply** $X_n \overset{\text{L}^p}{\to} X_0$

5. $X_n \overset{\text{P}}{\to} X_0$ **does not imply** $X_n \overset{\text{a.c.}}{\to} X_0$

6. $X_n \overset{\text{L}^p}{\to} X_0$ **does not imply** $X_n \overset{\text{a.c.}}{\to} X_0$

7. $X_n \overset{\text{d}}{\to} X_0$ **does not imply** $X_n \overset{P}{\to} X_0$.

Remark 9.4. Even though by (7) convergence in distribution does not imply convergence in probability for reasons given in Remark 9.3, there is one instance where this is so. It occurs when convergence in distribution is to a constant (degenerate). In **this case it does imply convergence in probability to that constant.** (see pp. 165 and 262)

9.2.1   Applications of Modes of Convergence

Proposition 9.2. Let $\{X_n : n \geq 0\}$ be a sequence of random variables defined on the probability space $(\Omega, \mathcal{A}, \mathcal{P})$ and suppose

$$X_n \overset{\text{P}}{\to} X_0.$$

If $\{Y_n : n \geq 1\}$ is another sequence such that it converges in probability to Y_0 and, if Y_0, X_0 are **equivalent** in the sense that they differ **only on a set of** $\mathcal{P}$-measure zero, then

$$|X_n - Y_n| \overset{\text{P}}{\to} 0. \tag{9.6}$$

See pp. 151–152.

---

[4]The implication in item (3) is easily established using Proposition 8.12, Chap. 8, (Generalized Chebyshev Inequality) as follows: define $Y_n = |X_n - X_0|^p$ and note that the Y-sequence consists of non-negative integrable rvs obeying $EY_n = s_n$, such that s_n converges to zero with n.

Proposition 9.3. Let $\{X_n, Y_n : n \geq 0\}$ be sequences of random variables defined on the probability space $(\Omega, \mathcal{A}, \mathcal{P})$ and suppose

$$Y_n \overset{d}{\to} Y_0, \quad |X_n - Y_n| \overset{P}{\to} 0.$$

Then

$$X_n \overset{d}{\to} Y_0. \tag{9.7}$$

See pp. 161–162.

Proposition 9.4. Let $\{X_n : n \geq 0\}$ be a sequence of random variables defined on the probability space $(\Omega, \mathcal{A}, \mathcal{P})$, and let

$$\phi : R \to R$$

be a $\mathcal{B}(R)-$**measurable function** whose discontinuities are contained in a set D of $\mathcal{P}$-measure zero. Then, provided $\phi(X_0)$ is well defined, the following is true:

i. $X_n \overset{P}{\to} X_0$ implies $\phi(X_n) \overset{P}{\to} \phi(X_0)$;

ii. $X_n \overset{a.c.}{\to} X_0$ implies $\phi(X_n) \overset{a.c.}{\to} \phi(X_0)$;

iii. $X_n \overset{d}{\to} X_0$ implies $\phi(X_n) \overset{d}{\to} \phi(X_0)$.

See pp. 144–145, pp. 147–148, pp. 242–243.

A more useful form of item iii is as follows:

Proposition 9.5. For the sequence of Proposition 9.4, let P_0 be the distribution of X_0 and define the transformed sequence

$$Y_n = \phi(X_n), \quad n \geq 0.$$

Then

$$Y_n \overset{d}{\to} \phi(X_0), \quad \text{whose distribution is} \quad P_0 \circ \phi^{-1}. \tag{9.8}$$

Proof: We show explicitly the form of the converged distribution of the Y-sequence. Let P_n and P_n^* be, respectively, the distributions induced by the elements of the $X-$ and $Y-$ sequences, and C an arbitrary set in $\mathcal{B}(R)$. If we can determine the value of $P_n^*(C)$, for arbitrary $C \in \mathcal{B}(R)$, we will have defined the distribution function of Y_n. But $Y_n \in C$ if and only if $X_n \in B = \phi^{-1}(C)$; now $X_n \in B$ with probability $P_n(B)$. Thus,

$$P_n^*(C) = P_n(B) = P_n[\phi^{-1}(C)] = P_n \circ \phi^{-1}(C),$$

where $P_n \circ \phi^{-1}$ is the **composition of** P_n and ϕ^{-1}. Since C is an arbitrary set in $\mathcal{B}(R)$, it follows

$$P_n^* = P_n \circ \phi^{-1},$$

and consequently

$$P_n \overset{w}{\to} P_0, \quad \textbf{implies} \quad P_n^* \overset{w}{\to} P_0 \circ \phi^{-1},$$

where the notation $\overset{w}{\to}$ denotes weak convergence of measures which is roughly comparable, in this context, to complete convergence of cumulative distribution functions explained above. Note, in addition, that

$$P_0 \circ \phi^{-1} \quad \textbf{is the distribution of} \quad Y_0 = \phi(X_0).$$

An interesting by-product of this is the following very useful result.

Proposition 9.6. Let A_n, a_n be a suitable random matrix and vector, respectively, **converging in probability to** A, a, and let ξ_n be a sequence of random vectors converging in distribution to ξ_0, then

$$\zeta_n = A_n \xi_n + a_n \overset{d}{\to} A\xi_0 + a. \tag{9.9}$$

See pp. 242–244.

9.3 Laws of Large Numbers and Central Limit Theorems

Let $\{X_n : n \geq 1\}$ be a sequence of random variables defined on the probability space $(\Omega, \mathcal{A}, \mathcal{P})$ and define

$$S_n = \sum_{i=1}^{n} X_i, \quad Q_n = \frac{S_n - a_n}{b_n},$$

where $b_n > 0$, for all n, $\lim_{n \to \infty} b_n = \infty$, and it, as well as the sequence a_n, are taken to be **non-random.** Both Laws of Large Numbers (LLN) and Central Limit Theorems (CLT) describe properties of the sequence Q_n as $n \to \infty$.

Definition 9.6 (Weak Law of Large Numbers (WLLN)). If $Q_n \overset{P}{\to} 0$ we say that the sequence obeys the WLLN.

Definition 9.7 (Strong Law of Large Numbers (SLLN)). If $Q_n \overset{a.c.}{\to} 0$, we say that the sequence obeys the SLLN.

Definition 9.8 (Central Limit Theorem (CLT)). If $Q_n \overset{d}{\to} \xi$ and ξ is a member of a well defined equivalence class of random variables, we say that the sequence obeys a CLT.

9.3.1 Criteria for Applicability, LLN

Let $\{X_n : n \geq 1\}$ be a sequence of random variables defined on the probability space $(\Omega, \mathcal{A}, \mathcal{P})$ and suppose the elements of the sequence are independent, identically distributed (i.i.d. or iid) with mean μ (actually we need $E|X_1| < \infty$, but existence of variance is not required). Taking

$$a_n = n\mu, \qquad b_n = n;$$

it can be shown that

$$Q_n = \frac{1}{n}\sum_{i=1}^{n}(X_i - \mu) \overset{a.ç.}{\to} 0,$$

and hence that it converges to zero in probability as well, so that it obeys both the WLLN and the SLLN. (See pp. 188–190)

For independent, not identically distributed random variables with finite variance we have Kolmogorov's criterion for a SLLN. Taking

$$Q_n = \frac{1}{b_n}\sum_{i=1}^{n}(X_i - E(X_n))$$

it can be shown that $Q_n \overset{a.ç.}{\to} 0$, **provided**

$$\sum_{n=1}^{\infty}\left(\frac{\text{var}(X_n)}{b_n^2}\right) < \infty.$$

Thus, for $b_n = n$, a sufficient condition for

$$Q_n = \frac{1}{n}\sum_{i=1}^{n}(X_i - E(X_n)) \overset{a.ç.}{\to} 0$$

is that

$$\sum_{n=1}^{\infty}\left(\frac{\text{var}(X_n)}{n^2}\right) < \infty.$$

Evidently this is satisfied when $\text{var}(X_n)$ is (uniformly) bounded, or more generally when $\text{var}(X_n) \sim cn^{\alpha}$, for $\alpha \in [0,1)$. (See pp. 186–188)

If the sequence above is merely a sequence of **uncorrelated** random variables with variance $\text{var}(X_n) \leq cn^{\alpha}$, and $\alpha \in [0, \frac{1}{2})$, then

$$Q_n = \frac{1}{n}\left(\sum_{i=1}^{n}[X_i - E(X_i)]\right) \overset{a.ç.}{\to} 0$$

See pp. 191–193.

9.3.2 Criteria for Applicability, CLT

Let $\{X_n : n \geq 1\}$ be a sequence of random variables defined on the probability space $(\Omega, \mathcal{A}, \mathcal{P})$ and suppose they are independent identically distributed (iid) with mean μ and variance σ^2, then

$$Q_n = \frac{\sum_{i=1}^{n}(X_i - \mu)}{\sqrt{n}} \xrightarrow{d} X \sim N(0, \sigma^2).$$

This is the most basic Central Limit Theorem.

See p. 264.

For the sequence above suppose that it is only one of **independent, not identically distributed** random variables, with mean μ_n and variance σ_{nn}, we have the following: put

$$z_n = \frac{S_n}{\sigma_n}, \quad S_n = \sum_{i=1}^{n}(X_i - \mu_i), \quad \sigma_n^2 = \sum_{i=1}^{n}\sigma_{ii}, \quad X_{in} = \frac{X_i - \mu_i}{\sigma_n},$$

and note that

$$z_n = \sum_{i=1}^{n} X_{in}, \quad \text{var}(X_{in}) = \frac{\sigma_{ii}}{\sigma_n^2} = \sigma_{in}^2.$$

If we denote the cdf of the X_i by F_i and the cdf of X_{in} by F_{in} we note that the mean of the last distribution is zero and its variance σ_{in}^2 which converges to zero with (the sample size) n. Moreover, define the **Lindeberg condition** by

$$\lim_{n\to\infty} W_n = 0, \quad W_n = \sum_{i=1}^{n} \int_{|\xi|>\frac{1}{r}} \xi^2 dF_{in}(\xi), \qquad (9.10)$$

for arbitrary integer r. The Lindeberg central limit theorem (CLT) then states

Proposition 9.7 (Lindeberg CLT). Under the conditions above

$$z_n \xrightarrow{d} \zeta \sim N(0, 1).$$

To see, roughly speaking, what the Lindeberg condition means, revert to the original variables and, for simplicity, assume the sequence has **zero means**. In this case the Lindeberg condition becomes

$$W_n = \frac{1}{\sigma_n^2} \sum_{i=1}^{n} \int_{|\xi|>\frac{\sigma_n}{r}} \xi^2 dF_i(\xi),$$

which says that as the sample increases the sum of the tails of the variance integral becomes negligible in comparison to the total variance.

Another way of displaying the Lindeberg CLT and the Lindeberg condition is to redefine X_{in} from

$$X_{in} = \frac{X_i - \mu_i}{\sigma_n} \quad \text{to} \quad X_{in} = \frac{X_i - \mu_i}{\sqrt{n}},$$

and put $z_n^* = \frac{S_n}{\sqrt{n}}$. This means that z_n^* is given by

$$z_n^* = \frac{S_n}{\sigma_n} \frac{\sigma_n}{\sqrt{n}} = z_n \frac{\sigma_n}{\sqrt{n}},$$

so that

$$z_n^* \overset{d}{\to} \zeta \sigma \sim N(0, \sigma^2), \quad \sigma^2 = \lim_{n \to \infty} \frac{\sigma_n^2}{n},$$

provided the latter exists, i.e. $\sigma^2 < \infty$. (See pp. 271–274)

When moments higher than the second are known to exist, it is possible to employ another condition for convergence; this is embodied in the CLT due to Liapounov.

Proposition 9.8 (Liapounov CLT). Let $\{X_n : n \geq 1\}$ be a sequence of independent non-identically distributed rvs defined on the probability space $(\Omega, \mathcal{A}, \mathcal{P})$, obeying $E(X_n) = \mu_n$, $|\mu_n| < \infty$, $E(X_n - \mu_n)^2 = \sigma_{nn}^2 < \infty$,[5] and for some constant $\delta > 0$, $E \, | \, Y_n - \mu_n \, |^{2+\delta} = \rho_{nn}^{2+\delta} < \infty$. Define, now,

$$z_n = \frac{S_n}{\sigma_n}, \quad S_n = \sum_{i=1}^{n} (X_i - \mu_i), \quad \sigma_n^2 = \sum_{i=1}^{n} \sigma_{ii}, \quad X_{in} = \frac{X_i - \mu_i}{\sigma_n},$$

and in addition,

$$\rho_n^{2+\delta} = \sum_{i=1}^{n} \rho_{in}^{2+\delta}, \quad \rho_{in}^{2+\delta} = E \, | \, X_{in} \, |^{2+\delta}.$$

Suppose further that σ_n^2 diverges to $+\infty$ with n. A sufficient condition for

$$z_n \overset{d}{\to} z \sim N(0, 1)$$

is that

$$\lim_{n \to \infty} \frac{\rho_n^{2+\delta}}{\sigma_n^{2+\delta}} = 0.$$

See pp. 275–276.

---

[5]Strictly speaking, this and the preceding are implied by the following condition, which states that the sequence of rvs in question possesses finite $(2 + \delta)$ th moments, for arbitrary $\delta > 0$. Moreover this last requirement implies the Lindeberg condition.

Martingale Difference (MD) Central Limit Theorem

First, we define what martingales and martingale differences are.

Definition 9.9. Let $\{X_n : n \geq 1\}$ be a sequence of random variables defined on the probability space $(\Omega, \mathcal{A}, \mathcal{P})$ and consider the sequence of (sub) σ-algebras, $\mathcal{A}_{n-1} \subseteq \mathcal{A}_n$, $n \in N$, where N is a subset of the integers in $(-\infty, \infty)$. The sequence $\{\mathcal{A}_n : n \in N\}$ is said to be a **stochastic basis** or a **filtration**. If, in the sequence above, X_n is $\mathcal{A}_n$-measurable, the sequence of pairs $\{(X_n, \mathcal{A}_n) : n \in N\}$ is said to be a **stochastic sequence**. If, in addition, $E|X_n| < \infty$, the sequence is said to be

  i. A **martingale** if $E(X_{n+1}|\mathcal{A}_n) = X_n$;

  ii. A **sub-martingale** if $E(X_{n+1}|\mathcal{A}_n) \geq X_n$;

  iii. A **super-martingale** if $E(X_{n+1}|\mathcal{A}_n) \leq X_n$;

  iv. A **martingale difference** if $E(X_{n+1}|\mathcal{A}_n) = 0$.

In the above, the notation $|\mathcal{A}_n$ means conditioning on the (sub) σ-algebra in question, or the random variables generating them.

Proposition 9.9 (Martingale Difference CLT). In the notation of the previous theorem, let $\{X_{in} : i \leq n\}$ be defined as

$$X_{in} = \frac{X_i - EX_i}{\sqrt{n}},$$

and suppose that, for each n, $\{(X_{in}, \mathcal{A}_{in}) : i \leq n\}$ is a martingale difference sequence satisfying the Lindeberg condition, i.e.[6]

$$\operatorname*{plim}_{n \to \infty} W_n = 0, \quad W_n = \sum_{i=1}^n E[X_{in}^2 I(|X_{in}| \geq \frac{1}{r})|\mathcal{A}_{i-1,n}],$$

where $\mathcal{A}_{i-1,n}$ is an element of the modified filtration induced by the sequence X_{in}. The following statements are true

  i. If $\sum_{i=1}^n E(X_{in}^2|\mathcal{A}_{i-1,n}) \xrightarrow{P} \sigma^2$, then $z_n = \sum_{i=1}^n X_{in}^2 \xrightarrow{d} \zeta \sim N(0, \sigma^2)$;

  ii. If $\sum_{i=1}^n X_{in}^2 \xrightarrow{P} \sigma^2$, then $z_n \xrightarrow{d} \zeta \sim N(0, \sigma^2)$.

See pp. 323–337.

---

[6]The function I in the equation below is the indicator function, which assumes the value 1, if $|X_{in}| \geq \frac{1}{r}$ and the value zero otherwise.

The reader may have noticed that we have given CLT only for scalar random variables and may have wondered why we did not provide such results for sequences of random **vectors**. In fact, such results may be subsumed under the discussion of scalar CLT as the following will make clear. First we note that the **characteristic function** (with parameter s) of a normal variable Y with mean μ and variance σ^2 is given by

$$\phi(s) = E e^{isY} = e^{is\mu - \frac{1}{2}s^2\sigma^2};$$

if Y is a **normal** random vector with mean vector μ and covariance matrix $\Sigma > 0$, its characteristic function, with parameter t, is given by

$$\phi(t) = E e^{it'Y} = e^{it'\mu - \frac{1}{2}t'\Sigma t}.$$

We now have

Proposition 9.10. Let X be a random vector, i.e.

$$X : \Omega \to R^m, \; EX = \mu, \; \text{Cov}(X) = \Sigma > 0.$$

If for **arbitrary** conformable vector λ

$$\lambda'X \sim N(\lambda'\mu, \; \lambda'\Sigma\lambda),$$

then

$$X \sim N(\mu, \Sigma).$$

Proof: For $y(\lambda) = \lambda'X$, with arbitrary conformable λ, $y(\lambda)$ is, by the premise of the proposition, normal with mean $\lambda'\mu$ and variance $\lambda'\Sigma\lambda$. Its characteristic function, with parameter s, is therefore given by

$$\phi(s) = e^{is\lambda'\mu - \frac{1}{2}s^2\lambda'\Sigma\lambda}.$$

But taking $t = s\lambda$, we find

$$\phi(s) = \phi(t) = E e^{it'X} = e^{it'\mu - \frac{1}{2}t'\Sigma t},$$

which is recognized as the characteristic function (with parameter t) of a multivariate normal with mean vector μ and covariance matrix Σ, which was to be proved.

Remark 9.5. The relevance of this to CLT is the following: if X_T is a zero mean (vector) sequence such that for arbitrary λ, $\lambda'X_T$ converges in distribution to a normal variable with mean zero and variance $\lambda'\Sigma\lambda$, then

$$X_T \xrightarrow{\text{d}} X \sim N(0, \Sigma). \tag{9.11}$$

9.4   Ergodicity and Applications

9.4.1   Preliminaries

Had we not dealt with time series models, the preceding would have been sufficient to deal with the estimation, properties of estimators and their limiting distribution in the context of the standard models dealt with in classical econometrics.

To see why this is so we note that the random variables (rvs) dealt with in time series models are far more complex that the error terms' probabilistic specification typically found in classical econometrics. In time series, we specify the structure and the white noise process that define them; the properties of the rvs are then to be derived from these specifications. For example in AR(n), we obtain its $MA(\infty)$ representation as

$$X(t) = \sum_{j=0}^{\infty} \psi_j u_{t-j}, \quad u \sim WN(0, \sigma^2), \tag{9.12}$$

with u sometimes $iid(0, \sigma^2)$. This is a far more complicated entity that the usual iid or uncorrelated specification (with mean zero and variance σ^2) we encounter in the general linear model or panel data models or even in the GLSEM so long as it is not dynamic.

Thus, in Eq. (9.12) we need to **prove** what the mean and variance are, **we cannot specify them arbitrarily**. We should also determine whether the time series is **weakly** or **strongly stationary**. For clarity we state these results as a proposition.

Proposition 9.11. Consider the (causal) $AR(n)$ sequence discussed above, whose $MA(\infty)$ representation appears in Eq. (9.12), with absolute convergence, i.e.

$$\sum_{j=0}^{\infty} |\psi_j| < \infty.$$

The following statements are true:

  i. The sequence has mean zero;

 ii. The sequence has constant variance;

iii. The sequence is **weakly stationary.**

Proof: For **causal** AR we require the roots of the characteristic polynomial (corresponding to the lag operator that defines the AR) be **greater than one** in absolute value, which enables the inversion of the polynomial lag operator.

In turn this enables the inversion of lag operators of the form $I - \lambda L$, with $|\lambda| < 1$, which is guaranteed since the latter is **the inverse of one of the roots of the characteristic polynomial**; this was dealt with extensively in the chapter on difference equations. But absolute convergence in Eq. (9.12), viz. that the series obeys

$$\sum_{j=0}^{\infty} |\psi_j| < \infty,$$

is also ensured by the causal property of the AR because all the parameters $|\lambda|^j$ converge to zero with j, and thus the series obeys the Cauchy criterion for convergence. This in turn guarantees that

$$\sum_{j=0}^{\infty} |\psi_j|^2 < \infty.$$

To prove i, note that

$$|EX(t)| \le \sum_{j=0}^{\infty} |\psi_j| E|u_{t-j}| = \sum_{j=0}^{\infty} |\psi_j| m < \infty. \tag{9.13}$$

This is so because $E|u_{t-j}| = Eu_{t-j}^+ + Eu_{t-j}^- = m$ and $Eu_{t-j} = Eu_{t-j}^+ - Eu_{t-j}^- = 0$; thus the expectations Eu_{t-j}^+, Eu_{t-j}^- are well defined. Moreover, for sufficiently large N, $|X(t) - \sum_{j=0}^{N} \psi_j u_{t-j}| < \delta$, for any pre-assigned $\delta > 0$; since $EX^N(t) = 0$, where $X^N(t) = \sum_{j=0}^{N} \psi_j u_{t-j}$, it follows that $EX(t) = 0$.

To prove ii,

$$\text{var}(X(t)) = EX^2(t) = \sum_{j=0}^{\infty} \sum_{s=0}^{\infty} \psi_j \psi_s Eu_{t-j}u_{t-s} = \sigma^2 \sum_{j=0}^{\infty} \psi_j^2 < \infty. \tag{9.14}$$

To prove iii, note that

$$c(t+h, t) = EX(t+h)X(t) = \sum_{j=0}^{\infty} \sum_{s=0}^{\infty} \psi_j \psi_s Eu_{t+h-j}u_{t-s}$$

$$= \sigma^2 \sum_{j=0}^{\infty} \psi_{j-h}\psi_j = c(-h).$$

If in the penultimate member of the equation above we made the association $j = s + h$, we would have obtained $c(h)$, for the last member. Consequently we obtain

$$c(t+h, t) = c(|h|), \tag{9.15}$$

which shows the existence of the auto-variance function as well as **weak stationarity for a causal AR.** Incidentally this would be true whether the u-sequence obeys

$$u \sim WN(0, \sigma^2), \quad \text{or} \quad u \sim iid(0, \sigma^2),$$

and also note that in the second case the time series is **strictly stationary** as well!

q.e.d.

In Chap. 6 (Proposition 6.1) we established that the auto-covariance function of a weakly stationary time series is **at least a positive semi-definite matrix.** An interesting by product is the following result.

Proposition 9.12. A real function defined on the integers is the auto-covariance function of a stationary time series if and only if it is even and non-negative definite (positive semi-definite).

For a proof see Theorem 1.5.1, p. 27 in Brockwell and Davis (1991),

In that proof it is shown that there exists a **strictly stationary Gaussian time series** with that auto-covariance function.

A casual reading of this result may lead the reader to conclude that the causal AR discussed in Proposition 9.11, **is also strictly stationary and Gaussian** because its auto-covariance function is even ($c(h) = c(-h)$) and positive semi-definite. Unfortunately this is not true; Proposition 9.12 only asserts the existence **of a** strictly stationary Gaussian time series with the auto-covariance function exhibited in Proposition 9.11, but this is not necessarily the causal AR which gave rise to that specific auto-covariance function.

9.4.2 Ergodicity

It would appear reasonable that we should be able to estimate such parameters through observations on the AR sequence. The problem is that, for a time series sequence

$$\{X(t) : \quad t \in T, \quad T \text{ a linear index set}\},$$

a realization is a series of observations $\{x_t : \quad x_t = X(\omega_t, t), \, t = 1, 2, \ldots, N\}$, where we have inserted the understood argument ω since $X(t)$ is a measurable function from $\Omega \to R$. What this means is that we have one observation per rv and, **without additional conditions we cannot assert that**

$$\frac{1}{N} \sum_{t=1}^{N} x_t \quad \text{converges in probability to} \quad EX(t). \tag{9.16}$$

It is at this stage that ergodic theory becomes relevant. In the following discussion we shall explain what ergodicity or the ergodic property means and provide the tools to determine that not only the usual mean function but also **the auto-covariance function estimated from realizations converge to the corresponding parameters not only in probability (WLLN) but with probability one (SLLN) as well.**[7] The discussion of the concept of ergodicity, its relevant results and its potential applications begins with the definition of **measure preserving transformations, MPT**.

Definition 9.10. Let $(\Omega, \mathcal{A}, \mathcal{P})$ be a probability space; the mapping (or transformation)

$$\mathcal{T}: \quad \Omega \to \Omega,$$

is said to be **measure preserving** if for $A \in \mathcal{A}$, $\mathcal{T}^{-1}(A) \in \mathcal{A}$, i.e. it is $\mathcal{A}$-measurable, and $P(\mathcal{T}^{-1}(A)) = P(A)$.

Lemma 9.1. If $\mathcal{T}$ is the transformation in Definition 9.10, minus the measure preserving property, and if $\mathcal{T}$ is viewed as a point transformation, i.e. for $\omega \in \Omega$, $\mathcal{T}^{-1}\omega \in \Omega$, and if, **in addition, it is one-to-one and onto, i.e. invertible**, then $\mathcal{T}$ is a MPT.

Proof: Let A be an arbitrary set in $\mathcal{A}$, and let $\omega \in A$ be arbitrary; then $\mathcal{T}\omega \in B = \mathcal{T}^{-1}A$, where $B \in \mathcal{A}$. Thus, given any $\omega \in A$, there exists a unique $\mathcal{T}\omega \in B$, and vice-versa. Thus, $P(A) = P(\mathcal{T}A)$, which shows $\mathcal{T}$ to be a MPT.

Definition 9.11. In the context of Definition 9.10 a set $A \in \mathcal{A}$, is said to be **invariant under** $\mathcal{T}$ if $\mathcal{T}A = A$.

Definition 9.12. In the context of Definition 9.10 a $\mathcal{A}-$ measurable function f is said to be **invariant**, if $f(\omega) = f(\mathcal{T}\omega)$.

Definition 9.13. In the context of Definition 9.11 the transformation $\mathcal{T}$ is said to be ergodic if the only $\mathcal{T}$-invariant sets in $\mathcal{A}$ have measure zero or one.

Definition 9.14. In the context of Definition 9.10, the orbit of a point $\omega \in \Omega$, under $\mathcal{T}$ is the sequence

$$\{\omega, \mathcal{T}\omega, \mathcal{T}^2\omega, \mathcal{T}^3\omega, \ldots\}.$$

---

[7]This discussion owes a great deal to Billingsley (1995), Shiryayev (1984), and Stout (1974) which contain a very lucid description of the concepts and issues involved in ergodicity, including the role played by Kolmogorov's extension theorem we discussed in Chap. 8 (Proposition 8.6), and Kolmogorov's Zero-One Law, Proposition 8.25.

We are now in a position to state the ergodic and ergodic related theorems that will enable us to deal with the properties of parameter estimators such as the mean and the auto-covariance or auto-correlation functions of **causal** time series.

Proposition 9.13 (Ergodic theorem). Let $(\Omega, \mathcal{A}, \mathcal{P})$ be a probability space and consider the MPT

$$\mathcal{T}: \quad \Omega \to \Omega,$$

and the $\mathcal{A}$-measurable and integrable function f. Then

$$\lim_{N \to \infty} \frac{1}{N} \sum_{k=1}^{N} f(\mathcal{T}^{k-1}(\omega)) = \hat{f}(\omega) \tag{9.17}$$

exists with probability 1, is invariant and integrable with $E\hat{f}(\omega) = f(\omega)$. Moreover, **if $\mathcal{T}$ is ergodic, $\hat{f} = f$ with probability 1**.

Proof: See Billingsley (1995, pp. 317–319) and Shiryayev (1984, pp. 379–385).

Another important result that has a direct bearing in the discussion of such issues in the context of time series is

Proposition 9.14. Let $\{X(t): \quad t \geq 1\}$ be stationary (and) ergodic, and ϕ a measurable function

$$\phi: \quad R^\infty \to (R, \mathcal{B}(R)), \text{ and define } Y(t) = \phi(X(t), X(t+1), X(t+2), \ldots).$$

Then $\{Y(t): \quad t \geq 1\}$ is stationary (and) ergodic.

For a proof and relevant discussion, see Stout (1974, pp. 182–185).

Strictly speaking Proposition 9.14, as stated, is not directly applicable to causal auto-regressions because there we need to have ϕ and $Y(t)$ defined, at least, as

$$Y(t) = \phi(X(t), X(t-1), X(t-2), \ldots), \quad t \geq 1,$$

in other words we need to deal with doubly infinite series. Since this is not a book on mathematics we shall not attempt an extension. Instead, we present an alternative result which corrects this deficiency.

Proposition 9.15. Let X be a stationary and ergodic (sequence); in particular if the X_n (X_n: $n = 0, \pm 1, \ldots$) are independent and identically distributed, (and if ϕ is measurable and time invariant) define the sequence

$$Y_n = \phi(\ldots, X_{n-1}, X_n, X_{n+1}, \ldots), \quad (n = 0, \pm 1, \pm 2, \pm 3, \ldots). \tag{9.18}$$

Then the sequence Y_n (as above) is stationary and ergodic.

For a discussion and proof of this result see Billingsley (1995, pp. 494–496). The expressions in parentheses are editorial comments inserted only for clarity and conformity with this author's notation and style.

Remark 9.6. While in the discussion above we have developed the tools to handle issues related to ergodicity, there is an important aspect of the convergence of ergodic series (i.e. estimators) that needs to be made clear. The methods of proof employed **do not allow us to determine rates of convergence.**

Remark 9.7. To properly employ the tools provided by the ergodic results above, it is useful to have a general conception of what ergodicity means beyond the formal mathematical definitions we provided. One view, interpreting the orbit of a point $\omega \in \Omega$ of Definition 9.14, is to think of it as passing, (over time) through nearly all sets of its σ-algebra (or at least the sub-σ-algebra $\sigma(X)$ generated by the elements of the sequence X); thus when operating with realizations over time we can capture moment related properties of the sequence $\{X(t): t \in T\}$ over the entire space; this may not be strictly accurate but it conveys the meaning and essence of the limit theorems.

Another useful view is to think of the individual components, $X(t)$, as rvs operating within a given probabilistic structure; thus a realization at "time" t is one of infinite possible outcomes that may have been generated by the random variable in question. Ergodicity in this context means, roughly speaking, that the probability structure remains constant over "time". Notice that this hints at "identical" distributions but not necessarily independence.

Remark 9.8. In some widely circulating textbooks on advanced econometrics, it is alleged that ergodicity means dealing essentially with nearly independent sequences or, substantially, the independence between elements of the sequence $\{X(t), X(s), t, s, \in T\}$ provided the indices are sufficiently far apart, i.e. that $|t - s|$ is sufficiently large.[8] **This is actually a confusion between two concepts, mixing and ergodicity.** Mixing, in the context of this discussion, and particularly in the context of an MPT, $\mathcal{T}$, means, for any two sets A, B in the σ-algebra of the probability space $(\Omega, \mathcal{A}, \mathcal{P})$

$$\lim_{n \to \infty} P(A \cap \mathcal{T}^{-n}B) = P(A)P(B); \tag{9.19}$$

---

[8]This is a property referred to under certain circumstances as *mixing*.

on the other hand $\mathcal{T}$ is ergodic if and only if[9]

$$\lim_{n\to\infty} \frac{1}{n} \sum_{k=1}^{n-1} P(A \cap \mathcal{T}^{-k}B) = P(A)P(B), \tag{9.20}$$

for any two sets A, B as in Eq. (9.19).

It is evident that the condition in Eq. (9.19) implies (is stronger than) the condition in Eq. (9.20), so that mixing implies ergodicity, **ergodicity does not imply mixing.**

9.4.3 Applications

Example 9.1. Consider the scalar (causal) AR(n)

$$X(t) = \sum_{j=1}^{n} \phi_j X(t-j) + \epsilon_t, \tag{9.21}$$

where the sequence ϵ_t is one of $iid(0, \sigma^2)$ rvs and thus stationary and ergodic. Since the $AR(n)$ of Eq. (9.21) is causal, its $M(\infty)$ representation is

$$X(t) = \sum_{i=0}^{\infty} \psi_i \epsilon_{t-i} = \psi(\epsilon_s, s \le t), \tag{9.22}$$

and the function ψ is both measurable and time invariant. Because of causality the right member of Eq. (9.22) converges absolutely and thus, by Proposition 9.15, the sequence $\{X(t): \ t \in T\}$ is stationary and ergodic.

Example 9.2. Consider now a realization of the sequence in Example 9.1, $\{x_k: \ k = 1, 2, 3, \ldots N\}$, and the mean realization function

$$\hat{\mu}_N = \frac{1}{N} \sum_{k=1}^{N} x_k. \tag{9.23}$$

The right member of Eq. (9.23) is measurable and integrable and its limit exists with probability 1, and moreover

$$\lim_{N\to\infty} \hat{\mu}_N = \lim_{N\to\infty} E\hat{\mu}_N = 0,$$

with probability 1 by Proposition 9.14, or Proposition 9.15.

---

[9]See Billingsley (1995), p. 325, who also gives a counterexample of an ergodic sequence which is not mixing.

Example 9.3. In the context of Example 9.2, consider the realization auto-covariance function (elements)

$$\hat{c}_N(h) = \frac{1}{N} \sum_{k=1}^{N} x_{k+h} x_k, \tag{9.24}$$

with the proviso that if for some index there is no corresponding observation that term is omitted from the sum. The function in the right member of Eq. (9.24) is both measurable and integrable; thus its limit exists with probability 1, by Proposition 9.13, and moreover

$$\lim_{N \to \infty} \hat{c}_N(h) = \sigma^2 \sum_{j=0}^{\infty} \psi_j^2,$$

with probability 1 by Proposition 9.14 or Proposition 9.15. In a similar way we can show that the estimators of (the elements of) the auto-correlation function converge to their corresponding parameters with probability 1.

Example 9.4. Consider, in the usual (econometric regression) context, the model

$$y_t = \lambda y_{t-1} + u_t, \quad |\lambda| < 1, \quad \text{where } u_t, \ t = 1, 2, 3, \ldots, T, \tag{9.25}$$

is iid and defined on the probability space $(\Omega, \mathcal{A}, \mathcal{P})$. The OLS estimator of λ is given by

$$\hat{\lambda} = \frac{\sum_{t=2}^{T} y_t y_{t-1}}{\sum_{t=2}^{T} y_{t-1}^2} = \lambda + \frac{\sum_{t=2}^{T} u_t y_{t-1}}{\sum_{t=2}^{T} y_{t-1}^2}.$$

Put

$$\hat{a}_T = \frac{\sum_{t=2}^{T} y_{t-1}^2}{T}, \quad \text{and note that}^{10} \quad \hat{a}_T \xrightarrow{\ P\ } \frac{\sigma^2}{(1 - \lambda^2)}.$$

Therefore we can write

$$\sqrt{T}(\hat{\lambda} - \lambda) \sim \frac{1 - \lambda^2}{\sigma^2} \sum_{t=2}^{T} X_{tT}, \quad X_{tT} = \frac{y_{t-1} u_t}{\sqrt{T}}.$$

Now define the (sub) σ-algebras $\mathcal{A}_{tT} = \sigma(X_{sT} : s \le t) \subseteq \mathcal{A}_{t+1,T}$, and note that X_{tT} is $\mathcal{A}_{tT}$-measurable. In addition,

$$E(X_{tT}|\mathcal{A}_{t-1,T}) = \frac{1}{\sqrt{T}} y_{t-1} E u_t = 0, \quad E X_{tT}^2 | \mathcal{A}_{t-1,T} = \frac{1}{T} y_{t-1}^2 E u_t^2 = \frac{1}{T} y_{t-1}^2 \sigma^2,$$

---

[10] Actually if we wished we could invoke the earlier discussion on ergodicity to argue that the convergence is actually with probability one, which of course implies convergence in probability.

so that $\{X_{tT}\}$ is a martingale difference (MD) sequence (by vi of Definition 9.9) and $\{A_{t,T}\}$ is a stochastic basis (also by Definition 9.9). Moreover,

$$\sum_{t=2}^{T} EX_{tT}^2 | A_{t-1,T} = \frac{1}{T} \sum_{t=2}^{T} y_{t-1}^2 \sigma^2 \overset{P}{\to} \frac{\sigma^4}{1-\lambda^2},$$

and the sequence obeys a Lindeberg condition.   Consequently, by i of Proposition 9.9

$$\sqrt{T}(\hat{\lambda} - \lambda) \overset{d}{\to} \zeta \sim N(0, 1 - \lambda^2).$$

Example 9.5. The previous example dealt with a rather simple case, that of a scalar AR(1). Here we shall deal with a more complex case that of a vector (r-element) AR(m). First some notation. Let

$$\{x_{t\cdot} : x_{t\cdot} = (x_{t1}, x_{t2}, \ldots, x_{tr}), \quad t = 1, 2, 3, \ldots N\}$$

be a realization of length N of the (causal) r-element vector $AR(m)$ time series sequence indexed on a linear index set. Define the (realization) matrices,

$$X = (x_{t\cdot}), \quad t = m+1, m+2, \ldots, N, \quad X_{-i} = (x_{t-i\cdot}), \quad t = m+1-i, m+2-i, \ldots, N-i, \tag{9.26}$$

and the $r \times r$ parameter matrices A_i, $i = 1, 2, 3, \ldots$, m, so that the system of observations can be represented as

$$X = \sum_{i=1}^{m} X_{-i} A_i + U, \quad U = (u_{t\cdot}), \quad t = m + 1, m + 2, \ldots, N, \tag{9.27}$$

where $u_{t\cdot}$ is a realization of an r-element $iid(0, \Sigma)$ sequence with $\Sigma > 0$.

Using the results of Chap. 4 to vectorize, we can represent Eq. (9.27) as

$$x = (I_r \otimes X^*)\text{vec}(A) + \text{vec}(U) = (I_r \otimes X^*)a + u, \quad \text{where}$$

$$x = \text{vec}(X), \quad X^* = (X_{-1}, X_{-2}, \ldots, X_{-m}),$$

$$a = \text{vec}(A), \quad A = (A_1', A_2', A_3', \ldots, A_m')', \quad u = \text{vec}(U). \tag{9.28}$$

Thus, the unknown parameters of the model, contained in the vector a, can be estimated by OLS methods as

$$\hat{a} = (I_r \otimes X^{*\prime}X^*)^{-1}(I_r \otimes X^{*\prime})x = a + (I_r \otimes X^{*\prime}X^*)^{-1}(I_r \otimes X^{*\prime})u. \tag{9.29}$$

By the discussion of ergodicity above

$$\left(\frac{I_r \otimes X^{*\prime}X^*}{N} \right) \overset{\text{a.c.}}{\to} I_r \otimes K, \tag{9.30}$$

where K is the block Toeplitz matrix (see Chap. 4)

$$K = \begin{bmatrix} K(0) & K(1) & K(2) & \cdots & \cdots & K(m-1) \\ K(1)' & K(0) & K(1) & \cdots & \cdots & K(m-2) \\ K(2)' & K(1)' & K(0) & \cdots & \cdots & K(m-3) \\ \vdots & \vdots & \vdots & \vdots & \vdots & \vdots \\ \vdots & \vdots & \vdots & \vdots & \vdots & \vdots \\ K(m-1)' & K(m-2)' & K(m-3)' & \cdots & \cdots & K(0) \end{bmatrix},$$

$$\frac{X'_{-i}X_{-j}}{N} \overset{\text{a.c.}}{\to} K(i-j), \quad K(j-i) = K'(i-j), \quad i,j = 1,2,3,\ldots,m. \quad (9.31)$$

Using Proposition 9.6 above we can, equivalently, deal with the problem of finding the limiting distribution of

$$\sqrt{N}(\hat{a} - a) = (I_r \otimes K^{-1})\frac{1}{\sqrt{N}}(I_r \otimes X^{*\prime})u, \quad \text{where}$$

$$(I_r \otimes X^{*\prime})u = \text{vec}(X^{*\prime}U),$$

$$X^{*\prime}U = \sum_{t=m+1}^{N} x^{*\prime}_{t\cdot} u_{t\cdot}. \quad (9.32)$$

We shall now show that the sequence above is a MD and thus apply the MD central limit theorem of Proposition 9.9, as we did in the simpler discussion above. To this end note that

$$x^{*\prime}_{t\cdot} = (x_{t-1\cdot}, x_{t-2\cdot}, x_{t-3\cdot}, \ldots, x_{t-m\cdot})', \quad u_{t\cdot} = (u_{t1}, u_{t2}, u_{t3}, \ldots, u_{tr}). \quad (9.33)$$

Thus, for the first summand of Eq. (9.32) we obtain

$$(x_{m\cdot}, x_{m-1\cdot}, x_{m-2\cdot}, \ldots, x_{1\cdot})' u_{m+1\cdot}.$$

Now, if we construct the sub-σ-algebra

$$\mathcal{A}_m = \sigma(u_{s\cdot} : \quad s \leq m),$$

the entities of the first summand obey

$$E[(x_{m\cdot}, x_{m-1\cdot}, x_{m-2\cdot}, \ldots, x_{1\cdot})' u_{m+1\cdot} | \mathcal{A}_m] = 0, \quad (9.34)$$

because the first factor is $\mathcal{A}_m$-measurable **and thus independent of** $u_{m+1\cdot}$.

Similarly, if we look at the second summand of Eq. (9.32) we obtain

$$(x_{m+1\cdot}, x_{m\cdot}, x_{m-1\cdot}, \ldots, x_{2\cdot})' u_{m+2\cdot},$$

whose first factor is $\mathcal{A}_{m+1}$ -measurable and thus independent of $u_{m+2.}$. Consequently the expectation of the second summand is zero and so on. Thus the sequence

$$\zeta_t = x_{t.}^{*\prime} u_{t.}, \quad t = m+1, m+2, m+3, \ldots \tag{9.35}$$

is a martingale difference sequence (MD sequence) which obeys i of Proposition 9.9 as well as the Lindeberg condition noted therein; the latter is true because of the causal property of the autoregression, which makes all parameters involved in the $MA(\infty)$ representation of the elements of the realization, less that one in absolute value. In addition

$$\sum_{t=m+1}^{N} \frac{1}{N} \left(E(I_r \otimes x_{t.}^{*\prime} u_{t.}' u_{t.} I_r \otimes x_{t.} | \mathcal{A}_{t-1}) \right) = \Sigma \otimes \frac{X^{*\prime} X^*}{N} \overset{\text{a.c.}}{\to} \Sigma \otimes K, \tag{9.36}$$

and the first absolute moment exists. Thus, applying Proposition 9.9 we conclude,

$$\sqrt{N}(\hat{a} - a) \overset{\text{d}}{\to} N(0, \Sigma \otimes K^{-1}). \tag{9.37}$$

Chapter 10

The General Linear Model

10.1 Introduction

In this chapter, we examine the General Linear Model (GLM), an important topic for econometrics and statistics, as well as other disciplines. The term general refers to the fact that there are no restrictions in the number of explanatory variables we may consider, the term linear refers to the manner in which **the parameters enter the model. It does not refer to the form of the variables.** This is often termed in the literature the **regression model**, and analysis of empirical results obtained from such models as **regression analysis**.

This is perhaps the most commonly used research approach by empirically oriented economists. We examine both its foundations, but also how the mathematical tools developed in earlier chapters are utilized in determining the properties of the parameter estimators[1] that it produces.

The basic model is

$$y_t = \sum_{i=0}^{n} x_{ti}\beta_i + u_t, \tag{10.1}$$

where y_t is an observation at "time" t on the phenomenon to be "explained" by the analysis; the x_{ti}, $i = 0, 1, 2, \ldots, n$ are observations on variables that the investigator asserts are important in explaining the behavior of the phenomenon in question; β_i, $i = 0, 1, 2, \ldots n$ are parameters i.e. they are fixed but **unknown** constants that modify the influence of the x's on y. In the language of econometrics, y is termed the **dependent** variable, while the x's are termed the **independent or explanatory** variables; in the language of statistics, they are often referred to, respectively, as the **regressand** and the **regressors**. The u's simply acknowledge that the enumerated variables

---

[1]The term estimator recurs very frequently in econometrics; just to fix its meaning in this chapter and others we define it as: an estimator is a function of the data only ($x's$ and $y's$), say $h(y, X)$ that **does not include unknown parameters**.

do not provide an exhaustive explanation for the behavior of the dependent variable; in the language of econometrics, they are typically referred to as the error term or the **structural errors**. The model stated in Eq. (10.1) is thus the data generating function for the data to be analysed. Contrary to the time series approach to data analysis, econometrics nearly always deals within a reactive context in which the behavior of the dependent variable is conditioned by what occurs in the economic environment beyond itself and its past history.

One generally has a set of observations (sample) over T periods, and the problem is to obtain estimators and carry out inference procedures (such as tests of significance, construction of confidence intervals, and the like) relative to the unknown parameters. Such procedures operate in a certain environment. Before we set forth the assumptions defining this environment, we need to establish some notation. Thus, collecting the explanatory variable observations in the $T \times n + 1$ matrix

$$X = (x_{ti}), \quad i = 0, 1, 2, \ldots, n, \quad t = 1, 2, 3, \ldots, T, \tag{10.2}$$

and further defining

$$\beta = (\beta_0, \beta_1, \ldots, \beta_n)', \quad y = (y_1, y_2, \ldots, y_T)', \quad u = (u_1, u_2, \ldots, u_T)', \tag{10.3}$$

the observations on the model may be written in the compact form

$$y = X\beta + u. \tag{10.4}$$

10.2   The GLM and the Least Squares Estimator

The assumptions that define the context of this discussion are the following:

i. The elements of the matrix X are **nonstochastic** and its columns are **linearly independent**. Moreover,

$$\lim_{T \to \infty} \frac{X'X}{T} = M_{xx} > 0 \quad \text{i.e. it is a positive definite matrix.}$$

Often, the explanatory variables may be considered random, **but independent of the structural errors of the model**. In such cases the regression analysis is carried out **conditionally on the x 's**.

ii. The errors, u_t, are independent, identically distributed (iid) random variables with mean zero and variance $0 < \sigma^2 < \infty$.

iii. In order to obtain a distribution theory, one often adds the assumption that the errors have the normal joint distribution with mean vector zero and covariance matrix $\sigma^2 I_T$, or more succinctly one writes

$$u \sim N(0, \sigma^2 I_T); \tag{10.5}$$

with increases in the size of the samples (data) available to econometricians over time, this assumption is not frequently employed in current applications, relying instead on central limit theorems (CLT) to provide the distribution theory required for inference.

The least squares estimator is obtained by the operation

$$\min_{\beta}(y - X\beta)'(y - X\beta),$$

which yields

$$\hat{\beta} = (X'X)^{-1}X'y. \tag{10.6}$$

The first question that arises is: how do we know that the inverse exists? Thus, how do we know that the estimator of β is uniquely defined? We answer that in the following proposition.

Proposition 10.1. If X obeys condition i, $X'X$ is positive definite, and thus invertible.

Proof: Since the columns of X are linearly independent, the only vector α such that $X\alpha = 0$ is the zero vector—see Proposition 2.61. Thus, for $\alpha \neq 0$, consider

$$\alpha' X' X \alpha = \gamma' \gamma = \sum_{j=0}^{n+1} \gamma_j^2 > 0, \qquad \gamma = X\alpha. \tag{10.7}$$

Hence, $X'X$ is positive definite and thus invertible—see Proposition 2.62.

The model also contains another unknown parameter, namely the common variance of the errors σ^2. Although the least squares procedure does not provide a particular way in which such a parameter is to be estimated, it seems intuitively reasonable that we should do so through the residual vector

$$\hat{u} = y - X\hat{\beta} = [I_T - X(X'X)^{-1}X']u = Nu. \tag{10.8}$$

First, we note that N is a **symmetric idempotent** matrix—for a definition of symmetric matrices, see Definition 2.4; for a definition of idempotent matrices, see Definition 2.8. It appears intuitively reasonable to think of $\hat{u}$ of Eq. (10.8) as an "estimator" of the unobservable error vector u; thus it is also natural that we should define an estimator of σ^2, based on the sum of squares $\hat{u}'\hat{u}$. We return to this topic in the next section.

10.3   Properties of the Estimators

The properties of the least squares, termed in econometrics the OLS (ordinary least squares), estimator are given below.

Proposition 10.2 (Gauss-Markov Theorem). In the context set up by the three conditions noted above, the OLS estimator of Eq. (10.6) is

   i. Unbiased,

   ii. Efficient **within the class of linear unbiased estimators.**

Proof: Substituting from Eq. (10.4), we find

$$\hat{\beta} = \beta + (X'X)^{-1}X'u, \tag{10.9}$$

and taking expectations, we establish, in view of conditions i and ii,

$$E\hat{\beta} = \beta + E(X'X)^{-1}X'u = \beta. \tag{10.10}$$

To prove efficiency, let $\tilde{\beta}$ be **any other** linear (in y) unbiased estimator. In view of linearity, we may write

$$\tilde{\beta} = Hy, \quad \text{where } H \text{ depends on } X \text{ only and not on } y. \tag{10.11}$$

Without loss of generality, we may write

$$H = (X'X)^{-1}X' + C. \tag{10.12}$$

Note further that since $\tilde{\beta} = Hy$ is an **unbiased** estimator of β, we have

$$HX\beta = \beta, \quad \text{which implies } HX = I_{n+1}, \quad \text{or, equivalently, } CX = 0. \tag{10.13}$$

It follows then immediately that

$$\mathrm{Cov}(\tilde{\beta}) = \sigma^2(X'X)^{-1} + \sigma^2 CC' \quad \text{or, equivalently,} \tag{10.14}$$

$$\mathrm{Cov}(\tilde{\beta}) - \mathrm{Cov}(\hat{\beta}) = \sigma^2 CC' \geq 0. \tag{10.15}$$

That the rightmost member of the equation is valid may be shown as follows: If $\tilde{\beta}$ is not trivially identical to the OLS estimator $\hat{\beta}$, the matrix C is of rank **greater** than zero. Let this rank be r. Thus, there exists at least one (non-null) vector α such that $C'\alpha \neq 0$. Consequently, $\alpha'CC'\alpha > 0$, which demonstrates the validity of the claim.

q.e.d.

Corollary 10.1. The OLS estimator of β is also consistent[2] in the sense that

$$\hat{\beta}_T \xrightarrow{\text{a.c.}} \beta,$$

provided $|x_t . x_t'.| < B$ uniformly.

Proof: From Sect. 9.3.1 use Kolmogorov's criterion with $b_n = T$. Then

$$\left| \sum_{t=1}^{T} \frac{\text{var}(x_t'.u_t)}{t^2} \right| \leq B\sigma^2 \sum_{t=1}^{T} \frac{1}{t^2},$$

which evidently converges. Hence $\hat{\beta} \xrightarrow{\text{a.c.}} \beta$, as claimed. If one does not wish to impose the uniform boundedness condition there are similar weaker conditions that allow the proof of convergence with probability 1. Alternatively, one may not add any further conditions on the explanatory variables, and prove convergence in quadratic mean, see Definition 9.4. With $p = 2$. This entails showing unbiasedness, already shown, and asymptotic vanishing of the estimator's variance. Thus, by Proposition 10.2 and condition ii,

$$\lim_{T \to \infty} E(\hat{\beta} - \beta)(\hat{\beta} - \beta)' = \sigma^2 \lim_{T \to \infty} \frac{1}{T} \left(\frac{X'X}{T} \right)^{-1} = 0. \tag{10.16}$$

q.e.d.

We now examine the properties of the estimator for σ^2, hinted at the end of the preceding section.

Proposition 10.3. Consider the sum of squares $\hat{u}'\hat{u}$; its expectation is given by

$$E\hat{u}'\hat{u} = (T - n - 1)\sigma^2. \tag{10.17}$$

Proof: Expanding the representation of the sum of squared residuals we find $\hat{u}'\hat{u} = u'Nu$. Hence

$$Eu'Nu = Etru'Nu = EtrNuu' = trNEuu' = \sigma^2 trN. \tag{10.18}$$

The first equality follows since $u'Nu$ is a scalar; the second follows since for all suitable matrices $trAB = trBA$ —see Proposition 2.16; the third equality

---

[2]The term consistent generally means that as the sample, T, tends to infinity the estimator converges to the parameter it seeks to estimate. Since the early development of econometrics it meant almost exclusively convergence in probability. This is the meaning we shall use in this and other chapters, i.e. an estimator is consistent for the parameter it seeks to estimate if it converges to it in any fashion that implies convergence in probability.

follows from the fact that X, and hence N, is a nonstochastic matrix; the last equality follows from condition ii that defines the context of this discussion. Thus, we need only find the trace of N. Since $\operatorname{tr}(A + B) = \operatorname{tr}A + \operatorname{tr}B$ —see Proposition 2.16—we conclude that

$$\operatorname{tr}N = \operatorname{tr}I_T - \operatorname{tr}X(X'X)^{-1}X' = T - \operatorname{tr}(X'X)^{-1}X'X$$

$$= T - \operatorname{tr}I_{n+1} = T - n - 1. \tag{10.19}$$

q.e.d.

Corollary 10.2. The unbiased OLS estimator for σ^2 is given by

$$\hat{\sigma}^2 = \frac{\hat{u}'\hat{u}}{T - n - 1}. \tag{10.20}$$

Proof: Evident from Proposition 10.3.

10.4 Distribution of OLS Estimators

10.4.1 A Digression on the Multivariate Normal

We begin this section by stating a few facts regarding the multivariate normal distribution. A random variable (vector) x having the multivariate normal distribution with mean (vector) μ and covariance matrix Σ is denoted by

$$x \sim N(\mu, \Sigma), \tag{10.21}$$

to be read x has the multivariate normal distribution with mean vector μ and covariance matrix $\Sigma > 0$.

Its moment generating function is given by

$$M_x(t) = Ee^{t'x} = e^{t'\mu + \frac{1}{2}t'\Sigma t}. \tag{10.22}$$

Generally, we deal with situations where $\Sigma > 0$, so that there are no linear dependencies among the elements of the vector x, which result in the singularity of the covariance matrix. We handle singular covariance matrix situations through the following convention.

Convention 10.1. Let the k-element vector ξ obey $\xi \sim N(\mu, \Sigma)$, such that $\Sigma > 0$, and suppose that y is an n-element vector ($k \leq n$) which has the representation

$$y = A\xi + b, \quad \operatorname{rank}(A) = k. \tag{10.23}$$

Then, we say that y has the distribution

$$y \sim N(\nu, \Psi), \quad \nu = A\mu + b, \quad \Psi = A\Sigma A'. \tag{10.24}$$

Note that in Eq. (10.24), Ψ **is singular**, but properties of y can be inferred from those of ξ which has a proper multivariate normal distribution. Certain properties of the (multivariate) normal that are easily derivable from its definition are:

i. Let $x \sim N(\mu, \Sigma)$, partition $x = \begin{pmatrix} x^{(1)} \\ x^{(2)} \end{pmatrix}$, such that $x^{(1)}$ has s elements and $x^{(2)}$ has $k - s$ elements. Partition μ and Σ conformably so that

$$\mu = \begin{pmatrix} \mu^{(1)} \\ \mu^{(2)} \end{pmatrix}, \quad \Sigma = \begin{bmatrix} \Sigma_{11} & \Sigma_{12} \\ \Sigma_{21} & \Sigma_{22} \end{bmatrix}. \tag{10.25}$$

Then, the **marginal distribution** of $x^{(i)}$, $i = 1, 2$, obeys

$$x^{(i)} \sim N(\mu^{(i)}, \Sigma_{ii}), \quad i = 1, 2. \tag{10.26}$$

The **conditional distribution** of $x^{(1)}$ given $x^{(2)}$ is given by

$$x^{(1)} | x^{(2)} \sim N(\mu^{(1)} + \Sigma_{12}\Sigma_{22}^{-1}(x^{(2)} - \mu^{(2)}), \ \Sigma_{11} - \Sigma_{12}\Sigma_{22}^{-1}\Sigma_{21}). \tag{10.27}$$

ii. Let the k-element random vector x obey $x \sim N(\mu, \Sigma)$, $\Sigma > 0$, and define $y = Bx + c$, where B is any conformable matrix; then

$$y \sim N(B\mu + c, \ B\Sigma B'). \tag{10.28}$$

iii. Let $x \sim N(\mu, \Sigma)$ and partition as in part i; $x^{(1)}$ and $x^{(2)}$ are **mutually independent** if and only if

$$\Sigma_{12} = \Sigma_{21}' = 0. \tag{10.29}$$

iv. We also have the sort of converse of ii, i.e. if x is as in ii, there exists a matrix C such that

$$y = C^{-1}(x - \mu) \sim N(0, I_k). \tag{10.30}$$

The proof of this is quite simple; by Proposition 2.15 there exist a nonsingular matrix C such that $\Sigma = CC'$; by ii, $y \sim N(0, C^{-1}\Sigma C'^{-1} = I_k)$.

v. An implication of iv is that

$$(x - \mu)'\Sigma^{-1}(x - \mu) = y'y = \sum_{i=1}^{k} y_i^2 \sim \chi_r^2, \tag{10.31}$$

because the y_i are iid $N(0, 1)$ whose squares have the χ^2 distribution. More about this distribution will be found immediately below.

In item iii, note that if joint normality **is not assumed** and we partition $x = \begin{pmatrix} x^{(1)} \\ x^{(2)} \end{pmatrix}$, such that $x^{(1)}$ has s elements and $x^{(2)}$ has $k - s$ elements, as above, then under the condition in iii, $x^{(1)}$ and $x^{(2)}$ **are still uncorrelated**, but **they are not necessarily independent.** Under normality uncorrelatedness implies independence; under any distribution independence always implies uncorrelatedness.

Other distributions, important in the GLM context are the chi-square, the t-(sometimes also termed the Student t) and the F-distributions.

The chi-square distribution with r degrees of freedom, denoted by χ_r^2, may be thought to be the distribution of **the sum of squares of** r **mutually independent normal variables with mean zero and variance one**; the t-distribution with r degrees of freedom, denoted by t_r, is defined as the distribution of the ratio

$$t_r = \frac{\xi}{\sqrt{\zeta/r}}, \quad \xi \sim N(0,1), \quad \zeta \sim \chi_r^2, \tag{10.32}$$

with ξ and ζ **mutually independent.**

The F-distribution with m and n degrees of freedom, denoted by $F_{m,n}$, is defined as the distribution of the ratio

$$F_{m,n} = \frac{\xi/m}{\zeta/n}, \quad \xi \sim \chi_m^2, \quad \zeta \sim \chi_n^2, \tag{10.33}$$

with ξ and ζ **mutually independent.**

Note that $F_{m,n} \neq F_{n,m}$. The precise relation between the two is given by $F_{m,n} = 1/F_{n,m}$.

10.4.2 Application to OLS Estimators

We now present a very important result.

Proposition 10.4. The OLS estimators $\hat{\beta}$ and $\hat{\sigma}^2$ are mutually independent.

Proof: Consider the $T + n + 1$-element vector, say $\phi = (\hat{\beta}', \hat{u}')'$. From the preceding discussion, we have the representation

$$\phi = Au + \begin{pmatrix} \beta \\ 0 \end{pmatrix}, \quad \text{where} \quad A = \begin{bmatrix} (X'X)^{-1}X' \\ N \end{bmatrix}. \tag{10.34}$$

From our discussion of the multivariate normal, we conclude

$$\phi \sim N(\nu, \Psi), \quad \text{where} \quad \nu = \begin{pmatrix} \beta \\ 0 \end{pmatrix}, \quad \Psi = \sigma^2 AA'. \tag{10.35}$$

But

$$AA' = \begin{bmatrix} (X'X)^{-1} & 0 \\ 0 & N \end{bmatrix}, \tag{10.36}$$

which shows that $\hat{\beta}$ and $\hat{u}$ are **uncorrelated** and hence, by the properties of the multivariate normal, they are **mutually independent**. Since $\hat{\sigma}^2$ depends only on $\hat{u}$ **and thus not on** $\hat{\beta}$, the conclusion of the proposition is evident.

q.e.d.

Corollary 10.3. Denote the vector of coefficients of the *bona fide* variables by β_* and the coefficient of the "fictitious variable" one (x_{t0}) by β_0 (the constant term), so that we have

$$\beta = (\beta_0, \beta_*')', \quad \beta_* = (\beta_1, \beta_2, \ldots, \beta_n)', \quad X = (e, X_1), \tag{10.37}$$

where e is a T-element column vector all of whose elements are unities. The following statements are true:

   i. $\hat{u} \sim N(0, \sigma^2 N)$;

  ii. $\hat{\beta} \sim N(\beta, \sigma^2 (X'X)^{-1})$;

 iii. $\hat{\beta}_* \sim N(\beta_*, \sigma^2 (X_1^{*'} X_1^*)^{-1}), \quad X_1^* = (I_T - ee'/T)X_1$.

Proof: The first two statements follow immediately from Proposition 10.3 and property i of the multivariate normal. The statement in iii also follows immediately from property i of the multivariate normal and the properties of the inverse of partitioned matrices; however, we also give an alternative proof because we will need a certain result in later discussion.

The first order conditions of the OLS estimator read

$$T\beta_0 + e'X_1\beta_* = e'y, \quad X_1'e\beta_0 + X_1'X_1\beta_* = X_1'y.$$

Solving by substitution, we obtain, from the first equation,

$$\hat{\beta}_0 = \bar{y} - \bar{x}_1 \hat{\beta}_*, \quad \bar{y} = \frac{e'y}{T}, \quad \bar{x}_1 = \frac{X_1'e}{T}, \tag{10.38}$$

and from the second equation

$$\hat{\beta}_* = \left[X_1' \left(I_T - \frac{ee'}{T} \right) X_1 \right]^{-1} \left[X_1' \left(I_T - \frac{ee'}{T} \right) y \right]; \tag{10.39}$$

substituting for $y = e\beta_0 + X_1\beta_* + u$, and noting that $(I_T - ee'/T)e = 0$, we find equivalently

$$\hat{\beta}_* = \beta_* + (X_1^{*'} X_1^*)^{-1} X_1^{*'} u. \tag{10.40}$$

The validity of statement iii then follows immediately from Property ii of the multivariate normal.

q.e.d.

Remark 10.1. The results given above ensure that tests of significance or other inference procedures may be carried out, even when the variance parameter, σ^2, is not known. As an example, consider the coefficient of correlation of multiple regression

$$R^2 = 1 - \frac{\hat{u}'\hat{u}}{(y - e\bar{y})'(y - e\bar{y})} = \frac{(y - e\bar{y})'(y - e\bar{y}) - \hat{u}'\hat{u}}{(y - e\bar{y})'(y - e\bar{y})}.$$

To clarify the role of matrix algebra in easily establishing the desired result, use the first order condition and the proof of Corollary 10.3 to establish

$$y - e\bar{y} = \hat{u} + \left(I_T - \frac{ee'}{T} \right) X_1 \hat{\beta}_*$$

$$(y - e\bar{y})'(y - e\bar{y}) - \hat{u}'\hat{u} = \hat{\beta}_*' X_1' \left(I_T - \frac{ee'}{T} \right) X_1 \hat{\beta}_*, \quad \text{and}$$

$$\frac{R^2}{1 - R^2} = \frac{\hat{\beta}_*' X_1' \left(I_T - \frac{ee'}{T} \right) X_1 \hat{\beta}_*}{\hat{u}'\hat{u}}.$$

It follows therefore that

$$\left(\frac{R^2}{1 - R^2} \right) \frac{T - n - 1}{n} = \left(\frac{\hat{\beta}_*' X_1^{*'} X_1^* \hat{\beta}_*/\sigma^2}{\hat{u}'\hat{u}/\sigma^2} \right) \frac{T - n - 1}{n}. \tag{10.41}$$

The relevance of this result in carrying out "significance tests" is as follows: first, note that since $\hat{\beta}_* \sim N(\beta_*, \sigma^2(X_1^{*'}X_1^*)^{-1})$ the numerator and denominator of the fraction are **mutually independent** by Proposition 10.3. From Proposition 2.15, we have that every positive definite matrix has a nonsingular decomposition, say AA'. Let

$$(X_1^{*'}X_1^*)^{-1} = AA'. \tag{10.42}$$

It follows from property ii of the multivariate normal that

$$\xi = \frac{A^{-1}(\hat{\beta}_* - \beta_*)}{\sqrt{\sigma^2}} \sim N(0, I_n). \tag{10.43}$$

This implies that **all** of the elements of the vector ξ are scalars, mutually independent (normal) random variables with mean zero and variance one. Hence, by the preceding discussion,

$$\xi'\xi = \frac{(\hat{\beta}'_* - \beta_*)X_1^{*'}X_1^*(\hat{\beta}_* - \beta_*)}{\sigma^2} \sim \chi_n^2. \tag{10.44}$$

Similarly, $\hat{u}'\hat{u} = u'Nu$ and N is a symmetric idempotent matrix of rank $T - n - 1$. As a symmetric and idempotent matrix, it has the representation

$$N = Q\Lambda Q' \qquad \Lambda = \begin{bmatrix} I_{T-n-1} & 0 \\ 0 & 0_{n+1} \end{bmatrix}, \tag{10.45}$$

where Q is an orthogonal matrix. To verify that, see Propositions 2.53 and 2.55.

Partition the matrix of characteristic vectors $Q = (Q_1, Q_2)$, so that Q_1 corresponds to the nonzero (unit) roots and note that

$$\frac{u'Nu}{\sigma^2} = \frac{u'Q_1Q_1'u}{\sigma^2}. \tag{10.46}$$

Put $\zeta = Q_1'u/\sqrt{\sigma^2}$ and note that by property ii of the multivariate normal

$$\zeta \sim N(0, Q_1'(Q_1Q_1')Q_1) = N(0, I_{T-n-1}). \tag{10.47}$$

It follows, therefore, by the definition of the chi-square distribution

$$\frac{u'Nu}{\sigma^2} \sim \chi_{T-n-1}^2. \tag{10.48}$$

Now, **under the null hypothesis,**

$$H_0 : \ \beta_* = 0$$

as against the alternative,

$$H_1 : \ \beta_* \neq 0,$$

the numerator of the fraction $(R^2/n)/[(1 - R^2)/T - n - 1]$ is chi-square distributed with n degrees of freedom. Hence, Eq. (10.41) may be used as a test statistic for the test of the hypothesis stated above, and its distribution is $F_{n,T-n-1}$.

Remark 10.2. The preceding remark enunciates a result that is much broader than appears at first sight. Let S_r be an $n \times r$, $r \leq n$, **selection** matrix;

this means that the columns of S_r are **mutually orthogonal** and in each column all elements are zero **except one which is unity**. This makes S_r of rank r. Note also that it is orthogonal only in the sense that $S_r'S_r = I_r$; on the other hand $S_rS_r' \neq I_n$. It is clear that if we are interested in testing the hypothesis, say, that $\beta_2 = \beta_3 = \beta_7 = \beta_{12} = 0$, we may define the selection matrix S_4 such that

$$S_4'\beta_* = (\beta_2, \beta_3, \beta_7, \beta_{12})'. \qquad (10.49)$$

Since $S_4'\hat{\beta}_* \sim N(S_4'\beta_*, \sigma^2\Psi_4)$, where

$$\Psi_4 = S_4'(X_1^{*'}X_1^*)^{-1}S_4, \qquad (10.50)$$

it follows, from the discussion of Remark 10.1 and property i of the multivariate normal distribution, that

$$\left(\frac{\hat{\beta}_*'S_4\Psi_4^{-1}S_4'\hat{\beta}_*}{\hat{u}'\hat{u}}\right)\left(\frac{T-n-1}{4}\right) \sim F_{4,T-n-1} \qquad (10.51)$$

is a suitable test statistic for testing the null hypothesis

$H_0:\ S_4'\beta_* = 0,$

as against the alternative,

$H_1:\ S_4'\beta_* \neq 0,$

and its distribution is $F_{4,T-n-1}$.

Finally, for $r = 1$—i.e. for the problem of testing a hypothesis on a **single** coefficient—we note that the preceding discussion implies that the appropriate test statistic and its distribution are given by

$$\tau^2 = \frac{\hat{\beta}_i^2/\text{Var}(\hat{\beta})_i}{\hat{u}'\hat{u}/\sigma^2(T-n-1)} \sim F_{1,T-n-1}, \qquad (10.52)$$

where $\text{var}(\hat{\beta}_i) = \sigma^2 q_{ii}$, and q_{ii} is the ith diagonal element of $(X_1^{*'}X_1^*)^{-1}$. Making the substitution $\hat{\sigma}^2 = \hat{u}'\hat{u}/T - n - 1$ and taking the square root in Eq. (10.52) we find

$$\tau = \frac{\hat{\beta}_i}{\sqrt{\hat{\sigma}^2 q_{ii}}} \sim \sqrt{F_{1,T-n-1}}, \qquad (10.53)$$

A close inspection indicates that τ of Eq. (10.53) is simply the usual t-ratio of regression analysis. Finally, also observe that

$$t_{T-n-1} = \sqrt{F_{1,T-n-1}}, \qquad (10.54)$$

or more generally the distribution of the **square root** of a variable that has the $F_{1,n}$ distribution is precisely the t_n distribution.

10.5 Nonstandard Errors

In this section, we take up issues that arise when the error terms do not obey the standard requirement $u \sim N(0, \sigma^2 I_T)$, but instead have the more general normal distribution

$$u \sim N(0, \Sigma), \quad \Sigma > 0. \tag{10.55}$$

Since Σ is $T \times T$, we cannot in practice obtain efficient estimators **unless** Σ is known. If it is, obtain the nonsingular decomposition and the transformed model, respectively,

$$\Sigma = AA', \quad w = Z\beta + v, \quad w = A^{-1}y, \quad Z = A^{-1}X, \quad v = A^{-1}u. \tag{10.56}$$

It may be verified that the transformed model obeys the standard conditions, and hence the OLS estimator

$$\hat{\beta} = (Z'Z)^{-1}Z'w, \quad \text{with} \quad \hat{\beta} \sim N(\beta, (Z'Z)^{-1}) = N(\beta, (X'\Sigma^{-1}X)^{-1}) \tag{10.57}$$

obeys the Gauss-Markov theorem and is thus efficient, within the class of linear unbiased estimators of β.

If Σ is **not known** the estimator in Eq. (10.57), termed in econometrics the **Aitken** estimator, is not available. However, (as $T \to \infty$), if Σ has only a **fixed finite number of distinct elements** which can be estimated consistently, say by $\tilde{\Sigma}$, the estimator in Eq. (10.57) **with Σ replaced by $\tilde{\Sigma}$, is feasible and is termed the Generalized Least Squares (GLS) estimator.**

If we estimate β by OLS methods, what are the properties of that estimator and how does it compare with the Aitken estimator? The OLS estimator evidently obeys

$$\tilde{\beta} = (X'X)^{-1}X'y = \beta + (X'X)^{-1}X'u. \tag{10.58}$$

From property ii of the multivariate normal, we easily obtain

$$\tilde{\beta} \sim N(\beta, \Psi), \tag{10.59}$$

where $\Psi = (X'X)^{-1}X'\Sigma X(X'X)^{-1}$. It is evident that the estimator is **unbiased**. Moreover, provided

$$\lim_{T \to \infty} \frac{X'\Sigma X}{T^2} = 0, \tag{10.60}$$

we have

$$\lim_{T \to \infty} \Psi = \lim_{T \to \infty} \frac{1}{T} \left(\frac{X'X}{T} \right)^{-1} \left(\frac{X'\Sigma X}{T} \right) \left(\frac{X'X}{T} \right) = 0, \tag{10.61}$$

which shows that the estimator is consistent in the mean square sense i.e. it converges to β **in mean square** and thus also in probability.

How does it compare with the Aitken estimator? To make the comparison, first express the Aitken estimator in the original notation, Thus,

$$\hat{\beta} = (X'\Sigma^{-1}X)^{-1}X'\Sigma^{-1}y, \quad \hat{\beta} \sim N(\beta, (X'\Sigma^{-1}X)^{-1}). \tag{10.62}$$

Because both estimators are normal with the same mean, the question of efficiency reduces to whether the difference between the two covariance matrices is positive or negative semi-definite or indefinite. For simplicity of notation let $\Sigma_{\tilde{\beta}}$, $\Sigma_{\hat{\beta}}$ be the covariance matrices of the OLS and Aitken estimators, respectively. If

i. $\Sigma_{\tilde{\beta}} - \Sigma_{\hat{\beta}} \geq 0$, the Aitken estimator is efficient relative to the OLS estimator;

ii. $\Sigma_{\tilde{\beta}} - \Sigma_{\hat{\beta}} \leq 0$, the OLS estimator is efficient relative to the Aitken estimator;

iii. Finally if $\Sigma_{\tilde{\beta}} - \Sigma_{\hat{\beta}}$ is an **indefinite matrix** i.e. it is neither positive nor negative (semi)definite, the two estimators cannot be ranked.

To tackle this issue directly, we consider the simultaneous decomposition of two positive definite matrices; see Proposition 2.64.

Consider the characteristic roots of $X(X'\Sigma^{-1}X)^{-1}X'$ **in the metric of** Σ i.e. consider the characteristic equation

$$|\lambda\Sigma - X(X'\Sigma^{-1}X)^{-1}X'| = 0. \tag{10.63}$$

The (characteristic) roots of the (polynomial) equation above are exactly those of

$$|\lambda I_T - X(X'\Sigma^{-1}X)^{-1}X'\Sigma^{-1}| = 0, \tag{10.64}$$

as the reader may easily verify by factoring out on the right Σ, and noting that $|\Sigma| \neq 0$. From Proposition 2.43, we have that the nonzero characteristic roots of Eq. (10.64) are exactly those of

$$|\mu I_{n+1} - (X'\Sigma^{-1}X)^{-1}X'\Sigma^{-1}X| = |\mu I_{n+1} - I_{n+1}| = 0. \tag{10.65}$$

We conclude, therefore, that Eq. (10.63) has $n+1$ unit roots and $T-n-1$ zero roots. By the simultaneous decomposition theorem, see Proposition 2.64, there exists a nonsingular matrix A such that

$$\Sigma = AA', \quad X(X'\Sigma^{-1}X)^{-1}X' = A\begin{bmatrix} 0_{T-n-1} & 0 \\ 0 & I_{n+1} \end{bmatrix}A'. \tag{10.66}$$

It follows, therefore, that

$$\Sigma - X(X'\Sigma^{-1}X)^{-1}X' = A \begin{bmatrix} I_{T-n-1} & 0 \\ 0 & 0_{n+1} \end{bmatrix} A' \geq 0. \qquad (10.67)$$

Pre- and post-multiplying by $(X'X)^{-1}X'$ and its transpose, respectively, we find

$$(X'X)^{-1}X'[\Sigma - X(X'\Sigma^{-1}X)^{-1}X']X(X'X)^{-1} = \Sigma_{\tilde{\beta}} - \Sigma_{\hat{\beta}} \geq 0, \qquad (10.68)$$

which shows that the Aitken estimator is efficient relative to the OLS estimator. The validity of the last inequality is established by the following argument: let B be a $T \times T$ positive semidefinite matrix of rank r, and let C be $T \times m$ of rank m, $m \leq T$; then, $C'BC \geq 0$. For a proof, we show that either $C'BC = 0$, or there exists at least one vector $\alpha \neq 0$, such that $\alpha'C'BC\alpha > 0$ and no vector η such that $\eta'C'BC\eta < 0$. Since C is of rank $m > 0$, its column space is of dimension m; if the column space of C is **contained in the null space of** B, $C'BC = 0$; if not, then there exists at least one vector $\gamma \neq 0$ in the column space of C such that $\gamma'B\gamma > 0$, **because** B **is positive semidefinite.** Let α be such that $\gamma = C\alpha$; the claim of the last inequality in Eq. (10.68) is thus valid. Moreover, no vector η can exist such that $\eta'C'BC\eta < 0$. This is so because $C\eta$ is **in the column space** of C **and** B is positive semidefinite.

Remark 10.3. We should also point out that there is an indirect proof of the relative inefficiency of the OLS estimator of β in the model above. We argued earlier that the Aitken estimator in a model with nonstandard errors is simply the OLS estimator in an appropriately transformed model and thus obeys the Gauss-Markov theorem. It follows, therefore, that the OLS estimator in the **untransformed model** cannot possibly obey the Gauss-Markov theorem and is thus **not efficient.**

We now take up another question of practical significance. If in the face of a general covariance matrix, $\Sigma > 0$, for the errors of a GLM we estimate parameters and their covariance matrix **as if the model had a scalar covariance matrix**, do the resulting test statistics have a tendency (on the average) to reject too frequently, or not frequently enough, relative to the situation when the correct covariance matrix is estimated.

If we pretend that $\Sigma = \sigma^2 I_T$, the covariance matrix of the OLS estimator of β is estimated as

$$\tilde{\sigma}^2(X'X)^{-1}, \quad \text{where} \quad \tilde{\sigma}^2 = \frac{1}{T-n-1}\tilde{u}'\tilde{u}. \qquad (10.69)$$

The question posed essentially asks whether the matrix

$$W = E\tilde{\sigma}^2(X'X)^{-1} - (X'X)^{-1}(X'\Sigma X)(X'X)^{-1} \qquad (10.70)$$

is positive semidefinite, negative semidefinite or indefinite. If it is positive semidefinite, the test statistics (t-ratios) will be understated and hence the hypotheses in question will tend to be **accepted** too frequently; if it is negative semi-definite, the test statistics will tend to be overstated, and hence the hypotheses in question will tend to be **rejected** too frequently. If it is indefinite, no such statements can be made.

First we obtain

$$E\tilde{\sigma}^2 = \frac{1}{T-n-1}Etruu'N = \frac{1}{T-n-1}tr\Sigma N = k_T. \qquad (10.71)$$

To determine the nature of W we need to put more structure in place. Because Σ is a positive definite symmetric matrix, we can write

$$\Sigma = Q\Lambda Q', \qquad (10.72)$$

where Q is the **orthogonal matrix** of the characteristic vectors and Λ is the **diagonal matrix** of the (positive) characteristic roots arranged in **decreasing order**, i.e. λ_1 is the largest root and λ_T is the smallest.

What we shall do is to show that there exist data matrices (X) such that W is positive semidefinite, and data matrices such that W is negative semidefinite. To determine the nature of W we must obtain a result that holds for **arbitrary** data matrix X. Establishing the validity of the preceding claim is equivalent to establishing that W is an **indefinite** matrix.

Evidently, the columns of Q can serve as a basis for the Euclidean space R^T; the columns of the matrix X lie in an $(n+1)$-dimensional subspace of R^T. Partition $Q = (Q_1, Q_*)$ such that Q_1 corresponds to the $n+1$ largest roots and **suppose** we may represent $X = Q_1A$, where A is **nonsingular**. This merely states that X lies in the subspace of R^T **spanned by the columns of** Q_1. In this context, we have a simpler expression for k_T of Eq. (10.71). In particular, we have

$$(T-n-1)k_T = trQ\Lambda Q'[I_T - Q_1A(A'A)^{-1}A'Q_1']$$

$$= tr\Lambda[I_T - (I_{n+1}, 0)'(I_{n+1}, 0)] = \sum_{j=n+2}^{T} \lambda_j, \qquad (10.73)$$

so that k_T is the **average** of the **smallest** $T-n-1$ roots. Since $X'X = A'A$, we obtain

$$W = k_T(A'A)^{-1} - A^{-1}Q_1'\Sigma Q_1A'^{-1} = A^{-1}[k_TI_{n+1} - \Lambda_1]A'^{-1}. \qquad (10.74)$$

Because k_T is the average of the $T-n-1$ smallest roots whereas Λ_1 contains the $n+1$ largest roots, we conclude that $k_TI_{n+1} - \Lambda_1 < 0$, and consequently

$$W < 0. \qquad (10.75)$$

But this means that the test statistics have a tendency to be larger relative to the case where we employ the correct covariance matrix; thus, hypotheses (that the underlying parameter is zero) would tend to be accepted **more frequently than appropriate.**

Next, suppose that X lies in the $(n+1)$-dimensional subspace of R^T spanned by the columns of Q_2, where $Q = (Q_*, Q_2)$, so that Q_2 corresponds to the $n+1$ **smallest roots.** Repeating the same construction as above we find that in this instance

$$(T - n - 1)k_T = \mathrm{tr}Q\Lambda Q'[I_T - Q_2 A(A'A)^{-1}A'Q_2']$$

$$= \mathrm{tr}\Lambda[I_T - (0, I_{n+1})'(0, I_{n+1})] = \sum_{j=1}^{T-n-1} \lambda_j, \quad (10.76)$$

so that k_T is the **average** of the **largest** $T - n - 1$ roots. Therefore, in this case, we have

$$W = k_T(A'A)^{-1} - A^{-1}Q_2'\Sigma Q_2 A'^{-1} = A^{-1}[k_T I_{n+1} - \Lambda_2]A'^{-1} > 0, \quad (10.77)$$

since k_T is the **average of the** $T - n - 1$ **largest roots** and Λ_2 contains along its diagonal the $n+1$ **smallest roots** of Σ.

Evidently, in this case, the test statistics are smaller than appropriate and, consequently, we tend to reject hypotheses **more frequently** than appropriate. The preceding shows that the matrix W is **indefinite.**

Finally, the argument given above is admissible because no restrictions are put on the matrix X; since it is arbitrary it can, in principle, lie in an $(n+1)$-dimensional subspace of R^T **spanned by any set of** $n+1$ **of the characteristic vectors** of Σ. Therefore, we must classify the matrix W as **indefinite**, when X is viewed as arbitrary. It need not be so for any **particular** matrix X. But this means that no statement can be made with confidence on the subject of whether using $\tilde{\sigma}^2(X'X)^{-1}$ as the covariance matrix of the OLS estimator leads to any systematic bias in accepting or rejecting hypotheses. Thus, nothing can be concluded beyond the fact that using the OLS estimator, and OLS based estimators of its covariance matrix, (when actually the covariance matrix of the structural error in non-scalar) is inappropriate, unreliable, and should be avoided for purposes of hypothesis testing.

10.6 Inference with Asymptotics

In this section we dispose of the somewhat unrealistic assertion that the structural errors are jointly normally distributed. Recall that we had made this assertion only in order to develop a distribution theory to be used in inferences regarding the parameters of the model.

Here we take advantage of the material developed in Chap. 9, to develop a distribution theory for the OLS estimators **based on their asymptotic or limiting distribution.** To this end return to the OLS estimator as exhibited in Eq. (10.6). Nothing much will change relative to the results we obtained above, but the results **will not hold for every sample size** T, but only for "large" T, although it is not entirely clear what large is. Strictly speaking, **limiting or asymptotic results hold precisely only at the limit, i.e. as** $T \to \infty$ **; but if the sample is large enough the distribution of the entity in question could be well approximated by the asymptotic or limiting distribution.** We may, if we wish, continue with the context of OLS estimation embodied in condition i above, or we may take the position that the explanatory variables are random and all analysis is carried **conditionally** on the observations in the data matrix X; in this case we replace the condition therein by $\text{plim}_{T \to \infty}(XX/T) = M_{xx} > 0$. We shall generally operate under the last condition. At any rate developing the expression in Eq. (10.6) we find

$$\hat{\beta}=(X'X)^{-1}X'y=\beta+(X'X)^{-1}X'u, \quad \text{or} \quad \sqrt{T}(\hat{\beta}-\beta)=\left(\frac{X'X}{T}\right)^{-1}\left(\frac{X'u}{\sqrt{T}}\right).$$
$$(10.78)$$

Applying Proposition 9.6, we find[3]

$$\sqrt{T}(\hat{\beta} - \beta) \sim M_{xx}^{-1}\frac{1}{\sqrt{T}}\sum_{t=1}^{T}x'_{t.}u_{t}. \qquad (10.79)$$

The sum in the right member above is the sum of independent non-identically distributed random **vectors** with mean zero and covariance matrix $\sigma^2(x'_{t.}x_{t.}/T)$, to which a CLT may be applied. Since in all our discussion of CLT we have used only **scalar random variables,** ostensibly none of these results can be employed in the current (vector) context. On the other hand **using the observation in Remark 9.4** let λ be an arbitrary conformable vector and consider

$$\zeta_T = \frac{1}{\sqrt{T}}\sum_{t=1}^{T}\lambda'x'_{t.}u_{t}. \qquad (10.80)$$

The rvs $\lambda'x'_{t.}u_{t}$, are independent non-identically distributed with mean zero and variance $\sigma^2\lambda'(x'_{t.}x_{t.}/T)\lambda$, which also obey the Lindeberg condition,

---

[3]In the context of Eq. (10.79) the notation $\sim$ is to be read "behaves like".

because of the conditions put on $X'X/T$. Consequently by Proposition 9.7 we conclude

$$\zeta_T \xrightarrow{\text{d}} N(0, \sigma^2 \lambda' M_{xx} \lambda), \text{ and by Remark 9.4} \quad \frac{1}{\sqrt{T}} \sum_{t=1}^{T} x'_{t\cdot} u_t \xrightarrow{\text{d}} N(0, \sigma^2 M_{xx}),$$

thus establishing

$$\sqrt{T}(\hat{\beta} - \beta) \xrightarrow{\text{d}} N(0, \sigma^2 M_{xx}^{-1}). \tag{10.81}$$

For large T practitioners often use the approximation

$$\sqrt{T}(\hat{\beta} - \beta) \approx N(0, \tilde{\sigma}^2 \tilde{M}_{xx}^{-1}), \quad \tilde{M}_{xx}^{-1} = \left(\frac{XX}{T}\right)^{-1}, \quad \tilde{\sigma}^2 = \hat{u}'\hat{u}/T.$$

To translate the procedures for inference tests developed earlier (where we had normal distributions for every sample size, T) to the asymptotic case (where normality holds only for large T) we shall not follow step by step what we had done earlier. Instead we shall introduce the so called **general linear hypothesis**,

H_0: $A\beta = a_0$
as against the alternative
H_1: $A\beta \neq a_0$,

where A is $k \times n+1$, $k \leq n+1$, rank(A) $= k$, and A, a_0 are respectively a matrix and vector **with known elements.**

Remark 10.4. Note that the formulation above, encompasses all the types of inference tests considered in the earlier context. For example, if we wish to test the hypothesis that $\beta_i = a_{(0),i} = 0$ simply take $a_0 = 0$ and A as consisting of a single row, all of whose elements are zero save the one corresponding to β_i, which is unity. If we wish to test the hypothesis that $\beta_2 = \beta_3 = \beta_7 = \beta_{10} = 0$, simply take $a_0 = 0$, and A consisting of four rows, all of whose elements are zero except, respectively, those corresponding to β_i, $i = 2, 3, 7, 10$. If we wish to duplicate the test based on the ratio $R^2/1 - R^2$, which tests the hypothesis that the coefficients of all *bona fide* variables are zero, i.e. $\beta^* = 0$,[4] take $A = (0, I_n)$, which is an $n \times n+1$ matrix. Thus, all tests involving **linear** restrictions on the parameters in β **are encompassed in the general linear hypothesis** $A\beta = a_0$.

To apply the asymptotic distribution we need to design tests based exclusively on that distribution. To this end consider

$$\tau^* = \sqrt{T}(A\hat{\beta} - a_0)'[\sigma^2 A M_{xx}^{-1} A']^{-1} \sqrt{T}(A\hat{\beta} - a_0) \sim \chi_k^2,$$

---

[4]Occasionally this test is referred to as a test of significance of R^2.

because of item v (Eq. (10.31)) given in connection with the multivariate normal above. Unfortunately, however, τ^* is **not a statistic** since it contains the unknown parameters σ^2 and M_{xx}, not specified by the null. From Proposition 9.6, however, we know that if we replace σ^2, M_{xx} by their consistent estimators, viz. $\hat{u}'\hat{u}/T$, $X'X/T$ respectively, the limiting distribution will be the same; thus, consider instead

$$
\begin{aligned}
\tau &= \sqrt{T}(A\hat{\beta} - a_0)'[\tilde{\sigma}^2 A \tilde{M}_{xx}^{-1} A']^{-1} \sqrt{T}(A\hat{\beta} - a_0) \\
&= \frac{T(A\hat{\beta} - a_0)'[A(X'X/T)^{-1}A']^{-1}(A\hat{\beta} - a_0)}{\hat{u}'\hat{u}/T} \xrightarrow{d} \chi_k^2.
\end{aligned}
$$

Finally, clearing of redundancies, we may write

$$
\tau = \frac{(A\hat{\beta} - a_0)'[A(X'X)^{-1}A]^{-1}(A\hat{\beta} - a_0)}{\hat{u}'\hat{u}/T} \xrightarrow{d} \chi_k^2, \tag{10.82}
$$

which **is a statistic** in that it does **not** contain unknown parameters not specified by the null hypothesis.

Remark 10.5. A careful examination will disclose that when applied to the test statistic obtained earlier when estimators were normally distributed for every sample size T, the statistic τ of Eq. (10.82) duplicates them precisely, except for the denominator in $\hat{u}'\hat{u}/T$, which is immaterial. This means that if one does what is usually done in evaluating regression results (when it is assumed that normality of estimators prevails for all sample sizes T) the test procedures will continue to be valid **when the sample size is large and one employs the limiting distribution of the estimators.** The only difference is that what is t-test in the earlier case is now a z-test (i.e. based on N(0,1)) and what was an F-test is now a chi square test.

10.7   Orthogonal Regressors

Suppose that the regressors of the GLM are **mutually orthogonal,** meaning that in the representation

$$
y = X\beta + u, \quad \text{we have } X'X = D = \text{diag}(d_1, d_2, \ldots, d_{n+1}).
$$

In this case, the elements of β, the regression coefficients, can be estimated *seriatim*, i.e.

$$
\hat{\beta}_i = (x_{\cdot i}'x_{\cdot i})^{-1}x_{\cdot i}'y, \quad i = 0, 1, 2, \ldots, n+1.
$$

This may be verified directly by computing the elements of

$$
\hat{\beta} = (X'X)^{-1}X'y = D^{-1}X'y.
$$

Although this is a rare occurrence in actual practice, nonetheless it points out an important feature of least squares. Suppose, for example, the model is written as

$$y = X_1\beta_1 + X_2\beta_2 + u, \tag{10.83}$$

where X_1 is $T \times m + 1$ and X_2 is $T \times k$, $k = n - m$ such that $X_1'X_2 = 0$.
By the argument given above, we can estimate

$$\hat{\beta}_1 = (X_1'X_1)^{-1}X_1'y, \quad \hat{\beta}_2 = (X_2'X_2)^{-1}X_2'y. \tag{10.84}$$

The least squares residuals are thus given by

$$\hat{u} = N_1 y + (N_2 - I_T)y, \quad N_i = I_T - X_i(X_i'X_i)^{-1}X_i', \quad i = 1, 2. \tag{10.85}$$

Even if the two sets of variables in X_1 and X_2 are **not** mutually orthogonal, we can use the preceding discussion to good advantage. For example, suppose we are not particularly interested in the coefficients of the variables in X_1 but wish to carry out tests on the coefficients of the variables in X_2. To do so, we need estimators of the coefficient vector β_2 as well as its covariance matrix. Oddly enough, we may accomplish this with a simple regression as follows. Rewrite the model as

$$y = X\beta + u = X_1\beta_1 + N_1 X_2\beta_2 + u^*, \tag{10.86}$$

$$u^* = (I_T - N_1)X_2\beta_2 + u.$$

Carrying out an OLS regression, we find

$$\hat{\beta} = [(X_1, X_2^*)'(X_1, X_2^*)]^{-1}(X_1, X_2^*)'y, \quad X_2^* = N_1 X_2.$$

Making a substitution (for y) from Eq. (10.86), we can express the estimator as

$$\hat{\beta} = \begin{pmatrix} \beta_1 \\ \beta_2 \end{pmatrix} + \begin{pmatrix} X_1'X_2\beta_2 \\ 0 \end{pmatrix} + \begin{pmatrix} (X_1'X_1)^{-1}X_1' \\ (X_2^{*'}X_2^*)^{-1}X_2^{*'} \end{pmatrix} u. \tag{10.87}$$

Note that the estimator for β_2 is **unbiased** but the estimator for β_1 is **not**. Because we are not interested in β_1, this does not present a problem. Next, compute the residuals from this regression, namely

$$\hat{u} = y - X_1\hat{\beta}_1 - X_2^*\hat{\beta}_2 = y - X_1\beta_1 - N_1 X_2\beta_2$$

$$- (I_T - N_1)X_2\beta_2 - [(I_T - N_1) + X_2^{*'}(X_2^{*'}X_2^*)^{-1}X_2^{*'}]u$$

$$= y - X_1\beta_1 - X_2\beta_2 - [(I_T - N_1) + X_2^{*'}(X_2^{*'}X_2^*)^{-1}X_2^{*'}]u$$

$$= [N_1 - N_1 X_2'(X_2' N_1 X_2)^{-1} X_2' N_1]u$$

$$= [I_T - X_2^{*'}(X_2^{*'} X_2^*)^{-1} X_2^{*'}]N_1 u. \tag{10.88}$$

To complete this facet of our discussion we must show that the estimator of β_2 as obtained in Eq. (10.87) and as obtained in Eq. (10.6) or Eq. (10.9) are identical; in addition, we must show that the residuals obtained using the estimator in Eq. (10.6) and those obtained from the estimator in Eq. (10.87) are identical.

To show the validity of the first claim, denote the OLS estimator as originally obtained in Eq. (10.6) by $\tilde{\beta}$ to distinguish it from the estimator examined in the current discussion. By Corollary 10.3, its distribution is given by

$$\tilde{\beta} \sim N(\beta, \sigma^2 B), \quad B = \begin{bmatrix} B_{11} & B_{12} \\ B_{21} & B_{22} \end{bmatrix} = (X'X)^{-1}. \tag{10.89}$$

By the property of the multivariate normal given in Eq. (10.26), the marginal distribution of $\tilde{\beta}_2$ is given by

$$\tilde{\beta}_2 \sim N(\beta_2, \sigma^2 B_{22}). \tag{10.90}$$

From Proposition 2.31, pertaining to the inverse of a partitioned matrix, we find that

$$B_{22} = [X_2'(I_T - X_1(X_1'X_1)^{-1}X_1')X_2]^{-1} = (X_2^{*'} X_2^*)^{-1}, \tag{10.91}$$

thus proving that $\hat{\beta}_2$ as exhibited in Eq. (10.87) of the preceding discussion is indeed the OLS estimator of the parameter β_2, since, evidently, $\tilde{\beta}_2$ of Eq. (10.90) has precisely the same distribution.

To show the validity of the second claim, requires us to show that

$$I_T - X(X'X)^{-1}X' = N_1 - X_2^*(X_2^{*'} X_2^*)^{-1} X_2^{*'}, \tag{10.92}$$

thus demonstrating that the residuals as obtained in Eq. (10.88) and as obtained in Eq. (10.8) are, in fact, identical.

The OLS residuals obtained from the estimator in Eq. (10.6) are given by

$$\tilde{u} = y - X\tilde{\beta} = [I_T - X(X'X)^{-1}X']u. \tag{10.93}$$

Using the notation $X = (X_1, X_2)$, as in the previous discussion, we find

$$I_T - X(X'X)^{-1}X' = I_T - [X_1 B_{11} X_1' + X_1 B_{12} X_2' + X_2 B_{21} X_1' + X_2 B_{22} X_2']. \tag{10.94}$$

From Proposition 2.3 and Corollary 5.5,

$$B_{11} = [X_1'X_1 - X_1'X_2(X_2'X_2)^{-1}X_2'X_1)]^{-1} = (X_1'X_1)^{-1}$$

$$+(X_1'X_1)^{-1}X_1'X_2(X_2'N_1X_2)^{-1}X_2'X_1(X_1'X_1)^{-1}$$

$$B_{12} = -(X_1'X_1)^{-1}X_1'X_2(X_2'N_1X_2)^{-1}, \quad B_{21} = B_{12}',$$

$$B_{22} = (X_2'N_1X_2)^{-1} = (X_2^{*'}X_2^{*})^{-1}. \tag{10.95}$$

Substituting the preceding expressions in the right member of Eq. (10.93), we can render the standard OLS residuals of Eq. (10.93) as

$$\tilde{u} = [N_1 - N_1X_2(X_2'N_1X_2)^{-1}X_2'N_1]u = [I_T - N_1X_2(X_2'N_1X_2)^{-1}X_2'N_1]N_1u. \tag{10.96}$$

This provides a constructive proof that the residuals from the regression of y on X_1 and N_1X_2 are precisely the same (numerically) as the residuals of the regression of y on X_1 and X_2.

Remark 10.6. If we had proved in this volume the projection theorem, the preceding argument would have been quite unnecessary. This is so because the OLS procedure involves the projection of the vector y on the subspace spanned by the columns of the matrix (X_1, X_2), which are, by assumption, linearly independent. Similarly, the regression of y on (X_1, N_1X_2) involves a **projection** of the vector y on the subspace spanned by the columns of (X_1, N_1X_2). But the latter is obtained by a Gram-Schmidt orthogonalization procedure on the columns of the matrix (X_1, X_2). Thus, the two matrices span **precisely** the same subspace. The projection theorem also states that any vector in a T-dimensional Euclidean space can be written **uniquely** as the sum of two vectors, one from the subspace spanned by the columns of the matrix in question and one from the **orthogonal complement** of that subspace. Because the subspaces in question are identical, **so are their orthogonal complements**. The component that lies in the orthogonal complement is simply **the vector of residuals** from the corresponding regression.

Remark 10.7. The results exhibited in Eqs. (10.6) and (10.87) imply the following computational equivalence. If we are not interested in β_1, and we merely wish to obtain the OLS estimator of β_2 in such a way that we can construct confidence intervals and test hypotheses regarding the parameters therein, we can operate exclusively with the model

$$N_1y = N_1X_2\beta_2 + N_1u, \quad \text{noting that} \quad N_1X_1 = 0. \tag{10.97}$$

From the standard theory of the GLM, the OLS estimator of β_2 in Eq. (10.97) is

$$\hat{\beta}_2 = (X_2^{*'} X_2^*)^{-1} X_2^{*'} N_1 y, \tag{10.98}$$

and the vector of residuals is given by

$$\hat{u} = [I_T - X_2^* (X_2^{*'} X_2^*)^{-1} X_2^{*'}] N_1 u, \tag{10.99}$$

both of which are identical to the results obtained from the regression of y on $(X_1, N_1 X_2)$, or on (X_1, X_2).

10.8   Multiple GLM

In this section we take up the case where one has to deal with a number of GLM, that are somehow related but not in any obvious way. The term, multiple GLM, is not standard; in the literature of econometrics the prevailing term is **Seemingly Unrelated Regressions (SUR)**. In some sense it is the intellectual precursor to Panel Data Models, a subject we shall take up in the next chapter.

This topic arose in early empirical research that dealt with disaggregated investment functions at the level of the firm. Thus, suppose a GLM is an appropriate formulation of the investment activity for a given firm, i, say

$$y_{t(i)} = x_{t.}^i \beta^{(i)} + u_{t(i)}. \tag{10.100}$$

Suppose further the investigator wishes to deal with a small but **fixed** number of firms, say m. The explanatory variables in $x_{t(i)}$ need not have anything in common with those in $x_{t(j)}, i \neq j$, although they may; the vectors of coefficients need not be the same for all firms and may indeed have little in common. However, by the nature of the economic environment **the error terms may be correlated across firms**, since they all operate in the same (macro) economic environment. We may write the observations on the ith firm as

$$y_{\cdot i} = X^i \beta_{\cdot i} + u_{\cdot i}, \quad i = 1, 2, \ldots m, \tag{10.101}$$

where $y_{\cdot i}$ is a T element column vector,[5] X^i is a $T \times k_i$ matrix of observations on the explanatory variables, $\beta_{\cdot i}$ is the k_i-element column vector of the regression parameters and $u_{\cdot i}$ is the T-element column vector of the errors. Giving effect to the observation that all firms operate in the same economic environment, we are prepared to assume that

$$\text{Cov}(u_{t(i)} u_{t(j)}) = \sigma_{ij} \neq 0, \tag{10.102}$$

for all t. All other standard conditions of the GLM continue in force, for each firm.

---

[5] The sample size is assumed to be the same for all firms.

We could estimate all m GLM *seriatim*, as we discussed above, obtain estimators and make inferences. If we did so, however, we would be ignoring the information, or condition, exhibited in Eq. (10.102), and this raises the question of whether what we are doing is optimal. To address this issue, write the system in Eq. (10.101) as

$$y = X^*\beta^* + u, \quad y = (y'_{.1}, y'_{.2}, \ldots, y'_{.m})', \quad u = (u'_{.1}, u'_{.2}, \ldots, u'_{.m})', \quad \text{where}$$
$$X^* = \text{diag}(X^1, X^2, \ldots, X^m), \quad \beta^* = (\beta'_{.1}, \beta'_{.2}, \ldots, \beta'_{.m})',$$
$$\tag{10.103}$$

and note that

$$\text{Cov}(uu') = (Eu_{.i}u'_{.j}) = (\sigma_{ij}I_T) = \Sigma \otimes I_T = \Psi. \tag{10.104}$$

Also y is an mT-element column vector as is u, X^* is an $mT \times k$ matrix, $k = \sum_{i=1}^{m} k_i$ and β^* is a k-element column vector.

In view of the fact that the system in Eq. (10.103) is formally a GLM, the efficient estimator of its parameters is the Aitken estimator when Ψ is known and, when not the generalized least squares (GLS) estimator. The latter is given by

$$\hat{\beta}^* = (X^{*\prime}\tilde{\Psi}^{-1}X^*)^{-1}X^{*\prime}\tilde{\Psi}^{-1}y = \beta^* + (X^{*\prime}\tilde{\Psi}^{-1}X^*)^{-1}X^{*\prime}\tilde{\Psi}^{-1}u, \tag{10.105}$$

where $\tilde{\Psi}$ is a consistent estimator of Ψ. Since the latter, an $mT \times mT$ matrix, contains a fixed number of parameters viz. the elements of the $m \times m$ symmetric matrix Σ, this estimator is **feasible**. Indeed, we can estimate *seriatim* each of the m GLM by least squares (OLS) obtain the residuals

$$\tilde{u}_{.i} = y_{.i} - X^i\tilde{\beta}_{.i}$$

and thus obtain the consistent estimator

$$\tilde{\Psi} = \tilde{\Sigma} \otimes I_T, \quad \tilde{\Sigma} = (\tilde{\sigma}_{ij}), \quad \text{where} \quad \tilde{\sigma}_{ij} = \frac{\tilde{u}'_{.i}\tilde{u}_{.j}}{T}, \quad i, j = 1, 2, \ldots, m. \tag{10.106}$$

Since

$$\tilde{\sigma}_{ij} = \frac{\tilde{u}'_{.i}\tilde{u}_{.j}}{T} = \frac{u'_{.i}u_{.j}}{T} + Q_T$$

$$Q_T = \frac{u'_{.i}(N^iN^j - N^i - N^j)u_{.j}}{T}, \quad N^i = X^i(X^{i\prime}X^i)^{-1}X^{i\prime}, \quad i = 1, 2, \ldots m,$$

it follows by the standard assumptions of the GLM that $\text{plim}_{T\to\infty} Q_T = 0$. It also follows from the standard assumptions of the GLM that $u'_{.i}u_{.j}/T$ obeys the SLLN, see Sect. 9.3.1 on the applicability of the latter to sequences of iid rvs with finite mean. Consequently,

$$\tilde{\sigma}_{ij} \xrightarrow{P} \sigma_{ij}, \quad \text{and thus} \quad \tilde{\Psi} \xrightarrow{P} \Psi. \tag{10.107}$$

Using **exactly the same arguments as above** we can establish the limiting (asymptotic) distribution of the GLS estimator as

$$\sqrt{T}(\hat{\beta}^* - \beta^*) \xrightarrow{d} N(0, \Phi), \quad \Phi = \plim_{T \to \infty} \left(\frac{X^{*\prime}\Psi^{-1}X^*}{T} \right)^{-1}. \tag{10.108}$$

In the discussion above, see Eq. (10.68), we have already shown, in the finite T case, that the GLS estimator is efficient relative to the OLS estimator.

Remark 10.8. Perhaps the development of the argument in this section will help explain the designation of such models as **Seemingly Unrelated Regressions (SUR)** and justify their special treatment.

Chapter 11

Panel Data Models

11.1 Introduction

The study of empirical models based on cross-section times series data dates back well into the early 1950s. However, the first modern attempt to consistently model the behavior of agents in such contexts can be traced to Balestra and Nerlove (1966), hereafter referred to as BN, who studied the demand for natural gas **by state** in the US, over the period 1957–1962.

The subsequent development of the subject, however, arose in the context of studies of human populations by "labor" economists, for want of a better term.

A sample of observations is said to be a **cross section**, if it refers to a number of economic agents **at the same period of time**, or reasonably close to the same period. For example the study of family budgets based on consumer expenditure surveys, or the study of the "determinants" of wages based on a survey of population, are examples of use of cross sections, or cross sectional samples. On the other hand a sample of observations on a single or a group of agents **over time** is said to be a time series sample. For example a study of aggregate consumption, or investment, or exports etc. over a period of time are instances of use of time series samples.[1] The term Panel (or Panel data) refers to a situation where (the same) agent, or a number of agents, are observed over multiple periods of time; for example if there are n individuals observed at time t and $t = 1, 2, \ldots, T$, then **for each t** we have information over the same n individuals. This is more generally referred to as a Balanced Panel, and the term Panel is also applied to situations in which, by attrition or otherwise, not all n individuals are observed over all T periods, but a

---

[1]Over the last 20 years or so the term time-series has acquired a very specialized meaning in econometrics connected to the study of non-stationary, particularly integrated, stochastic processes. However, our use of the term **time series** here typically does **not** refer to this meaning.

P.J. Dhrymes, *Mathematics for Econometrics*,
DOI 10.1007/978-1-4614-8145-4_11, © The Author 2013

substantial number is.   The term **repeated cross sections** or **pooled samples** typically refer to situations in which there may be substantial differences in the composition of the cross sections, although broadly conforming to the original specification(s).

The purpose of this chapter is to exposit in simple form the basic structure of these models, indicating clearly their relation to the general linear model broadly and in particular when n is fixed and T is large, to the "seemingly unrelated regressions" model (SUR) discussed in Chap. 10.

It does not aim to be an exhaustive treatment of the subject.

11.2   Notation

Because the study of panel data models operates in **three dimensions**, the number of agents, the number of periods and the number of variables (dependent and independent in the regression sense of the terms), care is required in devising a notation to represent the data. We adopt the following conventions. In Eq. (11.1) below we represent the typical econometric relationship estimated in a panel setting,

$$y_{ti} = x_{ti\cdot}\beta + \gamma_i + u_{ti}, \quad i = 1, 2, \ldots, n, \quad t = 1, 2, \ldots T, \qquad (11.1)$$

where n is the number of agents observed, so that it is the dimension of the cross section, T is the number of the periods of observation, so that it is the dimension of the time series, and $x_{ti\cdot}$ is **a k element row vector**, containing the independent or explanatory variables of the model. The term γ_i is an idiosyncratic, time invariant, entity that refers exclusively to agent i, and u_{ti} is the standard error term.   The term γ_i may be random (so that it is distributed over the agents through a time invariant distribution), or nonrandom, like a distinct constant term.[2]

Because of the three-dimensional nature of panel data, it is not possible to give a (meaningful) single matrix representation for the entire sample; we shall now define the entities in terms of which we shall represent the data.

$$Y = (y_{ti}), \; t = 1, 2, \ldots, T, \quad i = 1, 2, \ldots, n, \qquad (11.2)$$

is the matrix of the dependent variable. Its ith **column**,

$$y_{\cdot i} = (y_{1i}, y_{2i}, \ldots, y_{Ti})', \qquad (11.3)$$

contains **all** T observations on **the individual agent** i, while its tth **row**,

$$y_{t\cdot} = (y_{t1}, y_{t2}, \ldots, y_{tn}), \qquad (11.4)$$

---

[2]In many studies, allowing for this particular type of individual effect removes as much as 30–40 % of the observed variation in the dependent variable.

contains **all observations** on the n agents at time t, i.e. it represents the observations on the entire cross section at time t.

With the help of this notation, we may represent all observations in the sample in two ways. First, we can exhibit, *seriatim*, all observations on the first agent, then the second agent and so on. The second is to present *seriatim* all the observations on the first cross section (i.e. the relevant observations at time $t = 1$), then those of the second cross section and so on. To implement the first, let

$$X_i = (x_{ti\cdot}), \quad t = 1, 2, \ldots, T, \tag{11.5}$$

note that it is a $T \times k$ matrix and that

$$y_{\cdot i} = X_i \beta + \gamma_i e_T + u_{\cdot i}, \quad i = 1, 2, \ldots, n \tag{11.6}$$

represents[3] all (time series) observations, for the ith agent, and thus for $i = 1, 2, \ldots, n$, across all individuals; finally e_T is a T-element column vector, all of whose elements are one.

To implement the second, let

$$y_{t\cdot}' = X_{(t)} \beta + \gamma + u_{t\cdot}', \quad \gamma = (\gamma_1, \gamma_2, \ldots, \gamma_n)', \tag{11.7}$$

where

$$X_{(t)} = \begin{bmatrix} x_{t1\cdot} \\ x_{t2\cdot} \\ \vdots \\ x_{tn\cdot} \end{bmatrix}, \tag{11.8}$$

so that $X_{(t)}$ is a matrix of order $n \times k$, and Eq. (11.7) exhibits the observations **on all agents** at time t, i.e. it exhibits all the cross section observations at time t.

To round out the required notation we introduce the following two definitions, noted already in Chap. 4 as Convention 4.1.

Definition 11.1. Let A be a matrix of dimension $q \times s$, and define

$$a = \text{vec}(A) = \begin{bmatrix} a_{\cdot 1} \\ a_{\cdot 2} \\ \vdots \\ a_{\cdot s} \end{bmatrix}, \tag{11.9}$$

i.e. it is of dimension $qs \times 1$ (a column vector) that exhibits the **columns** of A, *seriatim*.

---

[3]Note that this formulation implies homogeneity of the coefficients (parameters) of all explanatory variables i.e. the $x's$, **across all individual agents**.

Definition 11.2. Let A be a matrix as in Definition 11.1 and define the operator *rvec* by

$$a^* = \text{rvec}(A)' = \begin{bmatrix} a'_{1.} \\ a'_{2.} \\ \vdots \\ a'_{q.} \end{bmatrix}, \tag{11.10}$$

i.e. it is of dimension $qs \times 1$ (a column vector) that exhibits **the transposed rows (so they become column vectors)** of A, *seriatim*.

Remark 11.1. Note that the two (column) vectors in Eqs. (11.9) and (11.10) **contain the same elements** differently arranged.

If we put

$$Y = (y_{ti}), \quad U = (u_{ti}), \quad \text{both being } T \times n \text{ matrices,} \tag{11.11}$$

we can represent **the entire sample** in two ways; the first is

$$y = \text{vec}(Y) = X\beta + \gamma \otimes e_T + u, \quad u = \text{vec}(U), \quad X = \begin{bmatrix} X_1 \\ X_2 \\ \vdots \\ X_n \end{bmatrix}, \quad \text{an } nT \times k \text{ matrix,}$$

$$\tag{11.12}$$

while the second is

$$y^* = \text{rvec}(Y)' = X^*\beta + e_T \otimes \gamma + u^*, \quad u^* = \text{rvec}(U)',$$

$$X^* = \begin{bmatrix} X_{(1)} \\ X_{(2)} \\ \vdots \\ X_{(T)} \end{bmatrix}, \quad \text{an } nT \times k \text{ matrix,} \tag{11.13}$$

and both of them "look" like a general linear model (GLM).

11.3   Interpretation and Assumptions

11.3.1   Interpretation

The interpretation of the models in Eqs. (11.12) and (11.13), depends on how we view the entities γ_i ; in the Balestra and Nerlove (BN) context, they relate to the **structural error** specification; thus, if we wish to estimate parameters by means of maximum likelihood (ML) methods, we need to state the distribution of the γ_i and u_{ti}, and obtain various moments such as, e.g. the covariance matrix, so as to be able to write down the likelihood function, if normal, or the generalized least squares objective function, if not. In particular, the conditional expectation of y_i given x is given by

$$E(y_{ti}|x_{ti}) = x_{ti}.\beta, \tag{11.14}$$

because BN **assume** $E(\gamma_i|x) = 0$, $E(u_i|x) = 0$, for all i, and it is clear that the γ_i **are not** part of the mean specification.

In the context of such models **as used in "labor" economics**, the entities γ_i are viewed as part of the "mean component" of the dependent variable, and in fact they are a crucial component of their structure. The γ_i are presumed to denote native ability, which is not observed or observable; if it were, then it would have entered as one of the components of x_{ti}. and its coefficient would have been one of the elements of β. For that reason it is entered as a single entity **denoting both** the "magnitude" of native ability as well as the manner of its impact on the dependent variable, i.e. its coefficient. From an operational point of view, in the context of the so called "within groups" estimation, it acts very much like a conventional regression **constant term**. A more sophisticated view holds the γ_i to be (proportional to) a random assignment of an **unobservable** property, such as ability, to agent i. In this context the entities γ_i, within the sample, are (proportional to) realizations of this random process.

An important consequence of this difference in interpretation between the BN and "labor" economics versions is that, in the first, the vector x_{ti}. is allowed to have a component which is **unity**, corresponding to the constant term of the regression, while in the second we assert

Convention 11.1. The vector β **does not contain** a constant term, i.e. the vector x_{ti}. does not contain a unit entry.

11.3.2 Assumptions

In the GLM we generally make three types of assumptions: (a) regarding the error process (b) regarding the explanatory or "independent" variables and (c) the relation between them.

In this literature, generally, not much is assumed about the explanatory variables, the vectors x_{ti}. The tendency is to consider the variables (y, x, γ, u) as being defined on a probability space, say $(\Omega, \mathcal{A}, \mathcal{P})$, and subject to some joint distribution from which we can derive marginal and conditional distributions. They are also taken to be square integrable. In part the justification is that papers in this literature are seldom concerned with limiting distributions and take the distribution of the vector (y, x, γ, u) to be jointly normal. We shall depart from this practice, thus assuming in particular:

i.

Assumption 11.1. The explanatory variables, $x_{ti\cdot}$ are a realization of a square integrable stochastic process, so that $Ex_{ti\cdot}$ and $\text{Cov}(x_{ti\cdot})$ are both finite and $A_{ij} = \text{plim}_{T\to\infty} \frac{X'_i X_j}{T}$, exist for all i, j and for $i = j$ A_{ii} is nonsingular.

ii.

Assumption 11.2. The vectors $u_{\cdot i}$ are generally taken to be strictly or weakly stationary in the sense of Definitions 6.5 and 6.6 (of Chap. 6); for example in the strictly stationary case we have that, for arbitrary q, s the joint distribution[4] of the elements of the sequence $\{u_{t\cdot}:t \in \mathcal{T}\}$ are square integrable and have the property[5]

$$f(u_{t\cdot}, u_{t+1\cdot}, \ldots, u_{t+q\cdot}) = f(u_{t+s\cdot}, u_{t+s+1\cdot}, \ldots, u_{t+s+q\cdot});$$

often the elements of the sequence are taken to be *iid* (independent identically distributed).

iii.

Assumption 11.3. The conditional densities of $u|x$ and $\gamma|x$ have the property $E(u_{t\cdot}|x = 0)$, and where appropriate $E(\gamma_i|x) = 0$) and, moreover, their conditional (on x) covariance matrices are not a function of x .

Remark 11.2. The conditions in Assumption 11.3, rely too heavily on the conventions of the GLM, especially as they relate to finite samples and requirements for the validity of the Gauss-Markov theorem. Unless considerably more structure is imposed, limiting distribution arguments would require that x and u be (statistically) independent.

Remark 11.3. The condition $E(u|x) = 0$ is a bit stronger than $\text{Cov}(x, u) = 0$, whether we use the same time frame for both, or we are also claiming this

---

[4]The notation $f(\cdot)$ is used generically to denote a density, which may be marginal, joint, or conditional as the context requires.

[5]The change in notation regarding the linear index set, here denoted by $\mathcal{T}$, while in the definition of stochastic sequences, in earlier chapters was usually denoted by T , is necessary because in this context we use T as the length of the time series observations.

property for u_t. and $x_{t'i}$, $t \neq t'$. To see this in the simplest possible context, let u, x be **scalar** random variables, each with marginal mean zero. Then,

$$\text{Cov}(u, x) = \int_{-\infty}^{\infty} \int_{-\infty}^{\infty} xuf(x, u)dxdu = \int_{-\infty}^{\infty} xf(x) \left(\int_{-\infty}^{\infty} uf(u|x)du \right) dx = 0$$

$$(11.15)$$

because

$$h(x) = \int_{-\infty}^{\infty} uf(u|x)du = E(u|x) = 0, \quad \text{by assumption.}$$

The converse, however, is not true, i.e. $\text{Cov}(u, x) = 0$ does not necessarily imply that $E(u|x) = 0$. This is so because

$$\text{Cov}(u, x) = \int_{-\infty}^{\infty} \int_{-\infty}^{\infty} xuf(x, u)dxdu = \int_{-\infty}^{\infty} xf(x) \left(\int_{-\infty}^{\infty} uf(u|x)du \right) dx$$

$$= \int_{-\infty}^{\infty} xh(x)f(x)dx = 0 \qquad (11.16)$$

does not necessarily imply that $h(x)$ is equal to zero!

11.4 Estimation

11.4.1 Within Groups Estimation

The estimation of panel data models involves basically the same procedures one employs in the estimation of general linear models, or systems of general linear models (SUR) studied in Chap. 10, with the added complication of the unobservable "ability", termed in this literature **unobserved heterogeneity**.

The term "within groups" is something of a misnomer and derives from an older usage in the context of analysis of variance, where one could define variance within groups and between groups; such considerations, however, are irrelevant in this case and the term survives only as a historical relic.

If we look at Eqs. (11.6) and (11.12) we see that even though our primary interest is the parameter β, estimation by simple methods is hampered by the presence of the unobservable γ_i, for the ith agent. If $T \geq 2$ we can, in fact, estimate β from **centered** data. We shall now consider the case where n is fixed and $T \to \infty$, thus, initially we deal with the case where the structural errors, the u's are *iid*.

Let $I_T - ee'/T$ be the (sample mean) centering matrix, Dhrymes (1978, p. 18ff.), where e is a column vector of T unities; we may transform Eq. (11.6) to obtain

$$\left(I_T - \frac{ee'}{T}\right) y_{\cdot i} = \left(I_T - \frac{ee'}{T}\right) X_i \beta + \left(I_T - \frac{ee'}{T}\right) u_{\cdot i}, \qquad (11.17)$$

because the centering matrix and e_T are mutually orthogonal, i.e.

$$\left(I_T - \frac{ee'}{T}\right) e_T = 0.$$

The centering operation produces observations centered on their respective means, for example

$$\left(I_T - \frac{ee'}{T}\right) y_{\cdot i} = y_{\cdot i} - e_T \bar{y}_i,$$

etc. While this operation has gotten rid of one problem, it has created another in that the covariance matrix of the error vector in Eq. (11.17) is no longer scalar, and if X_i contains a constant term,[6] it has reduced its rank by one. In fact, we also have another "problem" in that the sum of all equations therein is identically zero, indicating that the covariance matrix is **singular**. Note that the centering matrix is a **symmetric idempotent matrix** of rank $T - 1$, see Chap. 2. Hence, it has the decomposition

$$I_T - \frac{ee'}{T} = Q \begin{bmatrix} I_{T-1} & 0 \\ 0 & 0 \end{bmatrix} Q' = Q_1 Q_1', \qquad (11.18)$$

where Q is the orthogonal matrix of the characteristic vectors, and Q_1 is (its) the sub-matrix corresponding to the **nonzero (unit)** roots.

Now, using another old device from Dhrymes (1969), transform Eq. (11.17) to obtain

$$Q_1' \left(I_T - \frac{ee'}{T}\right) y_{\cdot i} = Q_1' \left(I_T - \frac{ee'}{T}\right) X_i \beta + Q_1' \left(I_T - \frac{ee'}{T}\right) u_{\cdot i}. \qquad (11.19)$$

Noting that $Q_1'(I_T - \frac{ee'}{T}) = Q_1'$, it is apparent that this entire operation could have been done ab initio; it was done in an extensive fashion only to clarify what is behind this transformation.

The virtue of this transformation is that it eliminates the parameter γ_i , by centering observations about sample means, and retains the property of the covariance matrix of the error as a **scalar** matrix of the form $\sigma_{ii} I_{T-1}$.

Since in the transformed context of Eq. (11.19), the entities therein obey the conditions for the Gauss-Markov theorem, it follows that, given X_i , the estimator

---

[6]Note that this was ruled out by Convention 11.1.

$$\hat{\beta} = [X_i'Q_1Q_1'X_i]^{-1}[X_i'Q_1Q_1'y_{\cdot i}] = \beta + [(X_i'Q_1Q_1'X_i]^{-1}[X_i'Q_1Q_1'u_{\cdot i}], \quad (11.20)$$

is the best linear unbiased estimator in the context of Eq. (11.19). But there are n such estimators and it is clear that from a system-wide point of view this is not the most efficient estimator.

If we consider the system-wide version of Eq. (11.19), we are dealing with

$$(I_n \otimes Q_1')y = (I_n \otimes Q_1')X\beta + (I_n \otimes Q_1')u, \quad (11.21)$$

because $(I_n \otimes Q_1')(\gamma \otimes e_T) = 0$. We also observe that

$$\text{Cov}(u) = \Sigma \otimes I_T, \text{ so that } \text{Cov}[(I_n \otimes Q_1')u] = \Sigma \otimes I_{T-1}. \quad (11.22)$$

Applying the same device as in Dhrymes (1969)[7] we pre-multiply by $\Sigma^{-(1/2)} \otimes I_{T-1}$ to obtain the final estimating form as

$$(\Sigma^{-1/2} \otimes Q_1)y = (\Sigma^{-1/2} \otimes Q_1)X\beta + (\Sigma^{-1/2} \otimes Q_1)u, \quad (11.23)$$

the covariance matrix of whose error term is $I_{n(T-1)}$. Thus, when Σ is known the least squares estimator in the context of the transformed model as in Eq. (11.23) is the optimal estimator. If Σ is **not known** but can be estimated consistently, the estimator above is the optimal estimator in the sense that it has the smallest variance within the class of consistent estimators conditionally on the "instruments" X.

As of now, we have not specified what Σ is, but we operated on the (implicit) assumption that it is the scalar matrix $\sigma^2 \otimes I_{nT}$. In the large T, fixed n context we have several (three) possibilities for

$$\Sigma = (\sigma_{ij}), \quad (11.24)$$

(a): no restrictions; (b): $\sigma_{ij} = 0, i \neq j$; and (c): the condition in (b) and in addition $\sigma_{ii} = \sigma^2$, $i = 1, 2, \ldots, n$, which is the condition implicitly adopted in the estimation above.

In case (c), we have the typical condition imposed when $T = 1$ and a least squares procedure is applied. With $T \geq 2$, however, this condition is **unnecessarily** restrictive and we can operate instead with condition (b), which preserves the traditional assumption of independence in the error terms corresponding to different agents. At the same time, from Eq. (11.26) below, we can estimate consistently the unknown elements of Σ, thus rendering the

---

[7]The motivation for this device is to transform the equation whose parameters we are interested in estimating so that, at least asymptotically, it obeys the conditions of a Gauss-Markov like theorem. This idea (*mutatis mutandis*) is behind all subsequent (literature) developments in optimal non-linear simultaneous equations and generalized method of moments (GMM) procedures.

estimator in Eq. (11.27) **feasible**. The same is true if we operate with conditions (a), except that here we are asserting that the error terms for different agents are not independent, which is not an assumption made in this literature. Thus, in all cases with fixed n and large T it is possible to estimate consistently the σ_{ij} and hence the estimator of Eq. (11.23) is **feasible**. This shows that in the unrestricted case panel data model's efficient estimation is simply SUR estimation.

To complete the estimation phase we need to obtain an estimator for γ_i. Noting that in Eq. (11.6) $\gamma_{\cdot i}$ is dealt with as if it were the constant term of a standard GLM (general linear model), we can estimate

$$\hat{\gamma}_i = \bar{y}_i - \bar{x}_i \hat{\beta}, \quad i = 1, 2, \ldots, n, \tag{11.25}$$

where the symbols with overbars denote the sample means of the corresponding vectors ($y_{\cdot i}$), or matrices (X_i).

To enable us to obtain the efficient estimator for the system as a whole we can estimate the covariance matrix of $u_{t\cdot}$ as

$$\hat{\sigma}_{ij} = \frac{1}{T}(y_{\cdot i} - \hat{\gamma}_i e_T - X_i\hat{\beta})'(y_{\cdot j} - \hat{\gamma}_j e_T - X_j\hat{\beta}), \tag{11.26}$$

so that the efficient estimator of β is given by

$$\tilde{\beta} = [X'(\hat{\Sigma}^{-1} \otimes Q_1 Q_1)X]^{-1}X'(\hat{\Sigma}^{-1} \otimes Q_1 Q_1')y \tag{11.27}$$

Remark 11.4. Evidently the estimator in Eq. (11.27) is the only relevant one. The one in Eq. (11.20) is a "first stage" which serves to explain the procedure and helps us estimate γ_i and σ_{ij}, $i, j = 1, 2, \ldots, n$. Notice also that the system estimator in Eq. (11.27), in the case of large T fixed n panels, should enable us to test for a more general form of heterogeneity. Thus, if we endow each agent, or groups of agents based on certain characteristics, with a **different parameter vector**, say, $\beta_{\cdot j}$, we may test the hypothesis that all such vectors are the same. Otherwise we take, arbitrarily, parameter homogeneity across agents or groups of agents as a **given** and, **even though the data permit it**, we do not test this hypothesis (assumption).

Consistency of the estimators in Eqs. (11.25)–(11.27) is immediate by virtue of the assumption that the x's form a square integrable process, Assumptions 11.1 and 11.3.

Not sufficiently precise assumptions have been made to ensure that the limiting distribution of these estimators exist, but with little additional loss of generality we can ensure that the Lindeberg condition holds, see

Proposition 9.7 (of Chap. 9). This will permit us to conclude that

$$\sqrt{T}(\tilde{\beta} - \beta) \xrightarrow{\text{d}} N(0, \Omega), \quad \Omega = \plim_{T \to \infty} \left(\frac{X'(\hat{\Sigma}^{-1} \otimes Q_1 Q_1')X}{T} \right)^{-1}. \quad (11.28)$$

With a little additional effort we can devise a test for the hypothesis

$H_0 : \gamma_j = \gamma_1$
as against the alternative $H_1 : \gamma_i \neq \gamma_j$, for $i \neq j$.

We now take up the case, in the context of T large n fixed, where the vector $u_t.$ is **not** *iid* but is either strictly stationary or an *AR* of some order. If the error process is square integrable this means that

$$\text{Cov}(u_t'.) = K(t, t), \quad \text{Cov}(u_t'., u_{t'}'.) = K(t, t'). \quad (11.29)$$

If the process is strictly (or even weakly) stationary $K(t, t) = K(0)$; if it is **covariance stationary**, $K(t, t') = K(\tau)$, $\tau = t - t'$, and $K(\tau) = K(-\tau)'$. Thus, in the popular first order **stable, i.e. causal** autoregression, AR(1), (which is strictly stationary as well as covariance stationary) when the ϵ process is $iid(0, \Sigma > 0)$, we find

$$u_t'. = R u_{t-1}'. + \epsilon_t'., \quad u_t'. = \sum_{j=0}^{\infty} R^j \epsilon_{t-j}'., \quad \sum_{j=0}^{\infty} \| R \|^j < \infty.[8] \quad (11.30)$$

Moreover,

$$\text{Cov}(u_t'.) = K_0 = \sum_{j=0}^{\infty} R^j \Sigma R'^j, \quad \text{Cov}(u_t'., u_{t-\tau}'.) = \sum_{i=0}^{\infty} \sum_{j=0}^{\infty} R^i E(\epsilon_t'., \epsilon_{t-\tau}.) R'^j$$

$$= \sum_{i=0}^{\infty} \sum_{j=0}^{\infty} R^i K(i - j) R'^j = R^\tau K_0, \quad (11.31)$$

for **positive** $\tau = i - j$, and $K_0 R'^{|\tau|}$, for **negative** τ. Thus, the covariance matrix of the vector u in Eq. (11.13), for the stable first order autoregression above, is given by

$$\text{Cov}(u) = \begin{bmatrix} K_0 & RK_0 & R^2 K_0 & \cdots & R^{T-1}K_0 \\ K_0 R' & K_0 & RK_0 & \cdots & R^{T-2}K_0 \\ \vdots & \vdots & \vdots & \cdots & \vdots \\ K_0 R'^{T-1} & K_0 R'^{T-2} & K_0 R'^{T-3} & \cdots & K_0 \end{bmatrix} = K. \quad (11.32)$$

---

[8] The symbol $\| R \|$ means the **norm** of the matrix R.

Remark 11.5. When we discussed causal ARs, Definition 6.12 of Chap. 6, and surrounding discussion, we did not enter into the specifics of the computations involved in demonstrating the validity of their $M(\infty)$ representation, except in the scalar case in scattered occasions. Let us do the complete calculation in this instance. The representation in Eq. (11.30), requires absolute convergence of the series. If R were a scalar, say r we would require that $\sum_{j=0}^{\infty} |r|^j < \infty$; when it is a matrix on the other hand, the analogous requirement is that $\sum_{j=0}^{\infty} \| R \|^j < \infty$, where $\| R \|$ is the **norm** of the matrix R. The general definition of a matrix norm is as follows.

Definition 11.3. Let A be a square matrix of dimension n. The **scalar** function $\| A \|$, is said to be a matrix norm if it obeys

    i. $\| A \| \geq 0$ and $\| A \| = 0$, if and only if $A = 0$;

    ii. For any scalar c, $\| cA \| = |c| \| A \|$;

    iii. If B is an arbitrary conformable matrix $\| A + B \| \leq \| A \| + \| B \|$, and

    iv. $\| AB \| \leq \| A \| \| B \|$

In view of the definition of the norm above consider again the matrix R; conceding a very slight degree of generality we shall assume that its characteristic roots are distinct,[9] so that we can write

$$R = S^{-1} \Lambda S, \quad \text{and thus} \quad \| R \| \leq \| S^{-1} \| \| S \| \| \Lambda \|,$$

where Λ, S are, respectively, the matrices of the characteristic roots and vectors of R. Moreover, in this representation we can rewrite Eq. (11.30) as

$$u'_{t\cdot} = \sum_{j=0}^{\infty} S^{-1} \Lambda^j S \epsilon'_{t-j\cdot}.$$

In this context we have to show that

$$\sum_{j=0}^{\infty} \| S^{-1} \| \| S \| \| \Lambda \|^j < \infty.$$

Since the norms of S and S^{-1} **are finite** we need deal only with the magnitude of the norm $\| \Lambda \|$ which is a term of an infinite series. To this end using specifically the L_2 norm for matrices

$$\| A \| = \max_{z'z=1} |Az|_2,$$

---

[9] See the discussion in Propositions 2.39 and 2.40 in Chap. 2.

and applying it to Λ we find

$$\| \Lambda \| = \max_{z'z=1} |\Lambda z|_2 = \max_{z'z=1} \left(\sum_{s=1}^{n} \lambda_s^2 z_s^2 \right)^{1/2} \leq |\lambda_*|,$$

where $|\lambda_*| = \max_s |\lambda_s|$. Consequently,

$$\sum_{j=0}^{\infty} \| R \|^j \leq \| S^{-1} \| \| S \| \sum_{j=0}^{\infty} \| \Lambda \|^j \leq \| S^{-1} \| \| S \| \sum_{j=0}^{\infty} |\lambda_*|^j$$

$$= \| S^{-1} \| \| S \| \frac{1}{1 - \lambda_*} < \infty,$$

because the $AR(1)$ sequence is causal and thus all roots of R, including the largest, are less than one in absolute value.
We shall now address the question of how to estimate the parameters of Eq. (11.21) when the error terms constitute an $AR(1)$ sequence with covariance matrix as in Eq. (11.32)

In Eqs. (11.12) and (11.13) we have given two representations for the display of the observations on the model in Eq. (11.1). This was not done frivolously. Rather, the representation in Eq. (11.12) is most convenient for eliminating the "unobserved heterogeneity" parameters, γ_i, while the representation in Eq. (11.13) is most convenient when the error term follows a more complicated distribution such as for example the $AR(1)$ considered just above; in addition, Eq. (11.13) is most convenient in obtaining the covariance matrix of the error vector, when the latter is not *iid* but has a more complicated probabilistic structure, such as e.g. in the $AR(1)$ case noted above, but it is extremely inconvenient for the "within groups" estimator because it is not simple to eliminate the vector γ from Eq. (11.13) in any meaningful way.

Thus, we shall proceed by using Eq. (11.12) to eliminate the unobserved heterogeneity parameters, and we shall use Eq. (11.13) to obtain the covariance matrix of the error as we did in Eq. (11.32); having done that we must find some way to rearrange the elements of that matrix so that it becomes the covariance matrix of the vector as exhibited in Eq. (11.12). Noting that

$$u = \text{vec}(U), \quad u^* = \text{rvec}(U)' \tag{11.33}$$

contain the same elements differently arranged, we find from Lemma 4.3 (Chap. 4) that there exists a permutation matrix P_{nT}, such that

$$u = P_{Tn} u^*, \quad \text{or} \quad u^* = P_{nT} u, \quad P_{Tn} = \begin{bmatrix} I_n \otimes e'_{.1} \\ I_n \otimes e'_{.2} \\ \vdots \\ I_n \otimes e'_{.T} \end{bmatrix}, \tag{11.34}$$

where $e_{\cdot t}$ is a T element **column vector** all of whose elements are zero except the t th, which is unity.

Returning to Eq. (11.21) we thus find

$$\text{Cov}[(I_n \otimes Q_1')u] = \Phi, \quad \text{where} \tag{11.35}$$

$$\Phi = S'KS, \quad S = \begin{bmatrix} I_n \otimes q_{1\cdot} \\ I_n \otimes q_{2\cdot} \\ \vdots \\ I_n \otimes q_{T-1\cdot} \end{bmatrix}, \tag{11.36}$$

and $q_{t\cdot}$ is the tth row of Q_1. Notice that the rows of Q_1 have $T-1$ elements and thus $I_n \otimes q_{t\cdot}$ is $n \times n(T-1)$, so that S is $nT \times n(T-1)$; thus the matrix Φ is $n(T-1) \times n(T-1)$ and nonsingular, as required for the definition of the optimal estimator. The latter may be obtained as the OLS estimator in the context of

$$\Phi^{-(1/2)}(I_n \otimes Q_1')y^* = \Phi^{-(1/2)}(I_n \otimes Q_1')X^*\beta + \Phi^{-(1/2)}(I_n \otimes Q_1')u^*. \tag{11.37}$$

Given a consistent estimator of K (and thus $\hat{\Phi} = S'\tilde{K}S$) the feasible estimator is

$$\tilde{\beta} = [X^{*\prime}(I_n \otimes Q_1)\hat{\Phi}^{-1}(I_n \otimes Q_1')X^*]^{-1}X^{*\prime}((I_n \otimes Q_1)\hat{\Phi}^{-1}(I_n \otimes Q_1')y^*. \tag{11.38}$$

By arguments analogous to those leading to Eq. (11.28) we conclude that

$$\sqrt{T}(\tilde{\beta} - \beta) \xrightarrow{d} N(0, \Phi^{-1}), \quad \Phi = \plim_{T \to \infty} S'\tilde{K}S, \tag{11.39}$$

where $\tilde{K}$ is the consistent estimator of K obtained by **regressing** $\hat{u}_{t\cdot}$ on $\hat{u}_{t-1\cdot}$, thus obtaining $\hat{R}$, and by estimating

$$\hat{K}_0 = \frac{1}{T} \sum_{t=1}^{T} \hat{u}_{t\cdot}' \hat{u}_{t\cdot}. \tag{11.40}$$

This procedure yields a **consistent estimator** because **we have assumed that the error process is a stable** $AR(1)$, and thus **strictly stationary.** If the error process is more complex and/or is not stated parametrically, we can employ, *mutatis mutandis* a similar procedure by estimating the entries in the (general form of the) matrix $K = [K(t, t')]$, using the sample covariances, $(t \neq t')$

$$\hat{K}(t, t') = \sum w(t, t')\hat{u}_{t\cdot}'\hat{u}_{t'\cdot}, \quad \hat{K}_0 = \frac{1}{T} \sum_{t=1}^{T} u_{t\cdot}' u_{t\cdot}, \tag{11.41}$$

provided stationarity is preserved, and where w is an appropriate weighting function, akin to the spectral windows (also known as kernels) employed when such problems are considered in the frequency domain.

When n is large and $T \geq 2$ is **fixed** our flexibility is considerably more circumscribed. This is so because, (a) no time series parameters can be consistently estimated, owing to the fixity of T, and (b) the cases

$$\Sigma = \text{diag}(\sigma_{11}, \sigma_{22}, \ldots, \sigma_{nn}), \quad \Sigma = (\sigma_{ij}), \quad \sigma_{ij} \neq 0, \text{ for } i \neq j, \qquad (11.42)$$

do not permit "efficient estimators", because the parameters in Σ cannot be consistently estimated. Thus we are reduced to assuming, implicitly, that $\Sigma = \sigma^2 I_n$ and obtaining OLS estimators. In the case of a **diagonal** covariance matrix in Eq. (11.42) we can consistently estimate the (limiting) covariance matrix of the estimator $\hat{\beta}$, by the same methods as in the GLM, thus permitting valid inference. The case of an unrestricted covariance matrix in Eq. (11.42), however, does not allow for consistent estimation of the (limiting) covariance matrix of the estimator $\hat{\beta}$, and thus valid inference in this case is not possible.

Chapter 12

GLSEM and TS Models

12.1 GLSEM

12.1.1 Preliminaries

In this chapter, we take up two important applications, involving simultaneous equation, AR, and ARMA models, which are very important in several disciplines, such as economics, other social sciences, engineering and statistics. We partially discussed such topics in Chap. 4, in the context of difference equations, since simultaneous equation and AR models involve the use of difference equations and it is important to establish the nature and the properties of their solutions.

We begin with simultaneous equation models, more formally the General Linear Structural Econometric Model (GLSEM).
The basic model is

$$y_t.B^* = x_t.C + u_t., \quad t = 1, 2, 3, \ldots, T, \tag{12.1}$$

where $y_t.$ is an observation at "time" t on the m-element vector of **endogenous variables**, $x_t.$ is an observation on the G-element vector of the **predetermined variables**, $u_t.$ is the m-element vector of the (unobservable) structural errors, and B^*, C are matrices of unknown parameters to be estimated. This model may be written compactly as

$$YB^* = XC + U, \tag{12.2}$$

where Y is $T \times m$, X is $T \times G$, U is $T \times m$, B^* is $m \times m$, and C is $G \times m$.

12.1.2 The Structural and Reduced Forms

Endogenous variables are those determined by the system whose description is given in Eq. (12.1); **exogenous** variables are those whose behavior is determined **outside** the system above; **predetermined** variables are those whose behavior is **not determined by the system of Eq. (12.1) at time** t. Thus, the class of predetermined variables consists of **lagged endogenous** and **exogenous** variables.

Remark 12.1. The class of predetermined variables has only tradition to recommend it. The distinction that is important in econometrics is between the variables that are **independent** of, or at least **uncorrelated** with, the structural error at time t and those that are not. When the structural errors are assumed to be independent, identically distributed (*iid*) as was invariably the case at the initial stages of the literature on the GLSEM, predetermined variables, such as lagged endogenous variables, were indeed independent of the error term at time t. If the error process, however, is asserted to be a moving average of length n, **exogenous** variables are still **independent** of the error at time t, **but not all** lagged endogenous variables are!

The standard assumptions under which we operate the benchmark (simplest) full model are:

i. The error process is one of *iid* random vectors with mean zero and covariance matrix $\Sigma > 0$.

ii. The predetermined variables of the model are

$$x_{t\cdot} = (y_{t-1\cdot}, y_{t-2\cdot}, \ldots, y_{t-k\cdot}, p_{t\cdot}), \quad p_{t\cdot} = (p_{t1}, p_{t2}, \ldots, p_{ts}),$$

where $p_{t\cdot}$ is the vector of the s **exogenous** variables, so that

$$G = mk + s.$$

iii. The **exogenous** variables are non-collinear, or more precisely if $P = (p_{t\cdot})$ is the matrix of observations on the s exogenous variables, the rank of P is s, and (at least)

$$\plim_{T\to\infty} \frac{P'P}{T} = M_{pp} > 0.$$

iv. The system is stable, an issue that was explored at length in Chap. 6. To see what this means, write the model with k lagged endogenous and s exogenous variables as

$$y_{t\cdot}B^* = \sum_{i=1}^{k} y_{t-i\cdot}C_i + p_{t\cdot}C_{k+1} + u_{t\cdot},$$

and note that this is a difference equation of order k and forcing function $p_t \cdot C_{k+1} + u_t$. From the discussion therein, we see that its characteristic equation is

$$\left| B^* - \sum_{i=1}^{k} C_i \xi^i \right| = 0.$$

Stability means that the roots of this characteristic equation lie outside the unit circle.

v. (Identification and normalization) The coefficient of the ith endogenous variable, y_i, in the ith equation is set to unity (normalization), and some coefficients of endogenous and predetermined variables are known a priori to be zero (identification). We generally employ the convention that in the ith equation there are $m_i + 1 \le m$ **endogenous** variables and $G_i \le G$ **predetermined** variables.

vi. (Uniqueness of equilibrium) The matrix B^* is **nonsingular**.

The model as exhibited in Eq. (12.1) is said to be the **structural form** of the model, and the equations listed therein are termed the **structural equations**. By extension, the parameters appearing therein are said to be the **structural parameters**.

In contradistinction to the structural form of the model is the **reduced form**, which is given by

$$Y = \sum_{i=1}^{k} Y_{-i} \Pi_i + P \Pi_{k+1} + V, \quad \Pi_i = C_i D, \quad V = UD, \quad D = B^{*-1},$$

$$C = \begin{pmatrix} C_1 \\ C_2 \\ \vdots \\ C_k \\ C_{k+1} \end{pmatrix}, \quad \Pi = \begin{pmatrix} \Pi_1 \\ \Pi_2 \\ \vdots \\ \Pi_k \\ \Pi_{k+1} \end{pmatrix}, \quad X = (Y_{-1}, Y_{-2}, \dots, Y_{-k}, P), \quad (12.3)$$

which is obtained by using assumptions iv and vi. The dimensions of the parameter matrices in Eq. (12.3) are: Π_i, $i = 1, 2, \dots, k$ which are $m \times m$; Π_{k+1} is $s \times m$, thus Π is $G \times m$, where G is as defined in assumption ii above; finally D is evidently $m \times m$. Notice, in particular, that in this notation Π_i is the coefficient matrix of the ith lag, i.e. the coefficient matrix of Y_{-i}, in the vector difference equation representation of reduced form of the GLSEM in Eq. (12.3).

The model as exhibited in Eq. (12.3) is said to be the **reduced form**, and the equations listed therein are termed the **reduced form equations**. By extension, the parameters appearing therein are said to be the **reduced form parameters**.

The structural form purports to describe the manner in which economic agents actually behave, and through the feedback matrix B^* allows one endogenous variable to influence another, i.e. to appear as an **explanatory** variable in the latter's "equation". By contrast, the reduced form nets out all such feedbacks and displays the endogenous variables as functions **only** of **predetermined** variables and the (reduced form) errors.

12.1.3 Inconsistency of Ordinary Least Squares

Employing the convention noted previously regarding normalization, we write the system as[1]

$$Y = YB + XC + U, \tag{12.4}$$

where B has **only zeros** on its diagonal. Further, imposing the identification convention, we may write the ith structural equation as

$$y_{\cdot i} = Y_i \beta_{\cdot i} + X_i \gamma_{\cdot i} + u_{\cdot i}, \quad i = 1, 2, 3, \ldots, m. \tag{12.5}$$

We shall now make clear the connection between the symbols here and the matrices Y, X, B^*, C, U through the use of **selection** matrices. The latter are permutations of (some of) the columns of an appropriate **identity** matrix. Denote by $e_{\cdot i}$ a column vector all of whose elements are zero except the ith, which is unity. The $m \times m$ **identity matrix** is denoted by $I_m = (e_{\cdot 1}, e_{\cdot 2}, \ldots, e_{\cdot m})$, where the $e_{\cdot i}$ are m-element **column** vectors. The $G \times G$ **identity matrix** is denoted by $I_G = (e_{\cdot 1}, e_{\cdot 2}, \ldots, e_{\cdot G})$, where, here, the $e_{\cdot i}$ are G-element **column** vectors all of whose elements are zero except the ith, which is unity.

Because Y_i has as its columns the T observations on the endogenous variables that appear as explanatory (right hand) variables, and similarly for X_i, there exist selection matrices L_{1i} and L_{2i}, which are of dimension $m \times m_i$ and $G \times G_i$, respectively, such that

$$Y_i = Y L_{1i}, \quad X_i = X L_{2i}, \tag{12.6}$$

and, moreover, $y_{\cdot i} = Y e_{\cdot i}$, $u_{\cdot i} = U e_{\cdot i}$. It is immediately apparent that the columns of B and C are related to the structural parameter vectors exhibited in Eq. (12.5) by

$$b_{\cdot i} = L_{1i} \beta_{\cdot i}, \quad c_{\cdot i} = L_{2i} \gamma_{\cdot i}. \tag{12.7}$$

---

[1]In econometrics, the least squares procedure is termed "Ordinary Least Squares" (OLS), for reasons that will become evident presently.

Since the columns of L_{1i} and L_{2i} are **orthonormal**, we also have the relationship

$$\beta_{\cdot i} = L_{1i}' b_{\cdot i}, \quad \gamma_{\cdot i} = L_{2i}' c_{\cdot i}. \tag{12.8}$$

Writing the ith structural equation even more compactly as

$$y_{\cdot i} = Z_i \delta_{\cdot i} + u_{\cdot i}, \quad Z_i = (Y_i, X_i), \quad \delta_{\cdot i} = (\beta_{\cdot i}', \gamma_{\cdot i}')', \tag{12.9}$$

the OLS estimator of the **structural** parameters is given by

$$\tilde{\delta}_{\cdot i} = (Z_i' Z_i)^{-1} Z_i' y_{\cdot i}. \tag{12.10}$$

Putting

$$L_i = \operatorname{diag}(L_{1i}, L_{2i}), \quad \text{we have} \quad \tilde{\delta}_{\cdot i} - \delta_{\cdot i} = (L_i' Z' Z L_i)^{-1} L_i' Z' U e_{\cdot i}. \tag{12.11}$$

To demonstrate the inconsistency of the OLS estimator of the structural parameters, we need to show that $L_i'(Z'Z/T)L_i$ converges, at least in probability, to a nonsingular matrix **and** that $L_i'(Z'U/T)e_{\cdot i}$ converges to a nonzero vector. By definition, $Z = (Y, X)$ and, using the reduced form $Y = X\Pi + V$, we find

$$\frac{1}{T}Z'Z = \frac{1}{T}\begin{bmatrix} \Pi'X'X\Pi + D'U'UD & \Pi'X'UD \\ D'U'X\Pi & X'X \end{bmatrix}$$

$$+ \frac{1}{T}\begin{bmatrix} \Pi'X'UD + D'U'X\Pi & D'U'X \\ X'UD & 0 \end{bmatrix}.$$

In view of assumptions i, ii, iii, and iv,

$$\frac{1}{T}X'U = \frac{1}{T}\sum_{t=1}^{T} x_{t\cdot}' u_{t\cdot} \xrightarrow{P} 0,$$

$$\frac{1}{T}U'U = \frac{1}{T}\sum_{t=1}^{T} u_{t\cdot}' u_{t\cdot} \xrightarrow{P} \Sigma,$$

$$\frac{1}{T}X'X = \frac{1}{T}\sum_{t=1}^{T} x_{t\cdot}' x_{t\cdot} \xrightarrow{P} M_{xx} > 0. \tag{12.12}$$

Consequently,

$$\operatorname*{plim}_{T\to\infty}(\tilde{\delta}_{\cdot i} - \delta_{\cdot i}) = \left(L_i' \begin{bmatrix} \Pi'M_{xx}\Pi + D'\Sigma D & \Pi'M_{xx} \\ M_{xx}\Pi & M_{xx} \end{bmatrix} L_i \right)^{-1} L_i' \begin{pmatrix} D'\Sigma \\ 0 \end{pmatrix} e_{\cdot i},$$

and it is evident that $D^{'}\sigma_{\cdot i}$ is not necessarily null.[2] The demonstration that the OLS estimator of the structural parameters is inconsistent will be complete if we show that the right hand matrix above is **nonsingular**, and that the vector multiplying it is non-null. For the first, write the matrix in question as

$$L_i^{'}(\Pi, I_G)^{'} M_{xx}(\Pi, I_G)L_i + L_i^{'}\begin{bmatrix} D^{'}\Sigma D & 0 \\ 0 & 0 \end{bmatrix}L_i,$$

and note that[3] $(\Pi, I_G)L_i = (\Pi_i, I_{G_i}) = S_i$ is a matrix of full (column) rank, if the ith structural equations is identified, see Dhrymes (1994, Proposition 5, p. 25).[4] Since S_i is $G \times (m_i + G_i)$ and the rank of M_{xx} is G, it follows that $S_i^{'}M_{xx}S_i$ is **nonsingular**. This is true because of Propositions 2.61 and 2.62 of Chap. 2. Thus, the matrix is **positive definite**; adding to it the **positive semi-definite** matrix with $L_{1i}^{'}D^{'}\Sigma DL_{1i}$ in the upper left block and zeros elsewhere still leaves us with a positive definite matrix, which is thus invertible.

12.2   Consistent Estimation: 2SLS and 3SLS

In this section we derive consistent estimators for the parameters of the GLSEM, beginning with 2SLS (two stage least squares) and extending it to 3SLS (three stage least squares).

The discussion proceeds in two main directions: (a) first we discuss the original approach suggested by Theil (1953, 1958), the inventor of 2SLS and 3SLS estimation, and (b) the approach suggested in Dhrymes (1969), which first transforms the GLSEM so that, asymptotically, it "looks" like a GLM and then represents 2SLS and 3SLS, respectively, as OLS and GLS estimation in this transformed context. In the second direction (b), we first use a special notation to make the result visibly obvious by inspection; in addition we also

---

[2] In the special case where D is **upper triangular** and Σ is **diagonal**, it may be shown that $L_{1i}^{'}D\sigma_{\cdot i} = 0$ and, **in this case only**, the OLS estimator of the structural parameters is indeed consistent. Such systems are termed **simply recursive**.

[3] In order to avoid excessive notational clutter we shall use, in the following discussion, the notation Π_i **to indicate not** $C_i D$ **as defined in Eq. (12.3), but to mean** ΠL_{1i} **, as well as in all discussions pertaining to the estimation, limiting distributions and tests of hypotheses. In this usage** Π_i **refers to the coefficients of** X **in the reduced form representation of** Y_i **, i.e. the endogenous variables that appear as explanatory variables in the** ith **structural equation. We shall use the definition** $C_i D$ **, of Eq. (12.3), only in connection with issues of forecasting from the GLSEM.**

[4] We shall discuss identification issues at some length at a later section.

express our results in an alternative more standard notation which is very useful in exploring other GLSEM issues relevant to estimation and testing.

The initial motivation for the 2SLS estimator, which is also responsible for the terminology "two stage least squares" thus distinguishing it from the "ordinary least squares" estimator, is as follows: Retracing our steps in the preceding discussion, we note that the last term in

$$\left(L_i' \begin{bmatrix} \Pi' M_{xx} \Pi + D' \Sigma D & \Pi' M_{xx} \\ M_{xx} \Pi & M_{xx} \end{bmatrix} L_i \right)^{-1} L_i' \begin{pmatrix} D' \Sigma \\ 0 \end{pmatrix} e_{\cdot i}$$

is simply the probability limit of $L_i' Z' U e_{\cdot i}/T$; examining the random component, we see that we are dealing with

$$\frac{1}{T} Z' U = \frac{1}{T} \begin{pmatrix} Y' U \\ X' U \end{pmatrix} = \frac{1}{T} \left[\begin{pmatrix} \Pi' X' U \\ X' U \end{pmatrix} + \begin{pmatrix} D' U' U \\ 0 \end{pmatrix} \right].$$

What is responsible for the inconsistency of OLS is the limit of the term $D' U' U/T$; that term corresponds to the covariance matrix between the structural errors and the error component of the **reduced form system** $Y = X\Pi + V$. If we were to eliminate V from consideration, the resulting "OLS" estimator would be consistent. Unfortunately, we do not know Π; we may, however, estimate it consistently by (ordinary) least squares as

$$\tilde{\Pi} = (X' X)^{-1} X' Y.$$

This is the "first" stage of (two stage) least squares; in the "second" stage, we regress $y_{\cdot i}$ on $\tilde{Z}_i = (\tilde{Y}, X) L_i$, where $\tilde{Y} = X\tilde{\Pi}$.

Noting that

$$(\tilde{Y}, X) L_i = X(\tilde{\Pi} L_{1i}, L_{2i}) = X\tilde{S}_i, \quad \tilde{S}_i = (\tilde{\Pi}, I_G) L_i, \quad L_i = \text{diag}(L_{1i}, L_{2i}),$$

we find that the "second" stage yields

$$\tilde{\delta}_{\cdot i(2SLS)} = \delta_{\cdot i} + (\tilde{S}_i' X' X \tilde{S}_i)^{-1} \tilde{S}_i' X' u_{\cdot i}. \tag{12.13}$$

The estimator in Eq. (12.13) **is the 2SLS estimator of the parameters in the *i*th structural equation of the GLSEM.** In principle this can be applied **to every equation of the GLSEM.** Consequently, by the previous discussion, it is evident that the 2SLS estimator of the structural parameters of a GLSEM is consistent.

12.2.1   An Alternative Derivation of 2SLS and 3SLS

By the assumptions made in this model, it is evident that

$$\text{rank}(X) = G, \quad a.c. \quad \text{i.e. with probability one.}$$

Hence (with probability one), there exists a nonsingular matrix, R, such that

$$X'X = RR'.$$

Consider now the transformation

$$R^{-1}X'y_{\cdot i} = R^{-1}X'Z_i\delta_{\cdot i} + R^{-1}X'u_{\cdot i}, \quad i = 1, 2, \ldots, m. \tag{12.14}$$

It is easily verified, in the transformed model, that

$$\plim_{T\to\infty} \frac{1}{T}Q'_{\cdot i}r_{\cdot i} = 0, \text{ where } Q_i = R^{-1}X'Z_i, \quad r_{\cdot i} = R^{-1}X'u_{\cdot i}.$$

In effect, this transformation has yielded a formulation in which the conditions responsible for making the OLS estimator in the General Linear Model **consistent** hold here as well, at least asymptotically, and in this formulation the 2SLS estimator is simply the OLS estimator (in the transformed context)!

12.2.2 Systemwide Estimation

As a matter of notation, put

$$w_{\cdot i} = R^{-1}X'y_{\cdot i} \quad Q_i = R'\tilde{S}_i, \quad \tilde{S}_i = (\tilde{\Pi}, I_G)L_i$$

$$r_{\cdot i} = R^{-1}X'u_{\cdot i} \quad Q^* = \text{diag}(Q_1, Q_2, \ldots, Q_m) = (I_m \otimes R')\tilde{S},$$

$$\delta = \begin{pmatrix} \delta_{\cdot 1} \\ \delta_{\cdot 2} \\ \vdots \\ \delta_{\cdot m} \end{pmatrix} \quad r = \begin{pmatrix} r_{\cdot 1} \\ r_{\cdot 2} \\ \vdots \\ r_{\cdot m} \end{pmatrix} \quad w = \begin{pmatrix} w_{\cdot 1} \\ w_{\cdot 2} \\ \vdots \\ w_{\cdot m} \end{pmatrix}$$

$$\tilde{S} = \text{diag}(\tilde{S}_1, \tilde{S}_2, \ldots, \tilde{S}_m) \quad a = \text{vec}(A) \quad A = \begin{pmatrix} B \\ C \end{pmatrix}, \tag{12.15}$$

and write the entire system more compactly as

$$w = Q^*\delta + r, \quad \text{Cov}(r) = \Sigma \otimes I_G = \Phi. \tag{12.16}$$

The 2SLS estimator in the context of this transformed model is the OLS estimator

$$\tilde{\delta}_{2SLS} = (Q^{*'}Q^*)^{-1}Q^{*'}w = \delta + (Q^{*'}Q^*)^{-1}Q^{*'}r, \tag{12.17}$$

while the 3SLS estimator is the generalized least squares (GLS), or **feasible Aitken**, estimator

$$\tilde{\delta}_{3SLS} = [Q^{*'}\tilde{\Phi}^{-1}Q^*]^{-1}Q^{*'}\tilde{\Phi}^{-1}w = \delta + [Q^{*'}\tilde{\Phi}^{-1}Q^*]^{-1}Q^{*'}\tilde{\Phi}^{-1}r. \tag{12.18}$$

The notation employed above was particularly well suited for showing that
**2SLS is OLS estimation and 3SLS is generalized least squares (GLS)
estimation in the context of the transformed model of Eq. (12.16)**.
However, it will not be particularly suitable for certain other purposes
in the discussion to follow.   Thus, we also present here the alternative
representations.[5]

$$(\tilde{\delta} - \delta)_{2SLS} = \left(\tilde{S}'(I_m \otimes \tilde{M}_{xx})\tilde{S}\right)^{-1} \frac{(I_m \otimes X')u}{T}, \quad \tilde{M}_{xx} = \frac{X'X}{T}$$

$$(\tilde{\delta} - \delta)_{3SLS} = \left(\tilde{S}'(\tilde{\Sigma}^{-1} \otimes \tilde{M}_{xx})\tilde{S}\right)^{-1} (\tilde{\Sigma}^{-1} \otimes I_G)\frac{(I_m \otimes X')u}{T} \qquad (12.19)$$

Consistency and Asymptotic Normality of 2SLS and 3SLS

We take up first the issue of consistency of the two estimators. To this end we
note that by the assumptions made above we have, at least, that

$$\frac{X'X}{T} \xrightarrow{P} M_{xx} > 0,$$

and thus by Proposition 2.64 there exists a non-singular $G \times G$ matrix $\bar{R}$
such that $\bar{R}\bar{R}' = M_{xx}$.
Moreover

$$\frac{Q_i'Q_i}{T} \quad = \quad \frac{Z_i'XR'^{-1}R^{-1}X'Z_i}{T} = \frac{L_i'\begin{pmatrix} Y' \\ X' \end{pmatrix} X(X'X)^{-1}X'(Y \quad X)L_i}{T}$$

$$\xrightarrow{P} \quad L_i'(\Pi, I_G)'M_{xx}(\Pi, I_G)L_i = S_i'M_{xx}S_i, \qquad (12.20)$$

because, at least, $(X'V/T) \xrightarrow{P} 0$, thus confirming the implications of
Eq. (12.19) about the consistency of 2SLS and 3SLS estimators.
    We shall now derive their limiting distribution. To this end consider

$$\sqrt{T}(\tilde{\delta} - \delta)_{2SLS} \quad = \quad \left(\frac{Q^{*'}Q^*}{T}\right)^{-1} \frac{Q^{*'}r}{\sqrt{T}}$$

$$\sqrt{T}(\tilde{\delta} - \delta)_{3SLS} \quad = \quad \left(\frac{Q^{*'}\tilde{\Phi}^{-1}Q^*}{T}\right)^{-1} \left(\frac{Q^{*'}\tilde{\Phi}^{-1}r}{\sqrt{T}}\right).$$

---

[5]Notice that, in this alternative representation, consistency of the 2SLS and 3SLS
estimators is obvious.

We now have recourse to Proposition 9.6 (of Chap. 9) in order to simplify certain derivations to follow. From previous definitions of symbols we find

$$\frac{Q^{*\prime}Q^*}{T} = L'\left[I_m \otimes \frac{Z'X}{T}\left(\frac{X'X}{T}\right)^{-1}\frac{X'Z}{T}\right]L \xrightarrow{P} S'(I_m \otimes M_{xx})S$$

$$\frac{Q^{*\prime}\tilde{\Phi}^{-1}Q^*}{T} \xrightarrow{P} S'(\Sigma^{-1} \otimes M_{xx})S \qquad (12.21)$$

$$\frac{Q^{*\prime}r}{\sqrt{T}} \sim S'\frac{(I_m \otimes X')u}{\sqrt{T}}, \quad \frac{Q^{*\prime}\tilde{\Phi}^{-1}r}{\sqrt{T}} \sim S'(\Sigma^{-1} \otimes I_G)\frac{(I_m \otimes X')u}{\sqrt{T}},$$

and we see that the limiting distribution problem **is the same whether we are dealing with 2SLS or 3SLS estimators since both involve the limiting distribution of** $(I_m \otimes X')u)/\sqrt{T}$. From Proposition 4.1 (Chap. 4) we see that

$$\frac{1}{\sqrt{T}}(I_m \otimes X')u = \frac{1}{\sqrt{T}}\text{vec}(X'U) = \frac{1}{\sqrt{T}}\sum_{t=1}^{T} x'_{t\cdot}u_{t\cdot}. \qquad (12.22)$$

But $\{x'_{t\cdot}u_{t\cdot}: t = 1, 2, 3, \ldots\}$ is a MD sequence because its first moment exists and $E(x'_{t\cdot}u_{t\cdot}|\mathcal{A}_{t-1}) = 0$, where

$$\mathcal{A}_{t-1} = \sigma(u_{q\cdot}: \quad q \le t - 1).$$

This is so because $x_{t\cdot}$ contains k lagged endogenous variables and s exogenous variables. The latter are independent of the structural errors, or the analysis is done **conditionally on the exogenous variables**, and the lagged endogenous variables are $\mathcal{A}_{t-1}$-measurable. Moreover, given the conditions on $X'X/T$ the sequence obeys a Lindeberg condition, and thus we may apply the MD CLT, Proposition 9.9 (of Chap. 9), with the Lindeberg condtion (Proposition 9.7) provided one of the two subsidiary conditions hold. To this end note that for

$$\zeta_T = \frac{1}{\sqrt{T}}(I_m \otimes X')u = \frac{1}{\sqrt{T}}\sum_{t=1}^{T}(I_m \otimes x'_{t\cdot})u'_{t\cdot}$$

$$\text{Cov}(\zeta_T) = \Sigma \otimes \frac{X'X}{T} \xrightarrow{P} \Sigma \otimes M_{xx}, \qquad (12.23)$$

where the covariance is calculated term by term conditioning on the appropriate sub-σ-algebra $\mathcal{A}_{t-1}$. Equation (12.23) satisfies condition i of the MD CLT. Consequently,

$$\zeta_T \xrightarrow{d} N(0, \Sigma \otimes M_{xx}).$$

Since

$$C_2 = (S^{*'}S^*)^{-1}S^{*'}\Phi S^*(S^{*'}S^*)^{-1}$$

$$C_3 = (S^{*'}\Phi^{-1}S^*)^{-1}S^{*'}\Phi^{-1}\Phi\Phi^{-1}S^*(S^{*'}\Phi^{-1}S^*)^{-1} = (S^{*'}\Phi^{-1}S^*)^{-1}$$

we conclude

$$\sqrt{T}(\tilde{\delta} - \delta)_{2SLS} \xrightarrow{\text{d}} N(0, C_2),$$

$$\sqrt{T}(\tilde{\delta} - \delta)_{3SLS} \xrightarrow{\text{d}} N(0, C_3).$$

We have therefore proved

Proposition 12.1. Under the conditions above the system-wide 2SLS and 3SLS estimators of the parameters of the GLSEM are consistent and, moreover, their limiting distributions are given, respectively, by

$$\sqrt{T}(\tilde{\delta} - \delta)_{(2SLS)} \xrightarrow{\text{d}} N(0, C_2), \quad \sqrt{T}(\tilde{\delta} - \delta)_{(3SLS)} \xrightarrow{\text{d}} N(0, C_3),$$

$$C_2 = (S^{*'}S^*)^{-1}S^{*'}\Phi S^*(S^{*'}S^*)^{-1},$$

$$C_3 = (S^{*'}\Phi^{-1}S^*)^{-1}, \quad \Phi = \Sigma \otimes I_G,$$

$$\frac{X'X}{T} \xrightarrow{\text{P}} M_{xx} = \bar{R}\bar{R}', \quad S_i = (\Pi, I_G)L_i, \quad S_i^* = \bar{R}'S_i = \bar{R}'(\Pi, I_G)L_i,$$

$$S^* = \text{diag}(S_1^*, S_2^*, S_3^*, \ldots, S_m^*). \tag{12.24}$$

In terms of the alternative notation the covariance matrices of the systemwide 2SLS and 3SLS are given by

$$C_2 = (S'(I_m \otimes M_{xx})S)^{-1}S'(\Sigma \otimes M_{xx})S(S'(I_m \otimes M_{xx})S)^{-1}$$

$$C_3 = (S'(\Sigma^{-1} \otimes M_{xx})S)^{-1}.$$

In Chap. 10, we showed, in the context of the GLM, that the OLS estimator is inefficient relative to the Aitken estimator, using a certain procedure. Although the same procedure may be used to show that the 3SLS estimator is efficient relative to the 2SLS estimator, we shall use an alternative argument.

Proposition 12.2. 3SLS are efficient relative to 2SLS estimators in the sense that $C_2 - C_3 \geq 0$.

Proof: By Proposition 2.66, (of Chap. 2), it will suffice to show that

$$C_3^{-1} - C_2^{-1} \geq 0.$$

Since

$$\Delta = C_3^{-1} - C_2^{-1} = S^{*'} P' [I_{mG} - H] P S^*,$$

where $H = P'^{-1} S^* (S^{*'} \Phi S^*)^{-1} S^{*'} P^{-1}$, $P'P = \Phi^{-1}$, we need only show that $I_{mG} - H \geq 0$. Consider the characteristic roots of the symmetric matrix H, which is of dimension mG. By Corollary 2.8, the roots of H, given by

$$|\lambda I_{mG} - H| = |\lambda I_{mG} - P'^{-1} S^* (S^{*'} \Phi S^*)^{-1} S^{*'} P^{-1}| = 0$$

consist of $mG - q$ zeros, where $q = \sum_{i=1}^{m} (m_i + G_i)$, and q unit roots because the non-zero roots of the characteristic polynomial above are precisely those of

$$0 = |\mu I_q - (S^{*'} \Phi S^*)^{-1} S^{*'} P^{-1} P'^{-1} S^*| = |\mu I_q - I_q| = 0,$$

which are q unities. Since H is symmetric, its characteristic vectors, i.e. the columns of the matrix J below, may be chosen to be orthonormal and thus

$$I_{mG} - H = JJ' - J \begin{bmatrix} I_q & 0 \\ 0 & 0 \end{bmatrix} J' = J \begin{bmatrix} 0 & 0 \\ 0 & I_{mG-q} \end{bmatrix} J' \geq 0.$$

But then

$$\Delta = S^{*'} P' J \begin{bmatrix} 0 & 0 \\ 0 & I_{mG-q} \end{bmatrix} J' P S^* \geq 0.$$

q.e.d.

12.3   Identification Issues

Because we are dealing with a system the identification of whose parameters depends (in part) on prior restrictions, it is desirable to examine the nature of such restrictions, to possibly test for their validity and elucidate the role they play in the estimation procedures we developed.

In the early development of this literature, see e.g. Anderson and Rubin (1949, 1950), the basic formulation of the identification problem was this: Given knowledge of the parameters of **the reduced form** can we determine uniquely, within normalization, the parameters of the **structural form** of the model? First note that Π is $G \times m$ and thus **it contains mG parameters**. The structural form of Eq. (12.2), $YB^* = XC + U$, contains $m^2 + mG$ parameters and the relation between the two is $\Pi B^* = C$; thus we have a system of mG equations in $m(m+G)$ unknowns, (the unknown elements of B^*

and C). Evidently such a system has no unique solution! It is this requirement that we expressed under item vi in the discussion below Remark 12.1. In and of itself, however, this is **inadequate** because it is only a necessary but not a sufficient condition for identification. To sharpen the focus, consider the ith structural equation, and suppose we wish to impose the normalization that the coefficient of y_{ti} in the ith equation is unity, i.e. $b_{ii}^* = 1$; we retain the previously imposed exclusion restrictions that only m_i jointly dependent (endogenous) and G_i predetermined variables appear as "explanatory" variables in its right member. This additional requirement (of normalization) necessitates a modification of the selection matrices defined in Eqs. (12.7) and (12.8). Thus, concentrating on the ith equation we find

$$Yb_{\cdot i}^* = (y_{\cdot i}, Y_i) \begin{pmatrix} 1 \\ -\beta_{\cdot i} \end{pmatrix} = Y(e_{\cdot i}, L_{1i}) \begin{pmatrix} 1 \\ -\beta_{\cdot i} \end{pmatrix}, \quad Xc_{\cdot i} = X_i \gamma_i = XL_{2i}\gamma_i$$

and thus

$$b_{\cdot i}^* = L_{1i}^* \begin{pmatrix} 1 \\ -\beta_{\cdot i} \end{pmatrix}, \quad L_{1i}^* = (e_{\cdot i}, L_{1i}), \quad c_{\cdot i} = L_{2i}\gamma_i. \tag{12.25}$$

Noting that, for the ith equation, the connection between the structural and reduced form parameters is

$$\Pi b_{\cdot i}^* = c_{\cdot i}, \quad \text{or} \quad 0 = \Pi b_{\cdot i}^* - c_{\cdot i}$$

and substituting from Eq. (12.25), results in

$$0 = \Pi L_{1i}^* \begin{pmatrix} 1 \\ -\beta_{\cdot i} \end{pmatrix} - L_{2i}\gamma_{\cdot i} = \pi_{\cdot i} - (\Pi, I_G)L_i \delta_{\cdot i}, \quad \text{or}$$

$$\pi_{\cdot i} = (\Pi, \quad I_G)L_i \delta_{\cdot i} = S_i \delta_{\cdot i}. \tag{12.26}$$

By the convention that there are $m_i + G_i$, explanatory variables in the ith equation, m_i jointly dependent and G_i predetermined variables, the system in Eq. (12.26) **contains G equations and $m_i + G_i$ unknowns;** thus, it has a unique solution only if the matrix of the system is of full rank. We have thus proved

Proposition 12.3. A necessary and sufficient condition for the ith equation of the GLSEM above to be identified is that

$$\text{rank}[(\Pi \quad I_G)L_i] = \text{rank}(S_i) = m_i + G_i. \tag{12.27}$$

Remark 12.2. The proof of identification above is in the spirit of Anderson and Rubin (1949, 1950), but its motivation and argumentation represents an appreciable modification of what is usually done in such discussions.

Remark 12.3. Notice that the **sufficient rank condition can only be asserted** prior to estimation; we can **only verify the order condition, which is merely necessary.**

Remark 12.4. In contrast to the Anderson-Rubin treatment of identifiability, the discussion above makes quite clear the role of the identification condition in estimating the parameters of the system. In particular observe from Eq. (12.13) and the surrounding discussion that $S_i = (\Pi, \quad I_G)L_i$, and its role is quite crucial in 2SLS and 3SLS estimation. In this connection also note that we can reassure ourselves *ipso facto* that **if the 2SLS or 3SLS estimator of some structural equation can be obtained, that equation is identifiable!**

The preceding suggests the following definitions.

Definition 12.1. The ith equation of a GLSEM is said to be identified if and only if it obeys the identification condition in Eq. (12.27). In practice, however, we can only check the **order condition.**

If every equation of a GLSEM is identified, we say that the system is identified.

Definition 12.2. If for the ith equation of a GLSEM, $G > m_i + G_i$, the equation is said to be **over-identified**; if $G < m_i + G_i$ it is said to be **under-identified**; if $G = m_i + G_i$ the equation is said to be **just-identified**. Thus, by the order condition the ith equation is identified if $G \geq m_i + G_i$. Otherwise it is said to be non-identifiable.

12.3.1   Tests of Over-identifying Restrictions

When an equation is over-identified, $G > m_i + G_i$, the "excess" restrictions we imposed (beyond normalization), $q = G - m_i - G_i$, are said to be the[6] over-identifying restrictions. We may wish to test for their validity, indeed for the validity of all restrictions.[7] This, however, cannot be done routinely with the formulation of the estimation problem as given above. In that context, all we **can do** it to re-estimate the model with **fewer** restrictions (i.e. more parameters) and then either test the extra parameters so estimated for

---

[6]Some readers may be confused by the terminology. To clarify, note that if $q > 0$, this means that we have specified that **more** parameters in B and C are zero than is allowable consistent with identification, i.e. relative to the number of restrictions that render the equation just identified. To test for their validity we need to allow for **fewer** restrictions, i.e. to **increase** m_i and/or G_i; thus we would then be estimating a model with **more** parameters than before!

[7]As we shall see later this is not possible.

significance, or engage in a likelihood ratio like procedure by comparing the sum of squared residuals (from 2SLS or 3SLS) in the case of fewer restrictions to that of the case with more restrictions. If this difference is "significant" we conclude that some of the over-identifying restriction may not be valid; if not, the over-identifying restrictions are valid. This is a very cumbersome procedure. Instead, we shall give a simpler alternative based on the method of Lagrange multipliers which, computationally, modifies the estimation problem as exposited above.

To this end, return to Eq. (12.14) and define

$$Q = R^{-1}X'Z, \tag{12.28}$$

so that the ith equation may be written as

$$w_{\cdot i} = Qa_{\cdot i} + r_{\cdot i}, \quad a_{\cdot i} = \begin{pmatrix} b_{\cdot i} \\ c_{\cdot i} \end{pmatrix}, \quad i = 1, 2, 3, \ldots m. \tag{12.29}$$

Letting L_{1i}^*, L_{2i}^* be the **complements**[8] of L_{1i}, L_{2i} of Eqs. (12.7) and (12.8) in I_m and I_G, respectively, we may express the prior restrictions as

$$L_i^{*'} a_{\cdot i} = 0, \quad L_i^* = \mathrm{diag}(L_{1i}^*, L_{2i}^*), \tag{12.30}$$

so that the 2SLS estimator of the unknown parameters in that equation may be obtained by the process

$$\min_{a_{\cdot i}} \frac{1}{T}(w_{\cdot i} - Qa_{\cdot i})'(w_{\cdot i} - Qa_{\cdot i}) \text{ subject to } L_i^{*'} a_{\cdot i} = 0.$$

Operating with the Lagrangian expression

$$F = \frac{1}{T}(w_{\cdot i} - Qa_{\cdot i})'(w_{\cdot i} - Qa_{\cdot i}) - 2\lambda' L_i^{*'} a_{\cdot i}, \tag{12.31}$$

we obtain, after some rearrangement, the first order conditions

$$\begin{bmatrix} \frac{Q'Q}{T} & L_i^* \\ L_i^{*'} & 0 \end{bmatrix} \begin{pmatrix} a_{\cdot i} \\ \lambda \end{pmatrix} = \begin{pmatrix} \frac{Q'w_{\cdot i}}{T} \\ 0 \end{pmatrix}, \tag{12.32}$$

where λ is the $q = G - m_i - G_i$-element vector of the Lagrange multipliers, corresponding to the "excess" restrictions beyond normalization. If we examine the first set of equations in Eq. (12.32)

$$\frac{Q'Q}{T}a_{\cdot i} + L_i^*\lambda = \frac{Q'w_{\cdot i}}{T} = \frac{Q'Q}{T}a_{\cdot i} + \frac{Q'r_{\cdot i}}{T},$$

---

[8]This means that L_{1i}^* is what remains after we eliminate form I_m the columns of L_{1i} and from I_G the columns of L_{2i}.

substitute therein $a._{\cdot i} = L_i \delta._{\cdot i}$, and pre-multiply by L_i' , we find, noting that $L_i' L_i^* = 0$, that the first set of equations in Eq. (12.32) reads

$$\frac{L_i' Q' Q L_i}{T} \tilde{\delta}._{\cdot i} = \frac{L_i' Q' Q L_i}{T} \delta._{\cdot i} + \frac{L_i' Q' r._{\cdot i}}{T}, \quad \text{or}$$

$$\frac{Q_i^{*'} Q_i^*}{T} (\tilde{\delta} - \delta._{\cdot i}) = \frac{Q_i^{*'} r._{\cdot i}}{T}, \quad Q_i^* = Q L_i,$$

whose solution is **identical**[9] with the ith sub-vector in Eq. (12.18), which exhibits the standard system-wide 2SLS estimators. Thus, in deriving the limiting distribution of the estimators in Eq. (12.32) we need not be concerned with that of $(\tilde{a}._{\cdot i} - a._{\cdot i})$, which contains a lot of zeros (the over-identifying restrictions) and whose limiting covariance matrix is thus singular. We concentrate instead on the more relevant limiting distribution of the Lagrange multipliers.

As a matter of notation, set $\tilde{K} = Q' Q / T$ and note that

$$K = \plim_{T \to \infty} \frac{Q' Q}{T} = \binom{\Pi'}{I_G} M_{xx}(\Pi, I_G).$$

Thus, K is an $(m + G) \times (m + G)$ matrix of rank at most G. It is shown in Dhrymes (1994, Proposition 6, p. 45) that the ith equation is **identified** if and only if $K + L_i^* L_i^{*'}$ is of full[10] rank (nonsingular).

Because in the matrix

$$\begin{bmatrix} K & L_i^* \\ L_i^{*'} & 0 \end{bmatrix} = \begin{bmatrix} A_{11} & A_{12} \\ A_{21} & A_{22} \end{bmatrix}$$

both diagonal blocks (A_{11} and A_{22}) are **singular**, we cannot use Proposition 2.30 to obtain an explicit representation for the 2SLS estimator of $a._{\cdot i}$, and the Lagrange multiplier λ. On the other hand, because the ith equation is identified the condition for invertibility in Proposition 2.31a is satisfied.

Using the latter we find[11]

$$\begin{bmatrix} A_{11} & A_{12} \\ A_{21} & A_{22} \end{bmatrix}^{-1} = \begin{bmatrix} B_{11} & B_{12} \\ B_{21} & B_{22} \end{bmatrix}$$

---

[9]Note, incidentally, that $S_i^{*'} S_i^*$ is the probability limit of $L_i' Q^{*'} Q^* L_i / T$.

[10]The symbol L_i^* in this discussion corresponds to the symbol $L_i^{*\circ}$ in Proposition 6.

[11]Noting that $V_{11}^{-1} = \left(\plim_{T \to \infty} \frac{Q' Q}{T} \right) + L_i^* L_i^{*'}$, and that $M_{xx} = \plim_{T \to \infty} \frac{X' X}{T}$, and their frequent use in the immediately ensuing discussion, we shall not, in our usage, distinguish between these probability limits and their definition for finite samples, letting the context provide the appropriate meaning. We do so only in order to avoid notational cluttering.

$$B_{11} = V_{11} - V_{11}L_i^*(L_i^{*\prime}V_{11}L_i^*)^{-1}L_i^{*\prime}V_{11}, \quad V_{11} = (K + L_i^*L_i^{*\prime})^{-1}$$

$$B_{12} = V_{11}L_i^*(L_i^{*\prime}V_{11}L_i^*)^{-1}$$

$$B_{21} = (L_i^{*\prime}V_{11}L_i^*)^{-1}L_i^{*\prime}V_{11}$$

$$B_{22} = I_q - (L_i^{*\prime}V_{11}L_i^*)^{-1}, \quad q = G - G_i - m_i. \tag{12.33}$$

Consequently, we can solve explicitly for these parameters. Doing so, and utilizing the restrictions imposed by the Lagrange multiplier, we obtain

$$\begin{pmatrix} \tilde{a}_{\cdot i} \\ \tilde{\lambda} \end{pmatrix} = \begin{pmatrix} a_{\cdot i} \\ 0 \end{pmatrix} + \begin{pmatrix} B_{11} \\ B_{21} \end{pmatrix} \frac{1}{T} Q' r_{\cdot i},$$

$$\tilde{K} = \frac{1}{T}Q'Q, \quad V_{11} = (\tilde{K} + L_i^*L_i^{*\prime})^{-1}, \quad B_{11} = V_{11} - V_{11}L_i^*(L_i^{*\prime}V_{11}L_i^*)^{-1}L_i^{*\prime}V_{11}$$

$$B_{21}(L_i^{*\prime}V_{11}L_i^*)^{-1}L_i^{*\prime}V_{11}. \tag{12.34}$$

Concentrating on the Lagrange multipliers we have

$$\sqrt{T}\tilde{\lambda} = B_{21}\frac{Q'r_{\cdot i}}{\sqrt{T}}. \tag{12.35}$$

Since

$$\frac{Q'r_{\cdot i}}{\sqrt{T}} = \frac{1}{\sqrt{T}}\frac{Z'X}{T}\left(\frac{X'X}{T}\right)^{-1}\frac{X'u_{\cdot i}}{\sqrt{T}} \sim \begin{pmatrix} \Pi' \\ I_G \end{pmatrix}\frac{X'u_{\cdot i}}{\sqrt{T}},$$

we need only establish the limiting distribution of the last term, which we had encountered repeatedly in previous discussions, albeit in a system-wide context. Thus[12]

$$\frac{Q'r_{\cdot i}}{\sqrt{T}} \xrightarrow{\text{d}} N(0,\Theta_i), \quad \Theta_i = \sigma_{ii}\begin{pmatrix} \Pi' \\ I_G \end{pmatrix} M_{xx}\begin{pmatrix} \Pi & I_g \end{pmatrix}. \tag{12.36}$$

Consequently,

$$\sqrt{T}\tilde{\lambda} \xrightarrow{\text{d}} N(0,\Psi_i), \quad \Psi_i = B_{21}\Theta_i B_{21}'.$$

The matrix $B_{21} = (L_i^{*\prime}V_{11}L_i^*)^{-1}L_i^{*\prime}V_{11}$, is of dimension $q \times m + G$ and rank q; the matrix Θ_i is of dimension $m + G \times m + G$ and rank G, so that Ψ_i, a $q \times q$ matrix, is non-singular provided $q \leq G$, which is true in the present case.

We have thus proved

---

[12]Note that Θ_i is an $m + G \times m + G$ matrix of rank G.

Proposition 12.4. The limiting distribution of the Lagrange multiplier estimator for the over-identified restrictions in the ith structural equation is given by

$$\sqrt{T}\tilde{\lambda} \xrightarrow{\mathrm{d}} N(0, \Psi_i), \tag{12.37}$$

$$\Psi_i = \sigma_{ii}(L_i^{*\prime}V_{11}L_i^*)^{-1}[L_i^{*\prime}V_{11}\begin{pmatrix} \Pi' \\ I_G \end{pmatrix} M_{xx}(\Pi \quad I_g)V_{11}L_i^*](L_i^{*\prime}V_{11}L_i^*)^{-1}.$$

The matrix in square brackets, above, is $q \times q$ of rank q, and the first and last matrix terms are non-singular of dimension $q \times q$; since $q = G - m_i - G_i \leq G$, the matrix Ψ_i is **non-singular**, as claimed.

Statistical Tests

Given the limiting distribution in Eq. (12.37) we can test for the validity of **all** the over-identifying restrictions by using Eq. (10.31) of Chap. 10. Let A be an arbitrary, nonrandom $s \times q$ matrix, $s \leq q$; then

$$\sqrt{T}A\tilde{\lambda} \xrightarrow{\mathrm{d}} N(0, A\Psi_i A'),$$

and the test statistic for testing the null hypothesis that (some) Lagrange multipliers are zero, i.e. that (some of) the restrictions are not binding **because they are true**, is given by

$$\tau = T\tilde{\lambda}'\tilde{\Psi}_i^{-1}\tilde{\lambda} \xrightarrow{\mathrm{d}} \chi^2_{\mathrm{rank}(A)}, \quad \text{where} \tag{12.38}$$

$$\tilde{\Psi}_i = \tilde{\sigma}_{ii}(L_i^{*\prime}V_{11}L_i^*)^{-1}[L_i^{*\prime}V_{11}\begin{pmatrix} \tilde{\Pi}' \\ I_G \end{pmatrix} M_{xx}(\tilde{\Pi} \quad I_G)V_{11}L_i^*](L_i^{*\prime}V_{11}L_i^*)^{-1};$$

$$\tilde{\Pi} = (X'X)^{-1}X'Y, \quad \tilde{\sigma}_{ii} = \frac{\tilde{u}_{\cdot i}'\tilde{u}_{\cdot i}}{T},$$

$\tilde{u}_{\cdot i}$ being the 2SLS **residuals** of the ith equation.

Remark 12.5. Notice that the Lagrange multiplier estimator is centered on zero; this is because the implicit or null hypothesis is that over-identifying restrictions **are not binding**, i.e. they are **in fact correct**. If we wish to test all of them, we take $A = I_q$, in which case the test statistic has a limiting distribution which is central chi-squared with q degrees of freedom; if we only wish to test a subset, or even a single one, we define A accordingly. Acceptance is to be interpreted that, at the specified level of significance, **no restriction is binding** i.e. all restrictions tested are in fact true.

Rejection is to be interpreted that, at the given level of significance, one or more of the restrictions are in fact **binding** i.e. that one or more may be false.

Remark 12.6. Notice that we could not test for the validity of all possible restrictions; one such possibility is that all elements of $a_{\cdot i}$ (beyond normalization) **are zero**. In this case there is **no** simultaneous equations model to estimate and we are asserting that each endogenous variable equals possibly a constant plus a zero mean error term. We can only test the validity of those restrictions **beyond the number necessary to attain identification for the ith structural equation.**

It is perhaps instructive to make the point more forcefully, by showing what would happen if we attempted to estimate the ith equation of a model by allowing the number of parameters therein to exceed G, i.e. if we take $m_i + G_i > G$. The reader should bear in mind that identification **depends on the rank of the matrix S_i, which crucially depends on the true parameters, not merely its dimension.** So the premise is: the ith equation is identified, i.e. in fact $m_i + G_i \leq G$, but through lack of proper information or otherwise we attempt to fit an equation where $m_i + G_i > G$. The first thing to note is that the matrix $\tilde{S}_i = (\tilde{\Pi}, I_G)L_i$ of Eq. (12.13) is of dimension $G \times (m_i + G_i)$ and the matrix $\tilde{S}_i' X' X \tilde{S}_i$ (also of Eq. (12.13)) is of dimension $(m_i + G_i) \times (m_i + G_i)$ and of rank **at most** $G < m_i + G_i$. Hence its inverse **does not exist** and thus the 2SLS estimator is not defined. Consequently, we can test **overidentifying restrictions**, but we cannot test all restrictions! So the null hypothesis in the full test exhibited above is best thought of as: the equation in question is overidentified versus the equation in question is just identified. An interesting by-product of the preceding discussion is

Proposition 12.5. When every equation in the system is **just-identified** the structural and reduced forms **are identical**, in the sense that there is a one-to-one relationship between them; thus they convey precisely the same information.

Proof: Consider again Eq. (12.13) and note that if the ith structural equation is just-identified this means that the matrix S_i is $G \times G$ and **non-singular**. Consequently, we can rewrite it as

$$(\tilde{\delta}_{\cdot i} - \delta_{\cdot i})_{2SLS} = S_i^{-1}(\tilde{\pi}_{\cdot i} - \pi_{\cdot i}), \quad i = 1, 2, \ldots, m. \tag{12.39}$$

Writing down the entire system, we have

$$\sqrt{T}(\tilde{\delta} - \delta)_{2SLS} = S^{-1}\sqrt{T}(\tilde{\pi} - \pi)_{un}$$

$$\pi = \text{vec}(\Pi), \quad S = \text{diag}(S_1, S_2, \ldots, S_m), \quad \delta = \begin{pmatrix} \delta_{\cdot 1} \\ \delta_{\cdot 2} \\ \vdots \\ \delta_{\cdot m} \end{pmatrix}.$$

$$\sqrt{T}(\tilde{\pi} - \pi)_{un} \xrightarrow{d} N(0, D'\Sigma D \otimes M_{xx}^{-1}). \tag{12.40}$$

It then follows immediately that in the case **where all equations in the system are just-identified**, the structural and reduced form coefficients are **linear transformations of each other and thus convey precisely the same information about the system.**

Other interesting implications of just identifiability of a structural system will be considered more systematically in a subsequent section.

12.3.2 An Alternative Approach to Identification

In this section,[13] we give applications relating to the conditions for the identifiability of the parameters of a simultaneous equations system.

We recall that the parameters of the ith equation are identified if and only if the matrix S_i, as defined in Eq. (12.13), is of **full rank**.

The method to be employed below is most easily exposited if we assume that the structural errors are jointly normally distributed; consequently, if we write the log-likelihood function, L, divided by the number of observations, T,

$$L_T(\theta) = -\frac{m}{2}\ln 2\pi - \frac{1}{2}\ln|B^{*'}B^*| - \frac{1}{2}\text{tr}\Sigma^{-1}A^{*'}\left(\frac{1}{T}Z'Z\right)A^* \tag{12.41}$$

we can obtain the maximum likelihood (ML) estimator, $\hat{\theta}_T$, which obeys

$$L_T(\hat{\theta}_T) \geq L_T(\theta), \quad \text{for all } \theta. \tag{12.42}$$

Because of Jensen's inequality, (Proposition 8.14 of Chap. 8), we may write

$$0 = \ln E_{\theta_0}\frac{L^*(\theta)}{L^*(\theta_0)} \leq E_{\theta_0}L_T(\theta) - E_{\theta_0}L_T(\theta_0), \tag{12.43}$$

---

[13]This section may be omitted, if desired, without any loss of ability to deal with the remainder of this volume.

where L^* denotes the likelihood function, **not** the log-likelihood function and E_{θ_0} indicates that the expectation is taken relative to the distribution function with parameter θ_0. Moreover, because we deal with the case where the errors are identically distributed, given some minimal additional conditions, we can show that

$$L_T(\theta) \overset{a.c.}{\to} \bar{L}(\theta, \theta_0), \quad L_T(\theta_0) \overset{a.c.}{\to} \bar{L}(\theta_0, \theta_0), \tag{12.44}$$

which are the limits of the expectations of $L_T(\theta)$ and $L_T(\theta_0)$, respectively. These limits may be determined as follows: since

$$L_T(\theta_0) = -\frac{m}{2} - \frac{1}{2}\ln|\Sigma_0| + \frac{1}{2}\ln|B_0^{*'} B_0^*| - \frac{1}{2}\text{tr}\Sigma_0^{-1}\left[A_0^{*'}\left(\frac{Z'Z}{T}\right)A_0^*\right],$$

and since

$$A_0^{*'}\left(\frac{Z'Z}{T}\right)A_0^* = \frac{1}{T}U'U,$$

by the strong law of large numbers (SLLN) for independent, identically distributed random vectors it follows that

$$A_0^{*'}\left(\frac{Z'Z}{T}\right)A_0^* \overset{a.c.}{\to} \Sigma_0.$$

For $A^{*'}(Z'Z/T)A^*$, we first write $ZA^* = ZA_0^* - Z(A_0^* - A^*)$ and note that

$$A^{*'}\frac{Z'Z}{T}A^* = A_0^{*'}\frac{Z'Z}{T}A_0^* + (A_0^* - A^*)'\frac{Z'Z}{T}(A_0^* - A^*)$$

$$-\frac{U'Z}{T}(A_0^* - A^*) - (A_0^* - A^*)'\frac{Z'U}{T}.$$

Again, by using an appropriate SLLN we can show that

$$\frac{Z'U}{T} \overset{a.c.}{\to} \begin{pmatrix} B_0^{*'-1}\Sigma_0 \\ 0 \end{pmatrix}$$

and

$$\frac{Z'Z}{T} \overset{a.c.}{\to} \begin{bmatrix} \Pi_0' M_{xx}\Pi_0 & \Pi_0' M_{xx} \\ M_{xx}\Pi_0 & M_{xx} \end{bmatrix} + \begin{bmatrix} \Omega_0 & 0 \\ 0 & 0 \end{bmatrix} = P_0 + \begin{bmatrix} \Omega_0 & 0 \\ 0 & 0 \end{bmatrix},$$

where

$$P_0 = (\Pi_0, I_G)' M_{xx}(\Pi_0, I_G). \tag{12.45}$$

Hence, after some manipulation, we find

$$(A_0^* - A^*)' \frac{Z'Z}{T}(A_0^* - A^*) \overset{\text{a.c.}}{\to} (A_0^* - A^*)' P_0 (A_0^* - A^*)$$

$$+ (B_0^* - B^*)' \Omega_0 (B_0^* - B^*)$$

$$(A_0^* - A^*)' \frac{Z'U}{T} \overset{\text{a.c.}}{\to} (B_0^* - B^*)' B_0^{*'-1} \Sigma_0, \qquad (12.46)$$

and we conclude that

$$\bar{L}(\theta_0, \theta_0) = -\frac{m}{2}\ln 2\pi - \frac{1}{2}\ln|\Sigma_0| + \frac{1}{2}|B_0^{*'}B_0^*| - \frac{1}{2}\mathrm{tr}\Sigma_0^{-1}\Sigma_0$$

$$= -\frac{m}{2}(\ln 2\pi + 1) - \frac{1}{2}\ln|\Omega_0|$$

$$\bar{L}(\theta, \theta_0) = -\frac{m}{2}\ln 2\pi - \frac{1}{2}\ln|\Sigma| + \frac{1}{2}\ln|B^{*'}B^*| - \frac{1}{2}\mathrm{tr}\Sigma^{-1}H,$$

$$H = (A_0^* - A^*)' P_0 (A_0^* - A^*) + B^{*'}\Omega_0 B^*. \qquad (12.47)$$

By definition, the ML estimator, $\hat{\theta}_T$, obeys

$$L_T(\hat{\theta}_T) \geq L_T(\theta), \quad \text{for all } \theta \in \Theta, \qquad (12.48)$$

where Θ is the admissible parameter space; moreover, it is assumed that Θ is compact (closed and bounded) and that the true parameter θ_0 is an **interior** point of Θ. A rigorous proof that the ML estimator is strongly consistent (i.e. that it converges with probability one to the true parameter point) under certain conditions is given in Dhrymes (1994), Chap. 5, in a more general context than the one employed here. It is also shown therein that it (the ML estimator) is a minimum contrast estimator with (asymptotic) contrast

$$K(\theta, \theta_0) = \bar{L}(\theta_0, \theta_0) - \bar{L}(\theta, \theta_0). \qquad (12.49)$$

The contrast[14] is a **nonnegative** function, which attains its global minimum at $\theta = \theta_0$; moreover, this minimizer is **unique**. In the theory of minimum contrast estimators, identification means the uniqueness of the minimizer in Eq. (12.49) above. We employ this framework to obtain the necessary and sufficient conditions for identification of the parameters of the GLSEM in the context of ML estimation.

---

[14]The expression in the left member of Eq. (12.49) is also referred to in the literature as the (asymptotic) Kullback information of θ_0 on θ.

Since Eq. (12.42) holds for all $\theta \in \Theta$, we have, in particular,

$$L_T(\hat{\theta}_T) \geq L_T(\theta_0). \tag{12.50}$$

Since $\hat{\theta}_T \in \Theta$, it must have a limit point, say θ^*; or, alternatively, we may argue that since $L_T(\theta)$ is bounded by an integrable function, the convergence is uniform, thus obtaining

$$L_T(\hat{\theta}_T) - L_T(\theta_0) \overset{\text{a.c.}}{\to} \bar{L}(\theta^*, \theta_0) - \bar{L}(\theta_0, \theta_0) \geq 0.$$

But the strict inequality cannot hold because of Jensen's inequality noted previously, so that we conclude

$$K(\theta^*, \theta_0) = 0. \tag{12.51}$$

Clearly, $\theta^* = \theta_0$ satisfies Eq. (12.51). Is this the only vector that does so? Denoting the elements of θ_0 by a zero subscript and those of θ^* without a subscript, we have

$$K(\theta^*, \theta_0) = -\frac{m}{2} - \frac{1}{2}\ln|\Omega_0| + \frac{1}{2}\ln|\Sigma| - \frac{1}{2}\ln|B^{*'}B^*| + \frac{1}{2}\text{tr}\Sigma^{-1}H. \tag{12.52}$$

Because in the context of the GLSEM the identification conditions are customarily stated in terms of the parameters in B^* and C, we eliminate Σ from Eq. (12.52) by partial minimization. Thus, solving

$$\frac{\partial K}{\partial \text{vec}(\Sigma^{-1})} = -\frac{1}{2}\text{vec}(\Sigma)' + \frac{1}{2}\text{vec}(H)' = 0, \tag{12.53}$$

and substituting $\Sigma = H$ in Eq. (12.52), we find the concentrated contrast, or Kullback information,

$$K^*(\Omega_0, B^*, C) = \frac{1}{2}\ln\left(\frac{|\Omega_0 + B^{*'-1}(A_0^* - A^*)'P_0(A_0^* - A^*)B^{*-1}|}{|\Omega_0|}\right). \tag{12.54}$$

Evidently, taking $A^* = A_0^*$ gives us the global minimum. But is it unique, or, otherwise put, is the system identifiable? The answer is evidently no. To see this note that P_0 is a square matrix of dimension $G+m$ but of rank G. Hence, there are m linearly independent vectors, say the columns of the matrix J such that $P_0 J = 0$. Let N be an arbitrary square matrix of dimension m and consider

$$A^* = A_0^* + JN, \quad \text{so that} \quad A^* - A_0^* = JN. \tag{12.55}$$

The matrix $A_0^* + JN$ also attains the global minimum since

$$(A^* - A_0^*)'P_0(A^* - A_0^*) = N'J'P_0JN = 0. \tag{12.56}$$

Thus, **we have no identification if no restrictions are placed on** B^* and C.

By the conventions of the GLSEM literature, we have at our disposal two tools: normalization and prior (zero) restrictions. Imposing normalization, we find $A^* - A_0^* = -(A - A_0)$, where $A = (B', C')'$. Since in Eq. (12.56) B^* must be a **nonsingular matrix**, the only task remaining is to determine what restrictions must be put on A so as to produce uniqueness in the matrix A that satisfies

$$\Psi = (A - A_0)' P_0 (A - A_0) = 0, \tag{12.57}$$

Because the matrix P_0 of Eq. (12.57) is **positive semi-definite**, by Corollary 2.14, the condition in Eq. (12.57) is equivalent to the condition

$$\text{tr}\Psi = \sum_{i=1}^{m} (a_{\cdot i} - a_{\cdot i}^0)' P_0 (a_{\cdot i} - a_{\cdot i}^0) = 0 \tag{12.58}$$

Recalling the selection matrices employed earlier, we have

$$a_{\cdot i} - a_{\cdot i}^0 = L_i(\delta_{\cdot i} - \delta_i^0), \quad i = 1, 2, 3, \ldots, m. \tag{12.59}$$

Therefore, we may rewrite Eq. (12.58) as

$$\text{tr}\Psi = (\delta - \delta^0)' L' (I_m \otimes P_0) L (\delta - \delta^0) = 0, \quad L = \text{diag}(L_1, L_2, \ldots, L_m). \tag{12.60}$$

The matrix $L'(I_m \otimes P_0)L$ is block diagonal and its ith diagonal block is

$$L_i'(\Pi, I_G)' \bar{M}_{xx}(\Pi, I_G) L_i = S_i' M_{xx} S_i, \quad i = 1, 2, 3, \ldots, m. \tag{12.61}$$

For the solution of Eq. (12.60) to be unique in δ, the matrices in Eq. (12.61) must be **nonsingular**. Because M_{xx} is nonsingular, it follows from Propositions 2.61 and 2.62 that the necessary and sufficient condition for identification is, in the context of this discussion,

$$\text{rank}(S_i) = m_i + G_i, \quad i = 1, 2, 3, \ldots, m, \tag{12.62}$$

which is the condition for identification usually obtained by conventional means.

12.4   The Structural VAR Model

The structural VAR Model (SVAR) was developed as an alternative to the GLSEM in areas of research where there was lack of a broad intellectual consensus on the type of prior restrictions that should be employed.

Thus, reflecting a desire to avoid possibly controversial prior restrictions, many researchers turned to autoregressive (AR) models of the type examined in Sect. 6.4 of Chap. 6. The models first proposed were of the form

$$y_{t.} = y_{t-1.}A_1 + y_{t-2.}A_2 + \ldots + y_{t-k.}A_k + u_{t.}, \tag{12.63}$$

where $y_{t.}$ is an m-element **row** vector containing the endogenous variables, A_i are $m \times m$ matrices containing the unknown parameters and $u_{t.}$ is an m-element **row** vector containing the error terms. Initially, the latter were assumed to be $iid(0, \Sigma > 0)$. However, it did not take long to realize that a model where an economic variable is determined **solely** by its past history is devoid of any economics and indeed obscures or totally overlooks the essentially interactive and reactive behavior of economic agents. The natural evolution was the SVAR model, which is of the form

$$y_{t.}A_0 = y_{t-1.}A_1 + y_{t-2.}A_2 + \ldots + y_{t-k.}A_k + \epsilon_{t.}, \tag{12.64}$$

where now the matrix A_0 introduces the reactive and interactive features of the economic universe and the new error terms, the ϵ' s, are asserted to be $iid(0, I_m)$. Moreover, it is asserted that A_0 is **non-singular**. The condition on the covariance of the errors is, in effect, an **identification condition**.

Remark 12.7. The VAR model, closely associated with the work of Sims (1980), was in fact first examined by Mann and Wald (1943) who dealt with the GLSEM as a set of stochastic difference equations, i.e. **without exogenous variables**, and obtained identification and estimation by standard means, i.e. by imposing exclusion restrictions. The later papers by Anderson and Rubin (1949, 1950) complemented Mann and Wald by introducing exogenous variables, and imposing exclusion restrictions; they also obtained estimators by maximum likelihood (ML), as did Mann and Wald.

The novelty of SVAR lies not so much in its formulation, but rather in the method of estimating its parameters, a topic to which we now turn.

The method is basically this: (a) we first obtain its **reduced** form, by multiplying on the right by A_0^{-1}; (b) having done so we obtain the matrix of residuals, say $\hat{U}$ and estimate the reduced form covariance matrix, say $\hat{\Sigma} = \hat{U}'\hat{U}/T$; (c) noting that the reduced form covariance matrix is $(A_0 A_0')^{-1}$, decompose $\hat{\Sigma}^{-1}$, **imposing prior restrictions on the elements of A_0, as needed**, to produce a **unique estimator thereof**.

Following this process, we obtain the reduced form

$$y_{t.} = \sum_{i=1}^{k} y_{t-i.}\Pi_i + u_{t.}, \quad \Pi_i = A_i A_0^{-1}, \quad u_{t.} = \epsilon_{t.}A_0^{-1}, \quad i = 1, 2, 3, \ldots, k.$$
$$\tag{12.65}$$

and we write the entire sample as

$$Y = Y^*\Pi^* + U, \quad Y^* = (\, Y_{-1} \quad Y_{-2} \quad \ldots \quad Y_{-k} \,), \quad \Pi^* = \begin{pmatrix} \Pi_1 \\ \Pi_2 \\ \vdots \\ \Pi_k \end{pmatrix}$$

$$y = (I_m \otimes Y^*)\pi^* + u, \ y = \mathrm{vec}(Y), \ \pi^* = \mathrm{vec}(\Pi^*), \ u = \mathrm{vec}(U). \quad (12.66)$$

We then obtain the OLS estimator of π^* and we estimate the covariance matrix, $\tilde{\Sigma}$ as

$$\tilde{\pi}^* = [I_m \otimes (Y^{*\prime}Y^*)^{-1}Y^{*\prime}]y = \pi^* + [I_m \otimes (Y^{*\prime}Y^*)^{-1}Y^{*\prime})]u,$$

$$\tilde{u} = [I_m \otimes (Y^{*\prime}Y^*)^{-1}Y^{*\prime})]u, \quad \tilde{\Sigma} = \frac{\tilde{U}'\tilde{U}}{T}, \quad \tilde{U} = \mathrm{mat}(\tilde{u}), \qquad (12.67)$$

where mat is the (re)matricizing operator that reverses the vec operator, i.e. if $u = \mathrm{vec}(U)$, then $U = \mathrm{mat}(u)$.

Since

$$\tilde{\Sigma}^{-1} = \tilde{A}_0 \tilde{A}_0',$$

the procedure relies on Proposition 2.62 or Corollary 2.15, of Chap. 2, to argue that the implied estimator of A_0 is that non-singular matrix that decomposes the positive definite matrix $\tilde{\Sigma}^{-1}$. The problem is that **Proposition 2.62 and/or Corollary 2.15 do not assert that such a matrix is unique.** Indeed, if $\tilde{A}_0$ serves that purpose, then so does $\tilde{A}_0 W$, where W **is any arbitrary (conformable) orthogonal matrix.** Thus, we have again **an identification problem that necessitates prior restrictions**, even though one of the chief motivations for the SVAR was to avoid "arbitrary" prior restrictions. But the situation here is even worse, in that the nature of the problem requires us to impose **just identifiability restrictions**; as we saw in a previous section, parameters estimated under conditions of just identifiability do not contain any additional information beyond that which is conveyed by the information utilized for obtaining them, in this case the estimator of the covariance matrix of **the reduced form, i.e. the simple VAR!**. The reason is as follows: the matrix $\tilde{\Sigma}$ contains $m(m+1)/2$ **distinct elements**; $\tilde{A}_0$ contains, potentially, m^2 distinct elements; an orthogonal matrix obeys $WW' = I_m$, which implies that $m(m-1)/2$ restrictions need to be imposed for uniqueness. Another way of putting this is to note that the proposed decomposition has to satisfy the equations

$$\tilde{\sigma}^{ij} = \tilde{a}_{i.}\tilde{a}'_{j.}, \quad i, j = 1, 2, \ldots, m, \tag{12.68}$$

where $a_{i.}$ is the ith **row** of A_0.

Because of symmetry $m(m-1)/2$ of the equations are redundant, so that there are in fact only $m(m+1)/2$ equations to determine m^2 unknowns. Even though the equations are **nonlinear**, it should be clear to the reader that a **unique solution cannot possibly exist, without imposition of restrictions on the elements of** A_0. We need at least $m(m-1)/2$ restrictions so that the unknown elements of A_0 are reduced to $m(m+1)/2$, the **number of distinct equations.** But this means that we render the system in Eq. (12.68) **just-identified.** One typical choice by those applying the SVAR model is to take A_0 to be **lower or upper triangular.**

Example 12.1. An example will, perhaps, illustrate best the problematic nature of the SVAR model. Suppose we have estimated

$$\left(\frac{\tilde{U}'\tilde{U}}{T}\right)^{-1} = \tilde{\Sigma}^{-1} = \begin{bmatrix} 145 & 2 & 72 \\ 2 & 13 & 15 \\ 72 & 15 & 61 \end{bmatrix}.$$

Following the procedure outlined above, we may obtain an estimator of A_0 using the triangular decomposition of positive definite matrices, i.e.

$$\tilde{A}_0 = \begin{bmatrix} \sqrt{145} & 0 & 0 \\ \frac{2\sqrt{145}}{145} & \frac{3\sqrt{30305}}{145} & 0 \\ \frac{72\sqrt{145}}{145} & \frac{677\sqrt{30305}}{30305} & \frac{46\sqrt{209}}{209} \end{bmatrix} = \begin{bmatrix} 12.04 & 0 & 0 \\ .17 & 3.60 & 0 \\ 5.98 & .006 & 3.18 \end{bmatrix}, \tag{12.69}$$

and it can, indeed, be verified that $\tilde{A}_0\tilde{A}'_0 = \tilde{\Sigma}^{-1}$. However it **can also be** verified that

$$\hat{A}_0 = \begin{bmatrix} 1 & 0 & 12 \\ 2 & 3 & 0 \\ 0 & 5 & 6 \end{bmatrix} \tag{12.70}$$

satisfies the condition

$$\tilde{\Sigma}^{-1} = \hat{A}_0\hat{A}'_0!$$

Needless to say, the **economics implied by** $\tilde{A}_0$ of Eq. (12.69) and $\hat{A}_0$ of Eq. (12.70) **are vastly different** but **the empirical basis of the two claims are identical,** viz. the inverse of the estimated covariance matrix of the reduced form errors.

12.5   Forecasting from GLSEM and TS Models

12.5.1   Forecasting with the GLSEM

Preliminaries

Given the discussion in previous sections it is relatively simple to use the estimated GLSEM for one step ahead forecasts. However, it should be noted that **the structural form cannot be used for forecasting even one step ahead.** This is so since in order to forecast one of the dependent variables, even one period beyond the sample, would require us to specify not only the **exogenous variables in that future period,** but also **all other jointly dependent (endogenous) variables as well!** Thus, **for forecasting purposes we can only use the reduced form.**

Definition 12.3. The structural and reduced forms of the GLSEM are given by Eqs. (12.1) and (12.3), respectively. The reduced form, whose representation is reproduced below,

$$Y = X\Pi + V, \quad \Pi = CD, \quad V = UD, \quad D = B^{*-1}$$

can be estimated in any one of at least three ways, in the context of this volume; one can use:

i. The unrestricted reduced form, reproduced just above, which ignores all restrictions, including normalization associated with the structural form, and estimate

$$\tilde{\Pi}_{un} \;=\; (X'X)^{-1}X'Y, \quad \text{or}$$

$$(\tilde{\pi} - \pi)_{un} \;=\; (D' \otimes I_G)(I_m \otimes (X'X)^{-1})(I_m \otimes X')u,$$

using the notation $\pi = \text{vec}(\Pi)$.

ii. The restricted reduced form induced by the 2SLS estimators of the structural form

$$\hat{\Pi}_{RRF(2SLS)} = \tilde{C}_{2SLS}\tilde{D}_{2SLS}, \quad \tilde{D}_{2SLS} = (I - \tilde{B}_{2SLS})^{-1}; \quad \text{and}$$

iii. The restricted reduced form induced by the 3SLS estimators of the structural form

$$\hat{\Pi}_{RRF(3SLS)} = \tilde{C}_{3SLS}\tilde{D}_{3SLS}, \quad \tilde{D}_{3SLS} = (I - \tilde{B}_{3SLS})^{-1}.$$

Before we proceed with the discussion of forecasting issues, however, we need to derive all relevant limiting distributions. The **unrestricted reduced form**, as noted above, is estimated by least squares. The restricted reduced form is estimated as $\tilde{C}\tilde{D}$, where $\tilde{C}$, $\tilde{D}$ contain the structural parameters estimated by 2SLS or 3SLS methods.

Thus, the restricted reduced form estimator is given by

$$\hat{\Pi} - \Pi = \tilde{C}\tilde{D} - CD = \tilde{C}\tilde{D} - \tilde{C}D + \tilde{C}D - CD$$

$$= \hat{\Pi}(\tilde{B} - B)D + (\tilde{C} - C)D = (\hat{\Pi}, I_G)\left(\begin{array}{c}\tilde{B} - B \\ \tilde{C} - C\end{array}\right)D, \quad D = (I - B)^{-1},$$

and we note that,[15] using the results in Proposition 4.4 of Chap. 4,

$$\mathrm{vec}\left(\begin{array}{c}\tilde{B} - B \\ \tilde{C} - C\end{array}\right) = L(\tilde{\delta} - \delta); \quad \text{thus}$$

$$\sqrt{T}(\tilde{\pi} - \pi)_{un} \sim (I_m \otimes M_{xx}^{-1})\frac{(I_m \otimes X')u}{\sqrt{T}} \xrightarrow{\mathrm{d}} N(0, \Phi_1),$$

$$\Phi_1 = (D' \otimes I_G)(\Sigma \otimes M_{xx}^{-1})(D \otimes I_G), \tag{12.71}$$

$$\sqrt{T}(\hat{\pi} - \pi)_{RRF(2SLS)} = (D' \otimes I_G)\hat{S}\sqrt{T}(\tilde{\delta} - \delta)_{2SLS}$$

$$\sqrt{T}(\hat{\pi} - \pi)_{RRF(3SLS)} = (D' \otimes I_G)\hat{S}\sqrt{T}(\tilde{\delta} - \delta)_{3SLS}$$

$$\sqrt{T}(\tilde{\delta} - \delta)_{2SLS} \xrightarrow{\mathrm{d}} N(0, C_2), \quad \sqrt{T}(\tilde{\delta} - \delta)_{3SLS} \xrightarrow{\mathrm{d}} N(0, C_3)$$

$$C_2 = (S'(I_m \otimes M_{xx})S)^{-1}S'(\Sigma \otimes M_{xx})S(S'(I_m \otimes M_{xx})S)^{-1}$$

$$\sqrt{T}(\hat{\pi} - \pi)_{RRF(2SLS)} \xrightarrow{\mathrm{d}} N(0, \Phi_2), \quad \Phi_2 = (D' \otimes I_G)SC_2S'(D \otimes I_G) \tag{12.72}$$

$$\sqrt{T}(\hat{\pi} - \pi)_{RRF(3SLS)} \xrightarrow{\mathrm{d}} N(0, \Phi_3), \quad \Phi_3 = (D' \otimes I_G)SC_3S'(D \otimes I_G), \tag{12.73}$$

$$C_3 = (S'(\Sigma^{-1} \otimes M_{xx})S)^{-1}.$$

This is an opportune time to examine the issue of the difference, if any, between the **restricted and unrestricted reduced form estimators,** as well as the differences between the 2SLS and 3SLS estimators, especially when the system is **just identified.** A relatively easy calculation will prove

---

[15]We also include below the result for the unrestricted reduced form for ease of reference.

Proposition 12.6. The following statements are true when all equations of the GLSEM are **just identified.**

i. The 2SLS and 3SLS are identical with probability one, in the sense that their limiting distributions are **identical;**

ii. RRF(2SLS) and RRF(3SLS) are identical with probability one, in the sense that they have the same limiting distribution;

iii. Unrestricted and all (both) restricted reduced form estimators are identical with probability one, in the sense that they all have the same limiting distribution.

Proof: Under the premise of the proposition S is a nonsingular matrix. Thus,

$$C_2 = C_3 = S^{-1}(\Sigma \otimes M_{xx}^{-1})S'^{-1}, \tag{12.74}$$

which proves i; it follows immediately that

$$\Phi_2 = \Phi_3 = D'\Sigma D \otimes M_{xx}^{-1}, \tag{12.75}$$

which proves ii; finally a comparison with Eq. (12.71) above proves iii.

q.e.d.

In the general case, i.e. when the equations of the GLSEM are **not just-identified** we have

Proposition 12.7. The following statements are true, when not all of the equations of the GLSEM are just-identified.

i. The 3SLS induced restricted reduced form, RRF(3SLS), estimator is efficient relative to the 2SLS induced restricted reduced form, RRF(2SLS), estimator;

ii. RRF(3SLS) is efficient relative to the unrestricted reduced form estimator;

iii. The RRF(2SLS) and unrestricted reduced form estimators cannot be ranked in the sense that the difference of the covariance matrices in their respective limiting distributions **is indefinite.**

Proof: To prove i we note that

$$\Phi_2 - \Phi_3 = (D' \otimes I_G)S(C_2 - C_3)S'(D \otimes I_G);$$

since the matrix $(D' \otimes I_G)S$ is of full column rank, and we had already shown in earlier sections that $C_2 - C_3 \geq 0$, it follows that $\Phi_2 - \Phi_3 \geq 0$. The proofs of ii and iii are too complex to produce here, but they are easily found in Dhrymes (1973) or Dhrymes (1994, Theorem 5, pp. 96ff).

Forecasting with the GLSEM

What are the properties of one-step-ahead forecasts obtained by the three alternatives? Since we generally do not make distributional assumptions beyond the existence of second moments and the like, we need to rely on asymptotic theory, which we developed in the previous section specifically for this purpose.

The one step ahead forecast presents no new problems. Thus,

$$\hat{y}_{T+1\cdot} = x_{T+1\cdot}\hat{\Pi}, \quad \text{or} \quad \hat{y}'_{T+1\cdot} = (I_m \otimes x_{T+1\cdot})\hat{\pi}$$

$$e'_{T+1\cdot} = y'_{T+1\cdot} - \hat{y}'_{T+1\cdot} = (I_m \otimes x_{T+1\cdot})(\hat{\pi} - \pi) + v'_{T+1\cdot}, \quad (12.76)$$

where the second equation above denotes the **forecast error**; the latter, even if conditioned on the actually observed predetermined variables and the specified or hypothesized exogenous variables,[16] contains two components or sources of uncertainty (randomness), one related to the reduced form error, ($v'_{T+1\cdot}$), and the other to the fact **do not know the structural form, we only have an estimate**. But, **given the assumptions embedded in the specification of the GLSEM, and conditionally on the predetermined and specified (future) exogenous variables**, these two sources are **independent**. Thus, if the initial sample is sufficiently large, we can approximate

$$(\hat{\pi} - \pi)_{RRF(2SLS)} \approx N(0, \Phi_2/T), \quad (\hat{\pi} - \pi)_{RRF(3SLS)} \approx N(0, \Phi_3/T), \quad (12.77)$$

respectively, for 2SLS and 3SLS based restricted reduced forms. Moreover from Eqs. (12.72) and (12.73) we find

$$\text{Cov}(e'_{T+1\cdot})_{RRF(2SLS)} \approx N(0, \Psi_2), \quad (12.78)$$

$$\Psi_2 = D'\Sigma D + (I_m \otimes x_{T+1\cdot})(D' \otimes I_G)S(C_2/T)$$
$$\times S'(D \otimes I_G)(I_m \otimes x'_{T+1\cdot})$$

$$\text{Cov}(e'_{T+1\cdot})_{RRF(3SLS)} \approx N(0, \Psi_3), \quad (12.79)$$

$$\Psi_3 = D'\Sigma D + (I_m \otimes x_{T+1\cdot})(D' \otimes I_G)S(C_3/T)$$
$$S'(D \otimes I_G)(I_m \otimes x'_{T+1\cdot})$$

---

[16]In the forecasting literature, at least the one found in public discussions, the specification of the exogenous variables is referred to as "scenarios".

The analogous expressions for the one step ahead **unrestricted reduced form** forecast is given by

$$\hat{y}_{T+1\cdot} \ = \ x_{T+1\cdot}\hat{\Pi}, \quad \text{or} \quad \hat{y}'_{T+1\cdot} = (I_m \otimes x_{T+1\cdot})\tilde{\pi}$$

$$(e'_{T+1\cdot})_{un} \ = \ y'_{T+1\cdot} - \hat{y}'_{T+1\cdot} = -(I_m \otimes x_{T+1\cdot})(\tilde{\pi} - \pi)_{un} + v'_{T+1\cdot},$$

$$\text{Cov}(e'_{T+1\cdot})_{un} \ \approx \ N(0, \Psi_1), \Psi_1 = D'\Sigma D$$

$$+(I_m \otimes x_{T+1\cdot})(D'\Sigma D \otimes M_{xx}^{-1}/T)(I_m \otimes x'_{T+1\cdot}), \text{or}$$

$$\Psi_1 \ = \ D'\Sigma D \left(1 + \frac{x_{T+1\cdot}M_{xx}^{-1}x'_{T+1\cdot}}{T}\right). \tag{12.80}$$

Remark 12.8. Since, as we have just seen, a one step ahead forecast is simply a **linear** function of the reduced form estimates, it is immediate that efficiency in estimation translates into efficiency of forecasts. Thus, forecasts through RRF(3SLS) are efficient relative to other forecasts in the sense of having a **smaller covariance matrix**. What, if anything, is different when we attempt a forecast involving more than one period ahead. We will elucidate the issues by examining the case of a three-step-ahead forecast, which is sufficiently complex to bring forth all the relevant issues and at the same time remains manageable.

When we did one step ahead forecast we used the standard notation

$$x_{T+1\cdot} = (y_{T\cdot}, y_{T-1\cdot}, \ldots, y_{T+1-k\cdot}, p_{T+1\cdot}),$$

and **all entries therein were actually observed, or specified as in the case of** $p_{T+1\cdot}$.

When we attempt, say, a two step ahead forecast, **we cannot specify** $x_{T+2\cdot}$ **in the usual way because** $y_{T+1\cdot}$ **is not observed**; thus we cannot proceed routinely. What is done in practice, is to **substitute for it the one step ahead forecast obtained earlier.** The same will be true if we had to do a three period ahead forecast, in which case the requisite predetermined variables $y_{T+1\cdot}$, $y_{T+2\cdot}$ are not observed; in their stead we use the one and two period ahead forecasts, $\hat{y}_{T+1\cdot}$, $\hat{y}_{T+2\cdot}$. This provides an **operational mechanism for actually carrying out the requisite forecasts** but complicates the analysis of the sources of randomness and makes more difficult than it need be, the derivation of the covariance matrix of the three period ahead forecast.

For the purpose of **analyzing these issues only** we shall follow a slightly different strategy. Let us take up first the simpler case of a two period ahead forecast. Put[17]

$$\hat{y}_{T+2\cdot} = x_{T+2\cdot}\hat{\Pi} - e_{T+1\cdot}\hat{\Pi}_1, \tag{12.81}$$

---

[17]In the following discussion Π_1 and Π_2 refer to these entities as defined in Eq. (12.3).

and note what is being done; first, x_{T+2}. contains the unobserved predetermined variable y_{T+1}. , which may be written as $y_{T+1}. = \hat{y}_{T+1}. + e_{T+1}.$, where the second term is as defined in Eq. (12.76). If we examine the nature of the matrix Π in Eq. (12.3), we see that we have corrected for the inaccuracies in $x_{T+2}.\hat{\Pi}$ by subtracting $e_{T+1}.\hat{\Pi}_1$. Thus, we may write

$$e_{T+2}. \quad = \quad y_{T+2}. - \hat{y}_{T+2}. = -x_{T+2}.(\hat{\Pi} - \Pi) + e_{T+1}.\hat{\Pi}_1 + v_{T+2}.$$

$$= \quad -x_{T+2}.(\hat{\Pi} - \Pi) - x_{T+1}.(\hat{\Pi} - \Pi)\hat{\Pi}_1 + v_{T+2}. + v_{T+1}.\hat{\Pi}_1, \quad \text{or}$$

$$e_{T+2}. \quad \approx \quad -x_{T+2}.(\hat{\Pi} - \Pi) - x_{T+1}.(\hat{\Pi} - \Pi)\Pi_1 + v_{T+2}. + v_{T+1}.\Pi_1, \quad (12.82)$$

and observe that the forecast error is now more complex than the one for the one period ahead forecast. Similarly, for the three period ahead forecast error we have

$$e_{T+3}. \quad = \quad y_{T+3}. - \hat{y}_{T+3}. = -x_{T+3}.(\hat{\Pi} - \Pi) + e_{T+1}.\hat{\Pi}_2 + e_{T+2}.\hat{\Pi}_1 + v_{T+3}., \quad \text{or}$$

$$e_{T+3}. \quad \approx \quad -x_{T+3}.(\hat{\Pi} - \Pi) - x_{T+1}.(\hat{\Pi} - \Pi)\Pi_2 - x_{T+2}.(\hat{\Pi} - \Pi)\Pi_1 + v_{T+3}.$$

$$+v_{T+2}.\Pi_1 + v_{T+1}.\Pi_2. \quad (12.83)$$

To see what the covariance matrix of the three periods ahead forecast is, approximately, consider

$$e'_{T+3}. \approx -(I_m \otimes x_{T+3}.)(\hat{\pi} - \pi) - (\Pi'_2 \otimes x_{T+1}.)(\hat{\pi} - \pi) \quad (12.84)$$

$$-(\Pi'_1 \otimes x_{T+2}.)(\hat{\pi} - \pi) + v'_{T+3}. + \Pi'_1 v'_{T+2}. + \Pi'_2 v'_{T+1}..$$

Thus, the approximate covariance matrix for the three periods ahead forecast, based on RRF(3SLS) is

$$\text{Cov}(e'_{T+3}.) \approx (I_m \otimes x_{T+3}.)\Psi_3(I_m \otimes x'_{T+3}.) + (\Pi'_1 \otimes x_{T+2}.)\Psi_3(\Pi_1 \otimes x'_{T+2}.)$$

$$+(\Pi'_2 \otimes x_{T+1}.)\Psi_3(\Pi_2 \otimes x'_{T+1}.) + \text{crossproducts} + V + \Pi'_1 V \Pi_1 + \Pi'_2 V \Pi_2,$$

$$(12.85)$$

and if we proceed to a four periods ahead forecast the first three terms of Eq. (12.85) will become four terms, as will the last three terms, with a corresponding increase in the number of cross products. Note that because we estimate Π on the basis of the first T observations the last three terms of Eq. (12.84), corresponding to the reduced form errors for periods $T + 1, T + 2, T + 3$ **are not involved in any cross products.**

12.5.2 Forecasting with Time Series Models (TSM)

Preliminaries

Forecasting with time series (TS) is not essentially different from what one does in econometrics with respect to the GLM, or the GLSEM, but it is not identical, reflecting the differences in the data and problems encountered in the statistics, engineering or physical sciences as compared to those typically found in economics. Time series for dynamic systems are long in the disciplines mentioned above, but not in the case of economics. The universe in which the data originate in the former tend to be governed by more stable relationships, but much less so in economics. This makes economic forecasting extremely difficult and it accounts for the greater emphasis in the specification of economic relationships than we find in time series modeling. In the latter we rely solely on the past history of the phenomenon studied to provide us with insight into its future development. A similar reliance in economics is rarely productive of insights, except possibly with so called high frequency data in certain financial markets.

In Chap. 6, we dealt with the basic concepts of time series and, in Example 9.5 (Chap. 9), we have also estimated and gave the limiting distribution of parameters in the case of multivariate $AR(m)$ sequences. It is highly recommended that the reader review these sections before proceeding.

In previous discussions we did not produce estimators and their limiting distributions in the case of $MA(n)$ or $ARMA(m, n)$, whether scalar or multivariate. Since this is not a volume on TS per se it would take us too far afield to discuss these issues in detail; and even though we shall only consider scalar $MA(n)$ and $ARMA(m, n)$ sequences, we shall provide sufficient information for an interested reader to pursue their generalization to multivariate sequences, if desired.

The plan is roughly as follows: (a) The main tool of analysis in TS is the autocovariance function; thus, we shall define and obtain the autocovariance generating function (AGF) of stochastic sequences, analogous to the moment generating function (MGF) the reader is no doubt familiar with in connection with random variables. This enables us to express the AGF in terms of the underlying parameters of the sequences in question. (b) The main tool of prediction, or forecasting, in time series relies on the concept of the **best linear predictor** which we shall derive, and finally (c) we shall provide the limiting distribution of an arbitrary size vector of autocovariance estimators.

The Autocovariance Generating Function (AGF)

Definition 12.4. Let $\{c(k): k \in \mathcal{T}\}$ [18] be an autocovariance function, or sequence, of a (stationary) time series. The autocovariance generating function, AGF, is generally defined by the series

$$AGF = \sum_{k=-\infty}^{\infty} c(k)z^k,$$

where z is a real or complex indeterminate. If we specify the stochastic sequence, say $\{X_t: t \in \mathcal{T}\}$, **the autocovariance function above is that of that specific time series** and for this reason we use the notation

$$AGF(X) = \sum_{k=-\infty}^{\infty} c(k)z^k; \quad \text{if the TS is \textbf{stationary} we have the}$$

representation

$$AGF(X) = c(0) + \sum_{k=-\infty}^{-1} c(k)z^k + \sum_{k=1}^{\infty} c(k)z^k, \quad \text{or}$$

$$AGF(X) = c(0) + \sum_{k=1}^{\infty} c(k)[z^k + z^{-k}]. \tag{12.86}$$

Proposition 12.8. The AGF of a (stationary) zero mean[19] sequence that is capable of having the representation

$$X_t = \sum_{j=-\infty}^{\infty} \psi_j u_{t-j}, \quad u \sim WN(0, \sigma^2), \quad \text{such that} \tag{12.87}$$

$$\sum_{j=-\infty}^{\infty} |\psi_j| z^j < \infty, \quad \text{for} \quad \kappa^{-1} < |z| < \kappa, \quad \kappa > 1.$$

is given by

$$AGF(X) = \sigma^2 \psi(z)\psi(z^{-1}), \quad \psi(z) = \sum_{j=-\infty}^{\infty} \psi_j z^j, \tag{12.88}$$

and analogously with $\psi(z^{-1})$.

---

[18]We use the symbol $\mathcal{T}$ to denote the linear index set to avoid confusion with T, which was used repeatedly in this chapter to denote the length of the sample.

[19]If the sequence does not have zero mean one uses, instead of X_t, $X_t - \mu$, where μ is the constant mean of the sequence, or $x_t - \bar{x}$ if we are using a realization to estimate autocovariances.

Proof: First, we note that under the conditions above, the X sequence is stationary and thus $c(k) = c(-k)$; second $c(k) = EX_{t+k}X_t$.[20] It follows then, form Eq. (12.87) and the surrounding discussion, that

$$AGF(X) = c(0) + \sigma^2 \sum_{k=-\infty}^{-1} \sum_{j=-\infty}^{\infty} \psi_{j-k}\psi_j z^k + \sigma^2 \sum_{k=1}^{\infty} \sum_{j=-\infty}^{\infty} \psi_{j+k}\psi_j z^k, \quad \text{or}$$

$$AGF(X) = c(0) + \sigma^2 \sum_{k=1}^{\infty} \sum_{j=-\infty}^{\infty} \psi_j \psi_{j+k}[z^k + z^{-k}]. \tag{12.89}$$

Next, note that

$$\psi(z)\psi(z^{-1}) = \sum_{j=-\infty}^{\infty} \sum_{s=-\infty}^{\infty} \psi_j \psi_s z^{j-s}, \tag{12.90}$$

with the notational convention that the positive power (j) corresponds to the index of the first component of $\psi_j\psi_s$, and the negative power (-s) corresponds to the index of the second component. It follows then that the power (of z) corresponding to $\psi_s\psi_j$ is $-(j-s)$. More generally, let $r = j - s$; then the terms $\psi_j\psi_s$ are attached to z^r and $\psi_s\psi_j$ are attached to z^{-r}. The two products are identical, and putting $j = r + s$, we have the pairings $\psi_s\psi_{s+r}[z^r + z^{-r}]$, for $r \geq 1$, and for $r = 0$ we have the term $\sum_{s=-\infty}^{\infty} \psi_s^2$; thus, we can write

$$\psi(z)\psi(z^{-1}) = \sum_{s=-\infty}^{\infty} \psi_s^2 + \sum_{r=1}^{\infty} \sum_{s=-\infty}^{\infty} \psi_s\psi_{s+r}[z^r + z^{-r}].$$

Since in Eq. (12.89), $c(0) = \sigma^2 \sum_{j=-\infty}^{\infty} \psi_j^2$, $AFG(X) = \sigma^2\psi(z)\psi(z^{-1})$, a relation valid for $\kappa^{-1} < |z| < \kappa$ $\kappa > 1$, the two expressions are identical.

q.e.d.

Example 12.2. Consider the $MA(n)$ sequence X_t; its autocovariance generating function as adapted from Eqs. (12.89) or (12.90) is

$$AGF(MA(n))(z) = \sigma^2 \left(\sum_{j=0}^{n} a_j^2 + \sum_{r=1}^{n} \sum_{j=0}^{n-r} a_j a_{j+r}[z^r + z^{-r}] \right) \quad \kappa^{-1} < |z| < \kappa,$$

---

[20]There is a convention in notation the reader may not be aware of, viz. the lag k is, by this convention, the difference between the index of the first factor, X_{t+k} and the second factor X_t. Thus the autocovariance at lag $-k$ is $EX_t X_{t+k}$ or, of course, $EX_{t-k}X_t$. Since the two terms under expectation commute it is evident that $c(k) = c(-k)$. This is of course obvious also by the very definition of stationary sequences. To avoid such minor confusions one generally defines $c(k) = EX_{t+|k|}X_t$ **whether k is positive or negative.**

with the understanding that $a_s = 0$ for $s \geq n+1$. Needless to say when n is **finite** the restriction on $|z|$ is irrelevant. It follows then immediately that, for autocovariances at positive lag k, we have

$$c(0) = \sigma^2 \sum_{j=0}^{n} a_j^2, \quad c(1) = \sigma^2 \sum_{j=0}^{n-1} a_j a_{j+1}, \quad c(2) = \sigma^2 \sum_{j=0}^{n-2} a_j a_{j+2} \quad (12.91)$$

$$c(3) = \sigma^2 \sum_{j=0}^{n-3} a_j a_{j+3}, \quad \text{and finally} \quad c(n) = \sigma^2 \sum_{j=0}^{n-n} a_j a_{j+n} = \sigma^2 a_0 a_n$$

$$c(n+s) = 0, \ s \geq 1, \quad a_0 = 1,$$

and since we have 11 equations in 11 unknowns we can solve them, even though the equations are nonlinear. This is facilitated by noting that

$$\left(\sum_{j=0}^{n} a_j \right)^2 = \sum_{j=0}^{n} a_j^2 + 2 \sum_{r=1}^{n} \sum_{j=0}^{n-r} a_j a_{j+r} = \frac{1}{\sigma^2} \left(c(0) + 2 \sum_{k=1}^{n} c(k) \right).$$

The reason we present examples like the one above is to facilitate the derivation of autocovariances at certain lags, **from the parameters (coefficients) that define the stochastic sequence in question. Thus, when such parameters can be estimated directly, as we did with the multivariate** $AR(m)$ **of Example 9.5 (Chap. 9), there is no need to derive the AGF for the scalar** $AR(m)$. But to be complete, we simply display the AGF of a scalar $AR(m)$; thus,

$$AGF(AR(m))(z) = \sigma^2 \frac{1}{b(z)b(z^{-1})}, \quad \kappa^{-1} < |z| < \kappa, \quad (12.92)$$

where $b(z) = \sum_{j=0}^{m} b_j z^j$, $b(z^{-1}) = \sum_{j=0}^{m} b_j z^{-j}$, $b_0 = 1$, and the b_j are the coefficients of the lag polynomial $b(L) = \sum_{j=0}^{m} b_j L^j$.

Finally, in the case of an $ARMA(m, n)$ sequence, we have not given a way of estimating its parameters (a) because it involves a much more complex procedure, whose derivation lies outside the scope of this volume, and (b) because it has a rather limited direct application in econometrics, where ARMA-like specifications generally correspond to **exogenous variables not to error sequences.** There are several algorithms for obtaining such estimates and the interested reader may find them in many undergraduate texts on TS or can find a more advanced and complete treatment in Brockwell and Davis (1991), Chap. 8. The $ARMA(m, n)$ AGF is given by

$$AGF(ARMA(m, n))(z) = \sigma^2 \frac{a(z)a(z^{-1})}{b(z)b(z^{-1})}. \quad (12.93)$$

Best Linear Predictor (BLP)

Definition 12.5. Let $\{X_t:\ t \in \mathcal{T}\}$ be a zero mean stationary time series indexed on the linear index set $\{\mathcal{T}:\ 0, \pm 1, \pm 2, \pm 3, \ldots\}$. The best linear predictor of $X_{t+h},\ h \geq 1$, given $\{X_1, X_2, \ldots, X_n\}$, is the function

$$\hat{X}_{t+h} = \sum_{j=1}^{n} \alpha_j X_{n+1-j}, \tag{12.94}$$

which **minimizes**

$$S = E(X_{n+h} - \sum_{j=1}^{n} \alpha_j X_{n+1-j})^2.$$

Some authors use the notation $\hat{X}_{t+h} = \mathcal{P}_n(X_{t+h})$, where $\mathcal{P}$ is the prediction operator; the subscript indicates the number of prior realizations used and the argument is the entity to be predicted.

The first order conditions are

$$\frac{\partial S}{\partial \alpha} = -2E(X_{n+h} - a' X_{(n)}) X'_{(n)} = 0, \quad X_{(n)} = \begin{pmatrix} X_n \\ X_{n-1} \\ \vdots \\ X_1 \end{pmatrix}. \tag{12.95}$$

Writing Eq. (12.95) more extensively, we obtain

$$EX_{n+h} X'_{(n)} = \alpha' E X_{(n)} X'_{(n)}, \quad \text{or} \quad \alpha = C_n(i-j) c_{nh}, \tag{12.96}$$

$$C_n(i-j) = (EX_{n+1-j} X_{n+1-i}), \quad i,j = 1, 2, \ldots, n, \tag{12.97}$$

$$c_{nh} = (\,EX_{n+h} X_n \quad EX_{n+h} X_{n-1} \quad \cdots \quad EX_{n+h} X_1\,)', \quad \text{or}$$

$$c_{nh} = (\,c(h) \quad c(h+1) \quad \cdots \quad c(h+n-1)\,)', \tag{12.98}$$

where $C_n(i-j)$, c_{nh} are, respectively, the matrix of autocovariances of $X_{(n)}$ and the vector of autocovariances between X_{t+h} and the elements of the vector $X_{(n)}$.

Remark 12.9. Notice the similarities and differences in the equations above, especially Eq. (12.96), relative to what we have in econometric applications of the GLM, as specified for example in Eqs. (10.1) through (10.4) of Chap. 10. In the latter, the equivalent of $C(i-j)$ would have been a set of observations on the variables in the vector $X_{(n)}$, say the matrix W, and $\text{plim}_{N \to \infty}(W'W/N)$ would converge to the equivalent of $C(i-j)$. In the notation of Chap. 10, it is the matrix $X'X/T$. In time series **we have only one observation (realization) on any element of the time series,** X_t. **Thus what we do in the GLM is not available to us in TS analysis.** Moreover, in the econometric applications of the GLM we are able to conduct analyses of the results, conditional on the equivalent of $X_{(n)}$. This is not done in the context of time series (TS) analysis. Notice, also that in the GLM context of Eqs. (10.1) through (10.4) $X'y/T$ is the equivalent of c_{nh} in the TS context. Finally, in the GLM the probabilistic properties of the estimator of the parameter vecor, i.e. the coefficient vector of the explanatory variables, which are the equivalent of $X_{(n)}$, depends chiefly on the error term (of the GLM), something that is not directly available in TS analysis. Instead, we rely on the expected values, $EX_{n+h}X_{(n)}$, $EX_{(n)}X'_{(n)}$, the essential parameters of the TS, viz. their variances and autocovariances which are evidently non-random. Thus, the "regression" coefficients, the elements of the vector α, are estimated **indirectly** through the **estimated autocovariances**.

These features make TS analysis somewhat easier to comprehend but make the determination of the probabilistic features of the forecasts more difficult to pin down. For this reason we shall not give as complete a coverage of the probabilistic properties of TS forecasts as we did in the case of forecasts of the far more complex GLSEM.

Estimation and Limiting Distribution of Autocovariances

To treat forecasting with time series symmetrically with our treatment of forecasting from the GLSEM, we need to provide estimators of the autocovariances as well as establish their limiting distributions. Since this is not a volume devoted to time series we shall limit our scope. First, we shall only deal the moving average (MA), autoregressive (AR) and autoregressive moving average (ARMA) sequences. Second, to facilitate and simplify the derivation of the relevant limiting distributions, we shall assert that the underlying u-sequence defining them, is $iid(0, \sigma^2)$, rather than the standard $WN(0, \sigma^2)$. Where appropriate, all sequences are taken to be **causal and invertible.** Second, **we shall only deal with scalar stochastic sequences, and we shall omit treatment of multivariate (vector) TS.**

Estimation of Autocovariances

In this section we deal, as we noted above, with **zero mean** stationary time series which are, where appropriate, **causal and invertible.** Estimation of autocovariances is the same whether we are dealing with $MA(n)$, $AR(m)$ or $ARMA(m, n)$ TS. The nature of the TS will play a role **only** if we wish to connect autocovariances with the basic parameters of the TS, i.e. $c(k)$ would consist of different parametric configurations depending on whether we are dealing with $MA(n)$, $AR(m)$ or $ARMA(m, n)$ TS. Generally, this is not an issue in forecasting and, at any rate, this representation is rather complex if the orders m, n are at all large, like more than 3 or 4. The reader can determine these relationships by examining the representation of such TS in their $MA(\infty)$ form.

Let $\{x_t:\ t = 1, 2, \ldots, N\}$ be a realization; for any integer k the kth order autocovariance, or the autocovariance at lag k, is estimated by[21]

$$\hat{c}(k) = \frac{\sum_1^{N-|k|} x_{t+|k|}x_t}{N}, \quad -N < k < N. \tag{12.99}$$

Since we do not specify distributional properties of the TS under consideration we do not have any results for their "small sample" distribution, beyond their expected values. Even asymptotically, to establish their limiting distribution, we need additional assumptions on the existence of moments (of the basic $iid(0, \sigma^2)$), which are stated in the proposition below.

Proposition 12.9. Let $\{X_t:\ t \in \mathcal{T}\}$ be a stochastic sequence that has the representation

$$X_t = \sum_{j=-\infty}^{\infty} \psi_j u_{t-j}, \quad \sum_{j=-\infty}^{\infty} |\psi_j| < \infty, \quad u_t \sim iid(0, \sigma^2) \tag{12.100}$$

and, moreover, suppose that[22]

$$Eu_t^4 = \eta\sigma^4, \quad \text{for some constant} \quad \eta. \tag{12.101}$$

Then for any integer $s \in \mathcal{T}$

$$\hat{c}_{(s)} = (\hat{c}(1), \hat{c}(2), \ldots, \hat{c}(s))', \quad \text{obeys}$$

$$\sqrt{N}(\hat{c}_{(s)} - c_{(s)}) \xrightarrow{d} N(0, V), \quad \text{where} \tag{12.102}$$

$$V = \left((\eta - 3)c(i)c(j) + \sum_{k=-\infty}^{\infty} c(k)c(k - i + j) \right.$$

$$\left. + c(k + j)c(k - i) \right)$$

$,j = 1, 2, \ldots, s$, and N is the length of the realization, where the $\hat{c}(k)$ are defined in Eq. (12.99).

---

[21]Even though dividing by N, instead of $N - k$, renders this estimator biased, it is the preferred practice in this literature because it preserves the positive semi-definiteness of the autocovariance matrix or function.

[22]Note that when the u sequence is **also** $N(0, \sigma^2)$, $Eu_t^4 = 3\sigma^4$!

Proof: See Brockwell and Davis (1991), Proposition 7.3.4, pp. 229ff.

Prediction Based on BLP

To make use of the results obtained in the earlier discussion of the BLP (best linear predictor) we need to adapt the results of Proposition 12.9 to handle the limiting distribution of the estimator of the BLP vector $\alpha = C_n^{-1}c_{nh}$, obtained from Eqs. (12.97) and (12.98); thus, we need to determine the limiting distribution of

$$\hat{\alpha} - \alpha = \hat{C}_n^{-1}\hat{c}_{nh} - \hat{C}_n^{-1}c_{nh} + \hat{C}_n^{-1}c_{nh} - C_n^{-1}c_{nh} \quad \text{or}$$

$$\hat{\alpha} - \alpha = \hat{C}_n^{-1}(\hat{c}_{nh} - c_{nh} - (\hat{C}_n - C_n)\alpha)$$

$$= \hat{C}_n^{-1}\left(\hat{c}_{nh} - c_{nh} - (\alpha' \otimes I_n)\text{vec}(\hat{C} - C)\right). \quad (12.103)$$

We note that $\hat{C} - C$ is a **symmetric Toeplitz matrix** with n distinct elements contained in the vector

$$\hat{\gamma} - \gamma = \begin{pmatrix} \hat{c}(0) - c(0) \\ \hat{c}(1) - c(1) \\ \vdots \\ \hat{c}(n-1) - c(n-1) \end{pmatrix}. \quad (12.104)$$

The vector c_{nh}, assuming $h < n$, contains the $n - h$ elements $(c(h)$, $c(h+1), \ldots, c(h+(n-1-h)))$, **which are also contained in** γ, and the h elements $c(n), c(n+1), \ldots, c(n+h-1)$ **which are not contained in** γ. To separate them create the $(n - h) \times n$ selection matrix F_1, such that $F_1\gamma = (c(h), c(h+1), \ldots, c(h+(n-1-h)))'$ and construct the matrix

$$F = \begin{bmatrix} F_1 & 0 \\ 0 & I_h \end{bmatrix}, \quad (12.105)$$

which is $n \times n + h$, and note that

$$c_{nh} = F\begin{pmatrix} \gamma \\ \delta \end{pmatrix}, \quad \delta = (c(n), c(n+1), \ldots, c(n+h-1))'. \quad (12.106)$$

Next, we note that $\text{vec}(C)$ is $n^2 \times 1$; because C is a symmetric Toeplitz matrix, by Eq. (4.25) of Chap. 4, as defined therein, there exists an $n^2 \times n$ matrix B_{ST}, such that

$$\text{vec}(C) = B_{ST}\gamma; \quad \text{thus} \quad \text{vec}(\hat{C} - C) = B_{ST}(\hat{\gamma} - \gamma). \quad (12.107)$$

Combining the two results we may thus rewrite Eq. (12.103)

$$\sqrt{N}(\hat{\alpha} - \alpha) \quad = \quad \hat{C}_n^{-1}\left[F - (\alpha' \otimes I_n)(B_{ST}, 0)\right]\sqrt{N}\begin{pmatrix}\hat{\gamma} - \gamma \\ \hat{\delta} - \delta\end{pmatrix} \quad \text{or}$$

$$\sqrt{N}(\hat{\alpha} - \alpha) \quad = \quad \hat{G}\sqrt{N}(\hat{\theta} - \theta), \quad \hat{\theta} - \theta = \begin{pmatrix}\hat{\gamma} - \gamma \\ \hat{\delta} - \delta\end{pmatrix} \qquad (12.108)$$

$$\hat{G} \quad = \quad \hat{C}_n^{-1}[F - (\alpha' \otimes I_n)(B_{ST}, 0)],$$

where N is the length of the realization on the basis of which the autocovariances in $\hat{\theta}$ have been obtained. Incidentally the change in notation from B_{ST} to $(B_{ST}, 0)$ is to accommodate the change from γ to $\theta = (\gamma', \delta')'$, so that the inserted zero is a matrix of dimension $n \times h$.

From Proposition 9.6 (of Chap. 9) the limiting distribution of $\sqrt{N}(\hat{\alpha} - \alpha)$ is a linear transformation, by the probability limit G of $\hat{G}$, of the limiting distribution of $\sqrt{N}(\hat{\theta} - \theta)$ as derived in Proposition 12.9 above.

We have thus proved

Proposition 12.10. Under the conditions of Proposition 12.9, the limiting distribution of the coefficient vector of the best linear predictor (BLP) obeys

$$\sqrt{N}(\hat{\alpha} - \alpha) \quad \overset{d}{\to} \quad N(0, GVG'), \quad \text{where} \qquad (12.109)$$

$$G \quad = \quad C_n^{-1}[F - (\alpha' \otimes I_n)(B_{ST}, 0)], \quad \text{and } V \text{ is as in Eq. (12.102)}.$$

Chapter 13

Asymptotic Expansions

13.1 Introduction

This chapter deals with situations in which we wish to approximate the limiting distribution of an estimator. As such it is different from other chapters in that it does not discuss topics in core econometrics and the ancillary mathematics needed to develop and fully understand them. Moreover, its purpose is different from that of the earlier (theoretical) chapters. Its aim is not only to introduce certain (additional) mathematical concepts but also to derive certain results that may prove useful for econometric applications involving hypothesis testing.

Suppose the distribution of an estimator cannot be obtained in closed form, but we can establish its limiting distribution i.e. the probability density function of the estimator **as the sample size increases to infinity**. This is possible for nearly all problems dealt with in econometrics and is the major approach followed in previous chapters. The question then arises as to whether this result is a useful approximation in making inferences, and/or testing hypotheses regarding underlying parameters, when the sample is "finite". For example, the reader may perhaps accept without question that **as an approximation** this is indeed quite useful if the sample size is 1,000 or higher. But is it useful if it is 500, or even 75? Although precise answers to these questions are not always possible, nonetheless such questions yield to mathematical analysis; in many instances it is possible to augment the limiting distribution in ways in which it may be rendered amenable, and better adapted, to the issues we have raised.

To be more precise, suppose the cumulative distribution function (cdf) for an estimator based on a sample of size T is given by F_T; suppose further that it may be shown that $F_T \to F$, at the points of continuity of F, where the notation is to be read: as $T \to \infty$, F_T converges to F, **pointwise**, at the points where F is continuous. This also means that the sequence of random

P.J. Dhrymes, *Mathematics for Econometrics*,
DOI 10.1007/978-1-4614-8145-4_13, © The Author 2013

vectors, say $\sqrt{T}(\hat{\theta}_T - \theta^*)$, converges as the sample size increases to infinity to a random vector having the distribution function F; in addition, it would also usually mean that $\hat{\theta}_T$ converges (in some form) as the sample size increases to infinity to θ^*, which is the true value of the parameter in question. Indeed these are the procedures we routinely employed in earlier chapters.

The nature and adequacy of the approximation is determined by the criterion that, using F yields useful results in making inferences regarding the parameter being estimated by $\hat{\theta}_T$, even when the sample is finite but large.

13.2 Preliminaries

Econometric estimation, and other forms of inference, takes place[1] in the context of a probability space (p.s.). The latter is the triplet (Ω, $\mathcal{A}$, $\mathcal{P}$), where Ω is the sample space, or the universe of discourse (frequently referred to as the set containing all possible outcomes of a "random" experiment, but best left as a logical primitive); $\mathcal{A}$ is a collection of subsets of Ω, referred to as the σ-algebra, and $\mathcal{P}$ is a probability measure defined over the elements of the σ-algebra.

A random variable, in the context of a probability space, is a measurable function, say

$$X : \Omega \longrightarrow R,$$

so that for each $\omega \in \Omega$, $X(\omega) \in R$. Measurability means the following: if by $\mathcal{B}(R)$ we denote the one-dimensional Borel field, i.e. the collection of all half open intervals on the real line, their complements, and countable unions, and if we pick an arbitrary element of this σ-algebra, say $B \in \mathcal{B}(R)$, then the **inverse image** of B is a set in $\mathcal{A}$. More precisely, if we put

$$A = \{\omega : X(\omega) \in B\} = X^{-1}(B), \quad \text{the inverse image of } B$$

then $A \in \mathcal{A}$. The function X is then said to be $\mathcal{A}$-measurable.

This framework helps us understand the basic aspects of random variables and demystifies them in that it shows them to be simply real valued measurable functions, like many other well known functions. It is also an excellent framework for understanding issues related to convergence of sequences of random variables.

---

[1] Nearly all concepts contained in this chapter are being dealt in much greater detail elsewhere in this volume. We undertake this discussion only in order to make this chapter as self contained as possible for those who are familiar with the broad aspects of probability theory and do not wish to go through the more detailed expositions above.

Often, however, such issues are not the primary focus, and we wish to concentrate on distributional aspects, i.e. on the probability with which the random variable takes on values in certain sets, particularly sets like $(-\infty, x)$. It is interesting that if we begin with the fundamental framework given earlier, we may convert that framework into one that involves only the real line, the one dimensional Borel field, and a distribution function induced by the random variable in question. Thus, let

$$X : \Omega \longrightarrow R,$$

be a measurable function according to the general framework just discussed. On the range space, R, define the Borel σ-algebra $\mathcal{B}(R)$, so that $(R, \mathcal{B}(R))$ is a **measurable space**. Moreover, for any $B \in \mathcal{B}(R)$, define its inverse image $A = X^{-1}(B)$. If we wish to concentrate on the distributional aspects of X, we need to define a function that gives us the probability with which it takes on values in certain sets (elements of $\mathcal{B}(R)$) as a function of these sets. Consequently, we may define the probability measure induced by X as follows:

$$P_x(B) = \mathcal{P}(A), \quad \text{for every} \ \ B \in \mathcal{B}(R).$$

It may be verified that P_x is a probability measure, i.e.

$$P_x : \mathcal{B}(R) \longrightarrow [0, 1],$$

obeys all other requirements for probability measures, and, moreover, that $(R, \mathcal{B}(R), P_x)$ is the smallest p.s. on which we can consider the random variable X. To make this perfectly analogous to the definition we gave to probability spaces earlier we adopt the convention that X is the **identity transformation**

$$X : R \longrightarrow R,$$

so that all information on X is embedded in the probability measure P_x, which is termed the **probability distribution function (pdf)**. Note that P_x is a **set** function, i.e. its arguments are sets in $\mathcal{B}(R)$. The reader is more likely to be familiar with it in a slightly different form; more precisely, one is more likely to be familiar with its counterpart, the **cumulative distribution function (cdf)** F_x. The connection between the two is rather simple. Take B to be the special set $B = (-\infty, b)$, for any $b \in R$. Then, with F_x viewed as a function of the point b, we have

$$F_x(b) = P_x(B), \quad B = (-\infty, b).$$

Thus, the cdf is a point function, which is derived from the **set function** P_x for the special sets above. If the derivative of F exists, it is generally denoted by f and is said to be the density function (of X).

Notice that the fundamental p.s. is considerably more informative regarding the random variable in question than the probability space it induces. In the former, we are given the nature of the function (random variable), as well as the general rule of assigning probabilities to the elements (sets) of the σ-algebra in $(\Omega, \ \mathcal{A}, \ \mathcal{P})$. In the latter, we do not know the nature of the random variable X; all we know is its cdf, say F. But **there are infinitely many variables** that may have F as their cdf. Thus, given only F, we generally cannot obtain uniquely the particular random variable from which we may have derived the cdf in the discussion above.

Finally, the precise context in which we shall discuss asymptotic expansions is this: $\hat{\theta}_T$ is an estimator of some unknown parameter θ, based on a sample of size T. The estimator is such that we can assert that $\sqrt{T}(\hat{\theta}_T - \theta)$, is expressible as

$$\sqrt{T}(\hat{\theta}_T - \theta) = Z_T = \frac{1}{\sqrt{T}} \sum_{t=1}^{T} z_t,$$

where z_t are random variables, and that Z_T obeys a central limit theorem, i.e. in the limit it converges to a random variable which is normal with mean zero and a certain variance, say ψ^2. Denote its limit normal density by $\phi(0, \psi^2)$, and denote the limit random variable by Z. We may then make the approximation to be

$$\hat{\theta}_T \approx \theta + \frac{1}{\sqrt{T}} Z,$$

which therefore has the approximate density $\phi(\theta, \psi^2/T)$.

The object of asymptotic expansion theory is to improve on this approximation, because as we know the relation above is "exact" only as $T \to \infty$. For some specific value of T it may be wildly off.

13.3   General Bounds on Approximations

Let $X_T = (1/b_T) \sum_{t=1}^{T} x_t$ be a sequence of random variables obeying a CLT, where b_T is a positive sequence tending to infinity. This means that we have the result that $X_T \overset{d}{\to} X$ (to be read X_T converges in distribution to a random variable having the distribution (cdf) of some random variable X). Let this be the normal cdf Φ, with mean zero and variance one.

The most basic results on approximation are the so called Berry-Esseen theorems that put a bound on the quantity

$$D = \sup_{x \in R} |F_T(x) - \Phi(x)|,$$

where F_T is the cdf of X_T. One such theorem is given in Dhrymes (1989, p. 276), for the case where the underlying sequence obeys

$$E|x_t|^{2+\delta} = \rho_{tt}^{2+\delta} < \infty, \quad \text{for} \quad \delta > 0,$$

in which case

$$D < C \left(\frac{\rho_T}{\sigma_T} \right)^{2+\delta}, \quad \rho_T^{2+\delta} = \sum_{t=1}^{T} \rho_{tt}^{2+\delta}, \quad \sigma_T^2 = \sum_{t=1}^{T} \sigma_{tt}^2, \quad \sigma_{tt}^2 = \text{Var}(x_t).$$

The constant C is independent of the distribution of the x's, but may depend on δ. Another, for the case of i.i.d. random variables with mean zero, variance one, bounded third moment, and normalizing sequence $b_T = \sqrt{T}$, is given in Kolassa (1997, p. 19), in which case we have

$$D < C\rho \frac{1}{\sqrt{T}}, \quad \rho = E|x_1|^3 = \rho < \infty,$$

and C may be taken to be 3. The interested reader may search the literature for results that are more suitable for the problem he faces.

This of course is a bound on the **maximal** difference in the relevant cdfs, and as such it may not be deemed satisfactory. For example if the sample size is $T = 100$ and $\rho = 3$, the maximal difference is .9; if $T = 625$ and $\rho = 3$, the maximal difference is .38. In either case, this is quite sizeable and is generally unacceptable for an approximation on which we wish to base inference procedures. Of course this is the **maximal** difference, and in some applications the actual difference may well be less; however, we have no way of establishing this in general. Thus, we are led to consider alternatives. These alternatives rely crucially on the connection between the cdf and the characteristic function of a random variable, a topic we elucidate in the discussion below.

13.4 Characteristic Functions (CF) and Moment Generating Functions (MGF)

We begin by recalling that the sth moment of a random variable, X, denoted by μ_s, is defined by

$$\mu_s = \int_{-\infty}^{\infty} x^s f(x) \, dx,$$

where f is the density function. In addition, we often deal with the so called **central** moments, defined by

$$\sigma_k = \int_{-\infty}^{\infty} (x - \mu_1)^k f(x) \, dx.$$

We note that $\sigma_1 = 0$ and that μ_1 is termed the **mean** of the random variable in question; similarly, σ_2 is usually denoted by σ^2 and is termed its **variance**.

We also remind the reader of the following important fact: If a r.v. X has a finite moment of order n, i.e. if $E|X|^n < \infty$, then it has finite moments of order $s \leq n$.

We now turn to the main topic of this section.

Definition 13.1. Let X be a random variable with cdf F, and let $t \in R$. Its **characteristic** function is given by

$$\psi_x(t) = Ee^{itX} = \int_{-\infty}^{\infty} e^{itx} dF(x)$$

where, we remind the reader, $e^{itx} = \cos tx + i \sin tx$, i is the imaginary unit obeying $i^2 = -1$, and the expression above is to be interpreted as a Riemann-Stieltjes integral. If the cdf (F) is differentiable, and thus the density f exists, the CF is given by

$$\psi_x(t) = Ee^{itX} = \int_{-\infty}^{\infty} e^{itx} f(x) dx,$$

and is thus the Fourier tranform of the density function.

Characteristic functions have the following properties.

i. $|\psi(t)| \leq \psi(0) = 1$.

ii. $\psi(t)$ is **uniformly continuous**.

iii. $\overline{\psi(t)} = \psi(-t)$.

iv. $\psi(t)$ is a **real valued function** if and only if the distribution of the r.v. X is **symmetric**, in the sense that $f(x) = f(-x)$.

v. If $E|X|^m < \infty$, the mth derivative of $\psi(t)$ exists and, moreover,

$$\mu_s = \frac{1}{i^s} \frac{d^m \psi(t)}{dt^m}\bigg|_{t=0}, \quad s = 1, 2, 3, \ldots, m,$$

where $\mu_s = EX^s$. Conversely, if X has finite moment of order s, $s = 1, 2, 3, \ldots, m$, its characteristic function has derivatives of order s and the relation in item v holds.

Definition 13.2. Let X be as in Definition 13.1 and $t \in R$; its **moment generating function (MGF)** is defined by

$$M_x(t) = Ee^{tX} = \int_{-\infty}^{\infty} e^{tx} dF(x) dx, \quad \text{or if } F \text{ is differentiable, by}$$

$$M_x(t) = Ee^{tX} = \int_{-\infty}^{\infty} e^{tx} f(x) dx.$$

Remark 13.1. Notice that since, in Definition 13.1, the entity $e^{itx} f(x)$ is **absolutely integrable**, the characteristic function **always** exists. In Definition 13.2, however, the integrand is $e^{tx} f(x)$; if X is a continuous random variable with range $[-\infty, \infty]$, the MGF need not exist. Whether it exists or not depends crucially on the behavior of

$$\lim_{x \to \infty} e^{tx} f(x).$$

We provide a number of examples.

Example 13.1. Let $X \sim N(\mu, \sigma^2)$, which is to be read: X is a random variable whose distribution is normal with mean μ and variance σ^2. In this case,

$$Ee^{tX} = \left(\frac{1}{\sqrt{2\pi\sigma^2}} \right) \int_{-\infty}^{\infty} e^{tx} e^{-(1/2\sigma^2)(x-\mu)^2} \, dx$$

$$= e^{t\mu + \frac{t\sigma^2}{2}} + \left(\frac{1}{\sqrt{2\pi\sigma^2}} \right) \int_{-\infty}^{\infty} e^{-(1/2\sigma^2)(x-\mu-t\sigma^2)^2} \, dx.$$

If X obeys the normal density,

$$\lim_{|x| \to \infty} e^{tx} f(x) = \lim_{|x| \to \infty} e^{t\mu + \frac{t\sigma^2}{2}} \left[e^{-(1/2\sigma^2)(x-\mu-t\sigma^2)^2} \right] = 0,$$

so that the necessary condition for the existence of the MGF is satisfied; moreover, as the example makes clear, the MGF actually exists and is given by

$$M_x(t) = e^{t\mu + (1/2)t^2\sigma^2}. \tag{13.1}$$

On the other hand, if X has the Cauchy or the (Student) t density, it has **no moment generating function** but it **has a characteristic function**. We illustrate this with the Cauchy distribution.

Example 13.2. Let X be a r.v. with the Cauchy distribution, whose density is $f(x) = 1/\pi(1 + x^2)$. That this is a density function is easily verified from

$$\frac{1}{\pi} \int_{-\infty}^{\infty} \frac{dx}{1 + x^2} dx = \frac{1}{\pi} \int_0^1 u^{(1/2)-1}(1 - u)^{(1/2)-1} du = \frac{\Gamma(1/2)\Gamma(1/2)}{\pi \Gamma(1)} = 1,$$

where $\Gamma(\cdot)$ is the gamma function given by

$$\Gamma(\alpha) = \int_0^{\infty} x^{\alpha-1} e^{-x} dx, \quad \text{with } \Gamma(1/2) = \sqrt{\pi}.$$

The result is obtained by noting that the integrand (the Cauchy density) is symmetric (about zero) and making the change in variable $u = 1/(1 + x^2)$.

Consider now

$$\lim_{|x| \to \infty} \frac{e^{tx}}{1 + x^2};$$

if $t > 0$ and $x > 0$, the limit of the expression is unbounded (infinite), because e^{tx} grows faster than x^2; also, for $t < 0$ and $x < 0$, the limit of the expression converges to ∞. Hence, the Cauchy distribution does not satisfy the necessary condition for the existence of a MGF. In fact, the MGF does not exist for the Cauchy distribution. On the other hand its CF exists and is given by

$$\psi_c(t) = Ee^{itX} = e^{|t|}. \tag{13.2}$$

Note that $\psi_c(t)$ is continuously differentiable for $t > 0$ and for $t < 0$. However, continuity fails for $t = 0$, since the derivative is e^t for $t > 0$ and its limit as $t \downarrow 0$ is 1; on the other hand for $t < 0$ the derivative is $-e^{-t}$ and its limit as $t \uparrow 0 = -1$. Indeed, its derivative at $t = 0$ **does not exist**.

Actually carrying out the integration above is too complex to present here.

The other distinctions between CF and MGF follow immediately from the fact that in CF the exponential involves *it*, whereas in the MGF the exponential involves only t. Thus, in property v of Definition 13.1, the formula should be amended to

$$\mu_s = \left. \frac{d^m \psi(t)}{dt^m} \right|_{t=0}, \quad s = 1, 2, 3, \ldots, m,$$

Remark 13.2. The usefulness of characteristic (and moment generating) functions derives from two very important facts. First, given a characteristic function, say ψ, there exists a unique cdf that corresponds to it, say F, which may be derived from it through the so called **inversion formula**, where a, b, are points of continuity of F,

$$F(b) - F(a) = \lim_{s \to \infty} \frac{1}{2\pi} \int_{-s}^{s} \frac{e^{-ita} - e^{-itb}}{it} \psi_x(t) dt.$$

For a proof of this see, for example, Dhrymes (1989, pp. 254–256).

Moreover, if the CF is **absolutely integrable**, i.e. if

$$\int_{-\infty}^{\infty} |\psi_x(t)| \, dt < \infty,$$

F is differentiable and the density is given by

$$f(x) = \frac{1}{2\pi} \int_{-\infty}^{\infty} e^{-itx} \psi_x(t) \, dt,$$

so that f and ψ are a pair of **Fourier transforms**.

Second, if $\{\psi_T : T = 1, 2, 3, \ldots\}$ is a sequence of characteristic functions converging pointwise to ψ, then the corresponding sequence of cdfs $\{F_T : T = 1, 2, 3, \ldots\}$ converges to F, which is the cdf corresponding to ψ.

Mathematically, the most complete exposition of the subject of asymptotic expansions can be made only in terms of the CF. However, because many students in economics may not be very familiar with complex analysis, we shall make an effort as much as possible to use the device of MGF.

13.5 CF, MGF, Moments, and Cumulants

13.5.1 CF, MGF, and the Existence of Moments

Let X be a r.v. that possesses moments of all orders.

The connection between the MGF and the moments of a r.v. is seen immediately from the Taylor series of the exponential. Specifically, we have

$$e^{tx} = \sum_{s=1}^{\infty} \frac{t^s x^s}{s!},$$

where the Taylor series converges on all compact subsets of $[-\infty, \infty]$. Therefore,

$$M(t) = E e^{tX} = \sum_{s=1}^{\infty} \frac{t^s}{s!} \mu_s, \qquad (13.3)$$

which helps to explain the term **moment generating** function, and incidentally shows the validity of part v of Definition 13.1 in the case of MGF.

The same situation prevails relative to the CF except that here, even if all moments fail to exist (be finite), the CF is still **defined**. The problem is that the formula given in part v of Definition 13.1 is valid **only** when the sth moment is known to exist. In fact, we may always write the CF in terms of the moments that **do** exist plus a remainder that approaches zero at a certain order. To make this point clear, we need certain additional notation.

Orders of Magnitude Notation

It will simplify the presentation if we introduce the o, O and o_p, O_p notation.

Definition 13.3. Let $h(x)$ be a (real valued) function, and let $x^* \in \bar{R}$, the extended real number system (including $\pm\infty$).

    i. We say that h is of order little oh, and denote this by $o(x)$, as $x \to x^*$, if and only if

$$\lim_{x \to x^*} \frac{h(x)}{x} = 0.$$

    This means then that h "grows" at a rate slower than x, i.e. grows at a **sublinear rate**.

    ii. We say that h is of order big oh, and denote this by $O(x)$, as $x \to x^*$, if and only if

$$\lim_{x \to x^*} \frac{h(x)}{x} \quad \text{exists and is finite.}$$

    This means then that h "grows" at the same rate as x, i.e. grows at a **linear rate**.

    iii. We say that h is of order little oh in probability, and denote this by $o_p(x)$, as $x \to x^*$, if and only if

$$\underset{x \to x^*}{\text{plim}} \, \frac{h(x)}{x} = 0.$$

    iv. We say that h is of order big oh in probability, and denote this by $O_p(x)$ as $x \to x^*$, if and only if

$$\underset{x \to x^*}{\text{plim}} \, \frac{h(x)}{x} \quad \text{exists and is finite.}$$

In such cases, we employ the notation

$$h(x) \sim o(x), \quad \text{or} \quad h(x) \sim O(x), \quad \text{or} \quad h(x) \sim o_p(x), \quad \text{or} \quad h(x) \sim O_p(x).$$

We illustrate the use of the first two concepts by means of an example.

Example 13.3. Consider the expansion

$$h_n(x) = n[e^{x/n} - 1] = x + \sum_{j=2}^{\infty} \frac{x^j}{j!n^{j-1}} = x + O(1/n)$$

$$= x + o(1/n^\alpha), \quad \alpha \in (0,1).$$

This is so because

$$\frac{1}{1/n} \sum_{j=2}^{\infty} \frac{x^j}{j!n^{j-1}} = \frac{x^2}{2!} + \sum_{j=3}^{\infty} \frac{x^j}{j!n^{j-2}},$$

whose limit as $n \to \infty$ is $x^2/2$. On the other hand,

$$\frac{1}{1/n^\alpha} \sum_{j=2}^{\infty} \frac{x^j}{j!n^{j-1}} = \sum_{j=2}^{\infty} \frac{x^j}{j!n^{j-1-\alpha}},$$

which converges to zero with n, in view of the fact that for any $j = 2, 3, \dots$ the denominators of the series contain a term of the form n^β, for $\beta > 0$.

Existence of Moments and Derivatives of CF

In this section, we give the important result that connects the existence of moments with the existence of derivatives of the CF, and conversely. Notice that, in the statement of the result, the first part of the proposition is more general than the second.

Proposition 13.1. Let X be a random variable having a finite moment of order $n + \delta$, for some $\delta \in [0, 1]$. Then, its characteristic function has continuous derivatives up to order n and, moreover,

$$\psi_x(t) = \sum_{j=0}^{n} \frac{(it)^j}{j!} \mu_j + o(|t|^n). \tag{13.4}$$

Conversely, if $\psi_x^{(2k)}(0)$ exists and is finite for $k = 1, 2, 3, \dots$, then

$$EX^{2k} < \infty, \quad \text{for} \quad k = 1, 2, 3, \dots. \tag{13.5}$$

Proof: A complete proof is given in Chow and Teicher (1988, pp. 277–279); here we give only a proof of the first part of the proposition relating to the (partial) expansion of the CF as a complex polynomial whose coefficients are moments of the r.v. in question.

Use integration by parts to find

$$\int_0^y (y-x)^n e^{ix}\, dx = \frac{y^{n+1}}{n+1} + \frac{i}{n+1}\int_0^y (y-x)^{n+1} e^{ix}\, dx, \qquad (13.6)$$

which, evidently, holds for all $n \geq 1$. Next, we need to verify that it holds for $n = 0$ as well. More precisely, we must verify that

$$\int_0^y e^{ix}\, dx = y + i\int_0^y (y-x)e^{ix}\, dx.$$

Again using integration by parts, we note that

$$\int_0^y (y-x)e^{ix}\, dx = y\int_0^y e^{ix}\, dx - \int_0^y xe^{ix}\, dx = \frac{1}{i}y[e^{iy}-1] - \int_0^y xe^{ix}\, dx$$

$$\int_0^y xe^{ix}\, dx = \frac{1}{i}xe^{ix}\Big|_0^y - \frac{1}{i}\int_0^y e^{ix}\, dx = \frac{1}{i}ye^{iy} - \frac{1}{i}\int_0^y e^{ix}\, dx.$$

Substituting the last member of the second equation in the last member of the first and collecting terms, we find

$$\int_0^y e^{ix}\, dx = y + i\int_0^y (y-x)e^{ix}\, dx; \quad \text{moreover,}$$

$$e^{iy} = 1 + i\int_0^y e^{ix}\, dx, \qquad (13.7)$$

thus verifying the claim.

Applying repeatedly integration by parts to the integral in the right member of the last equation in Eq. (13.7) we find

$$e^{iy} = \sum_{j=0}^n \frac{(iy)^j}{j!} + \frac{i^{n+1}}{n!}\int_0^y (y-x)^n e^{ix}\, dx. \qquad (13.8)$$

Since

$$\left| \frac{i^{n+1}}{n!}\int_0^y (y-x)^n e^{ix}\, dx \right| \leq \frac{1}{(n+1)!}|y|^{n+1},$$

we have the bound

$$\left| e^{iy} - \sum_{j=0}^n \frac{(iy)^j}{j!} \right| \leq \frac{1}{(n+1)!}|y|^{n+1}. \qquad (13.9)$$

The difficulty with using this bound is that it involves the existence of the $(n+1)$ st moment, something we have **not** assumed. To obtain another bound that involves only the nth moment, return to Eq. (13.6) and reverse it to obtain

$$\int_0^y (y-x)^n e^{ix}\, dx = \frac{n}{i}\left(\int_0^y (y-x)^{n-1} e^{ix}\, dx - \frac{y^n}{n}\right)$$

$$= \frac{n}{i}\left(\int_0^y (y-x)^{n-1}[e^{ix}-1]\, dx\right).$$

Substituting in Eq. (13.8), we have the alternative representation

$$e^{iy} = \sum_{j=0}^n \frac{(iy)^j}{j!} + \frac{i^n}{(n-1)!}\int_0^y (y-x)^{n-1}[e^{ix}-1]\, dx. \tag{13.10}$$

Since

$$\left|\frac{i^n}{(n-1)!}\int_0^y (y-x)^{n-1}[e^{ix}-1]\, dx\right| \le 2\frac{|y|^n}{n!},$$

we have the alternative bound

$$\left|e^{iy} - \sum_{j=0}^n \frac{(iy)^j}{j!}\right| \le 2\frac{|y|^n}{n!}. \tag{13.11}$$

Replacing iy by itX and taking expectations, we have

$$\psi_x(t) = Ee^{itX}$$

$$\left|\psi_x(t) - \sum_{j=0}^n \frac{(it)^j}{j!}\mu_j\right| \le |t|^n \min\left[|t|\frac{1}{(n+1)!}E|X|^{n+1},\ \frac{2}{n!}E|X|^n\right], \tag{13.12}$$

which provide sharp bounds for the remainder.[2]

Since by previous discussion $\mu_s = \left.\frac{\partial^s \psi_x(t)}{\partial t^s}\right|_{t=0}$, expanding the characteristic function about $t=0$ we obtain, from Taylor's theorem,

$$\psi_x(t) = \sum_{j=0}^n \frac{(it)^j}{j!}\mu_j + o(|t|^n).$$

---

[2]Evidently, this is so only when moments of all orders exist. If, for example, the $(n+1)$ st moment does not exist, only the second bound is relevant.

13.5.2   Cumulants and Cumulant Generating Functions (CGF)

In dealing with asymptotic expansions one typically deals not with CF's or MGF's but with **cumulant generating functions** (CGF). This construct has no independent or distinctive significance; it is simply a more convenient representation of the MGF given the purpose served in this context. We begin with the fundamental definition.

Definition 13.4. Let X be a r.v. defined on the probability space (Ω, $\mathcal{A}$, $\mathcal{P}$) and let $M_x(t)$ be its MGF, assuming that one exists. The **cumulant generating function** is defined by

$$K_x(t) = \log M_x(t). \tag{13.13}$$

It follows immediately that, if X, Y are mutually independent r.v., and $Z = aX + b$ then

$$K_{x+y}(t) = K_x(t) + K_y(t), \quad K_z(t) = K_{ax+b}(t) = K_x(ta) + bt. \tag{13.14}$$

Remark 13.3. The relations in Eq. (13.14) attest very strongly to the usefulness and convenience of the CGF.

Definition 13.5. The jth cumulant of a r.v. X, as above, is given by

$$\kappa_j^{(x)} = \left.\frac{d^j K(t)}{dt^j}\right|_{t=0}.$$

The following result is immediate.

Proposition 13.2. Let X, Y, Z be as in Definition 7.4. Then,

$$\kappa_j^{x+y} = \kappa_j^x + \kappa_j^y, \quad \kappa_j^{ax+b} = a\kappa_j^x + b. \tag{13.15}$$

As we noted earlier, cumulants and CGF's have no distinctive significance because they are conceptually simple creatures of moments and MGF's; nonetheless, as the order of the cumulant increases its relationship to the underlying moments becomes increasingly complex.

There are at least two general ways in which we can establish a relationship between cumulants and moments. First, we may obtain the Taylor series expansion of the CGF about $t = 0$, thus producing the relation

$$K(t) = \sum_{j=0}^{\infty} \kappa_j \left(\frac{t^j}{j!} \right);$$ (13.16)

recalling the definition of the CGF, we observe that

$$e^{\sum_{j=0}^{\infty} \left(\kappa_j \frac{t^j}{j!} \right)} = M(t) = \sum_{j=0}^{\infty} \left(\mu_j \frac{t^j}{j!} \right).$$ (13.17)

The leftmost member of the equation may also be written as

$$e^{\sum_{j=0}^{\infty} \kappa_j \left(\frac{t^j}{j!} \right)} = \prod_{j=0}^{\infty} e^{\kappa_j \frac{t^j}{j!}}$$ (13.18)

and, moreover,

$$e^{\kappa_j \frac{t^j}{j!}} = \sum_{s=0}^{\infty} \left(\frac{(\kappa_j t^j / j!)^s}{s!} \right).$$ (13.19)

Next, expand every term in this fashion and multiply together the resulting power series in the leftmost member of Eq. (13.17); if we equate powers of t on both sides of that equation, we find the desired relationship.

The alternative approach is to find the relationship through the operation

$$\kappa_j = \left. \frac{d^j K(t)}{dt^j} \right|_{t=0} = \left. \frac{d^j \log M(t)}{dt^j} \right|_{t=0}.$$ (13.20)

In either case, this is exposited in great detail in Sect. 3.14, volume I of Kendall and Stuart (1963) as well as Kendall et al. (1987), hereafter, respectively, KS and KSO.

We illustrate the problem by giving a number of such relationships. More exhaustive equivalences are given in KS and KSO as noted.

Cumulants		Moments	
κ_0	$= 0$	μ_0	$= 1$
κ_1	$= \mu_1$	μ_1	$= \kappa_1$
κ_2	$= \mu_2 - \mu_1^2$	μ_2	$= \kappa_2 + \kappa_1^2$
κ_3	$= \mu_3 - 3\mu_1\mu_2 + 2\mu_1^3$	μ_3	$= \kappa_3 + 3\kappa_1\kappa_2 + \kappa_1^3$
κ_4	$= \mu_4 - 4\mu_1\mu_3 - 3\mu_2^2 + 12\mu_1^2\mu_2 - 6\mu_1^4$	μ_4	$= \kappa_4 + 4\kappa_1\kappa_3 + 3\kappa_2^2 + 6\kappa_1^2\kappa_2 + \kappa_1^4$

The general formula for obtaining higher order moments or cumulants (in terms of each other) is given in KSO as

$$\mu_r = \sum_{m=0}^{r} \left[\sum_{\pi} \prod_{s=1}^{m} \left(\frac{\kappa_{p_s}}{p_s!} \right)^{\pi_s} \right] \frac{r!}{\prod_{s=0}^{m} \pi_s},$$ (13.21)

where the second summation runs over all values of π_s, such that

$$\sum_{s=1}^{m} p_s \pi_s = m. \tag{13.22}$$

Although the coefficients of the various cumulants appear to be hopelessly complex, in fact we can establish the relations

$$\frac{\partial \mu_r}{\partial \kappa_1} = r \mu_{r-1},$$

$$\frac{\partial \mu_r}{\partial \kappa_j} = \binom{r}{j} \mu_{r-j}, \quad j = 2, 3, \ldots, r-1. \tag{13.23}$$

The analogous connection for cumulants is given by

$$\kappa_r = r! \sum_{m=0}^{r} \left[\sum_{\pi,\nu} \prod_{s=1}^{m} \left(\frac{\mu_{p_s}}{p_s!} \right)^{\pi_s} \right] \frac{(-1)^{\nu-1}(\nu-1)!}{\prod_{s=1}^{m}(\pi_s)!}, \tag{13.24}$$

where the second summation runs over all possible values of π 's and ν 's such that

$$\sum_{s=1}^{m} p_s \pi_s = m, \quad \text{and} \quad \sum_{s=1}^{m} \pi_s. \tag{13.25}$$

13.6   Series Approximation

The basic notion underlying this approach is roughly the following: Suppose it is desired to approximate the cdf F_y by some function of the cdf F_x so that the approximation is (nearly) as close as desired. One way to proceed is to obtain the CF of the two cdf's, say $\phi_y(t)$, $\phi_x(t)$, respectively, perform the desired operation in terms of these two entities, and then invert the result to obtain the desired cdf approximation. To minimize unnecessary arguments regarding the existence of logarithms for complex entities, we assume that both cdf's have **moment generating functions** and proceed accordingly.

Thus, let $M_y(t)$, $M_x(t)$, respectively, be the MGF's and consider the ratio

$$\psi(t) = \frac{M_y(t)}{M_x(t)} = \sum_{j=0}^{\infty} b_j \frac{t^j}{j!}, \tag{13.26}$$

assuming that the ratio has a Taylor series expansion about $t = 0$. It follows then that formally

$$M_y(t) = M_x(t) \left[\sum_{j=0}^{\infty} b_j \frac{t^j}{j!} \right]. \tag{13.27}$$

The preceding is then a precise representation of the MGF of F_y in terms of the MGF of F_x **and the unknown parameters** b_j. It is also evident that the more parameters we retain in our representation, the closer the approximation will be. That this is a solution to the problem can be made entirely transparent even to a reader with modest mathematical sophistication if we assume that the density functions exist, so that we are dealing with two random variables, Y and X, with the respective density functions f_y and f_x. Because the CF's and the densities are pairs of Fourier transforms, i.e.

$$\phi_x(t) = \int_{-\infty}^{\infty} e^{it\xi} f_x(\xi)\, d\xi, \quad f_x(\xi) = \frac{1}{2\pi} \int_{-\infty}^{\infty} e^{-it\xi} \phi_x(t)\, dt, \tag{13.28}$$

it follows that the CF uniquely determines the density function and vice-versa.

Bibliography

Anderson, T.W. and H. Rubin (1949), Estimation of the Parameters of a Single Equation in a Complete System of Stochastic Equations, *Annals of Mathematical Statistics*, pp. 46–63.

Anderson, T.W. and H. Rubin (1950), The Asymptotic Properties of Estimates of Parameters of in a Complete System of Stochastic Equations, *Annals of Mathematical Statistics*, pp. 570–582.

Balestra, P., & Nerlove, M. (1966). Pooling cross section time series data in the estimation of a dynamic model: The demand for natural gas. *Econometrica, 34*, 585–612.

Bellman, R. G. (1960). *Introduction to matrix analysis.* New York: McGraw-Hill.

Billingsley, P. (1968). *Convergence of probability measures.* New York: Wiley.

Billingsley, P. (1995). *Probability and measure* (3rd ed.). New York: Wiley.

Brockwell, P. J., & Davis, R. A. (1991). *Time series: Theory and methods* (2nd ed.). New York: Springer-Verlag.

Chow, Y. S., & Teicher, H. (1988). *Probability theory* (2nd ed.). New York: Springer-Verlag.

Dhrymes, P. J. (1969). Alternative asymptotic tests of significance and related aspects of 2SLS and 3SLS estimated parameters. *Review of Economic Studies, 36*, 213–226.

Dhrymes, P. J. (1970). *Econometrics: Statistical foundations and applications.* New York: Harper and Row; also (1974). New York: Springer-Verlag.

Dhrymes, P. J. (1973). Restricted and Unrestricted Reduced Forms: Asymptotic Distributions and Relative Efficiencies, *Econometrica*, vol. 41, pp. 119–134.

Dhrymes, P. J. (1978). *Introductory economics*. New York: Springer-Verlag.

Dhrymes, P.J. (1982) *Distributed Lags: Problems of Estmation and Formulation (corrected edition)*. Amsterdam: North Holland

Dhrymes, P. J. (1989). *Topics in advanced econometrics: Probability foundations*. New York: Springer-Verlag.

Dhrymes, P. J. (1994). *Topics in advanced econometrics: Volume II linear and nonlinear simultaneous equations*. New York: Springer-Verlag.

Hadley, G. (1961). *Linear algebra*. Reading: Addison-Wesley.

Kendall, M. G., & Stuart, A. (1963). *The advanced theory of statistics*. London: Charles Griffin.

Kendall M. G., Stuart, A., & Ord, J. K. (1987). *Kendall's advanced theory of statistics*. New York: Oxford University Press.

Kolassa, J. E. (1997). *Series approximation methods in statistics* (2nd ed.). New York: Springer-Verlag.

Sims, C.A. (1980). Macroeconomics and Reality, *Econometrica*, vol. 48, pp.1–48.

Shiryayev, A. N. (1984). *Probability*. New York: Springer-Verlag.

Stout, W. F. (1974). *Almost sure convergence*. New York: Academic.

Theil, H. (1953). *Estimation and Simultaneous Correlation in Complete Equation Systems, mimeograph*. The Hague: Central Plan Bureau.

Theil, H. (1958). *Economic Forecasts and Policy*, Amsterdam: North Holland.

Index

CPSIA information can be obtained at www.ICGtesting.com
Printed in the USA
LVOW05s1234211214

419818LV00002B/3/P